An Integrative Approach to Successional Dynamics

Tempo and Mode of Vegetation Change

Much of what is considered conventional wisdom about succession isn't as clear cut as it is generally believed. Yet, the importance of succession in ecology is undisputed, since it offers a real insight into the dynamics and structure of all plant communities.

Part monograph and part conceptual treatise, *An Integrative Approach to Successional Dynamics* presents a unifying conceptual framework for dynamic plant communities and uses a unique long-term data set to explore the utility of that framework.

The 14 chapters, each written in a non-technical style and accompanied by numerous illustrations and examples, cover diverse aspects of succession, including: community, population and disturbance dynamics, diversity, community assembly, heterogeneity, functional ecology, and biological invasion. This unique text will be a great source of reference for researchers and graduate students in ecology and plant biology, and others with an interest in the subject.

Scott J. Meiners is a professor in the Department of Biological Sciences of Eastern Illinois University where he teaches Plant Ecology, Introductory Botany, and a graduate course in Biostatistics. His research interests focus on the dynamics of regenerating communities using forest, grassland, and successional systems, as well as the dynamics of stream fish communities and sustainable agriculture. Since 2001, he has led the Buell–Small Succession Study, the longest continuous study of post-agricultural vegetation dynamics.

Steward T. A. Pickett, a Distinguished Senior Scientist at the Cary Institute of Ecosystem Studies, in Millbrook, New York, is an expert in the ecology of plants, vegetation dynamics and natural disturbance. His contributions to succession are in the realm of both theory and empirical mechanistic studies. He also directs the Baltimore Ecosystem Study, Long-Term Ecological Research program. He has edited and authored books on ecological heterogeneity, humans as components of ecosystems, conservation, the linkage of ecology and urban design, the philosophy of ecology and ecological ethics.

Mary L. Cadenasso is a professor in the Department of Plant Sciences at the University of California, Davis. She received a National Science Foundation Career award and was recently named a Chancellor's Fellow. Her research interests span landscape, ecosystem, and plant ecology and focus on determining how the spatial heterogeneity of a system is linked to ecosystem functions and associated changes. Her work has been widely published in more than 50 peer reviewed journal articles, 25 book chapters and two books.

An Integrative Approach to Successional Dynamics

Tempo and Mode of Vegetation Change

SCOTT J. MEINERS
Eastern Illinois University, USA

STEWARD T. A. PICKETT
Cary Institute of Ecosystem Studies, NY, USA

MARY L. CADENASSO
University of California, Davis, USA

CAMBRIDGE
UNIVERSITY PRESS

University Printing House, Cambridge CB2 8BS, United Kingdom

Cambridge University Press is part of the University of Cambridge.

It furthers the University's mission by disseminating knowledge in the pursuit of education, learning and research at the highest international levels of excellence.

www.cambridge.org
Information on this title: www.cambridge.org/9780521116428

© S. J. Meiners, S. T. A. Pickett and M. L. Cadenasso 2015

This publication is in copyright. Subject to statutory exception and to the provisions of relevant collective licensing agreements, no reproduction of any part may take place without the written permission of Cambridge University Press.

First published 2015

A catalog record for this publication is available from the British Library

Library of Congress Cataloging in Publication data
Meiners, Scott J., 1970– author.
An integrative approach to successional dynamics: tempo and mode of vegetation change / Scott J. Meiners, Eastern Illinois University, USA, Steward T. A. Pickett, Cary Institute of Ecosystem Studies, NY, USA, Mary L. Cadenasso, University of California, Davis, USA.
    pages cm
Includes bibliographical references and index.
ISBN 978-0-521-11642-8 (alk. paper)
1. Vegetation dynamics. 2. Plant succession. 3. Plant communities.
I. Pickett, Steward T., 1950– author. II. Cadenasso, Mary L., author. III. Title.
QK910.M45 2015
581.7′18–dc23
                                                                    2014035278

ISBN 978-0-521-11642-8 Hardback

Cambridge University Press has no responsibility for the persistence or accuracy of URLs for external or third-party internet websites referred to in this publication, and does not guarantee that any content on such websites is, or will remain, accurate or appropriate.

Contents

	Nou wè nan kote nou kanpe	page vii
1	Goals, concepts and definitions	1
Part 1	**The conceptual background and development of succession**	**13**
2	History and context of the Buell–Small Succession Study	15
3	Succession theory	30
4	Conceptual frameworks and integration: drivers and theory	48
Part 2	**Successional patterns in the BSS data**	**65**
5	Community patterns and dynamics	67
6	Dynamics of populations through succession	87
7	Impacts of drought and other disturbances on succession	111
8	Dynamics of diversity	131
Part 3	**Integrative themes**	**149**
9	Convergence and community assembly	151
10	Successional equivalence of native and non-native species	169
11	Heterogeneity in dynamic systems	189
12	Functional ecology of community dynamics	211
Part 4	**Synthesis**	**235**
13	Succession, habitat management and restoration	237
14	Where we stand: lessons and opportunities	248
	Appendix 1	256
	Appendix 2	261
	References	268
	Index	300

Nou wè nan kote nou kanpe

One of the things that we have endeavored to do in this book is pull in quotes that capture the rich history of ecological thought on succession and related topics. To start this book off, we could have easily referred to Newton's "Standing on the shoulders of giants" quote in homage to the wealth of information that we have benefited from. As this quote is used entirely too often, we have instead opted for something a little different. "Nou wè nan kote nou kanpe" is a Haitian saying that translates to "We see from where we stand." If we have been able to offer any real insights in this book, it is because of the foundational work of many, many others. In particular, we owe very much to Drs. Helen Buell (1901–1995), Murray Buell (1905–1975) and John Small (1900–1977). They started the study that is the basis for much of this publication. Without their vision and commitment, none of this would have been possible. Their enthusiasm for ecology and the closeness of the three of them is captured in a comment from John Small's remembrance of Murray's life – "Hundreds will attest that they were quite a team! To the extent that the field of ecology, physically and theoretically, is the better for them, that is the way they wanted it." (Small, 1975).

We also need to thank decades' worth of researchers, most of whom volunteered their time and helped in the collection of this massive data set. We have attempted to assemble a list of these people from the original data sheets and have included this as an appendix to this text. We apologize for any whom we have missed.

1 Goals, concepts and definitions

> "Every one [sic] has heard that when an American forest is cut down, a very different vegetation springs up; but it has been observed that ancient Indian ruins in the Southern United States, which must formerly have been cleared of trees, now display the same beautiful diversity and proportions of kinds as in the surrounding virgin forests. What a struggle must have gone on during the long centuries between the several kinds of trees, each annually scattering its seeds by the thousand; what war between insect and insect – between insects, snails and other animals with birds and beasts of prey – all striving to increase, all feeding on each other, or on the trees, their seeds and seedlings, or on the other plants which first clothed the ground and thus checked the growth of the trees."
>
> Charles Robert Darwin, *The Origin of Species by Means of Natural Selection* (1859)

As has often been noted, you can go back to the writings of Darwin and see most of modern biology contained within his keen observations. Succession is clearly no different. Within his text we see a pluralism of mechanisms; invoking the action of competition, consumers and seed dispersal. We also see an early example of one of the major conflicts in the development of succession – the idea that communities will repeat themselves in time, returning to some pre-disturbance state. Of course Darwin is not particularly well known for his work in succession, but rather his views on population ecology. If we look back over the history of succession, we can generate a "Who's Who" list of ecological thinkers from the last century – Braun, Connell, Cowles, Clements, Gleason, Grime, Keever, Odum, Oosting and Whittaker among many, many others who have all contributed to our understanding of succession.

Based on this long history of work, shouldn't we have a pretty good handle on everything by now? Perhaps we should, but in preparing this text we have learned that there is much still left to be done. Much of what is considered conventional wisdom regarding succession isn't as clear cut as we would like. Ecologists have been very good at examining community dynamics from a one-driver approach and have built a strong body of theory that supports the role of those drivers. Where we have failed is in embracing the multiplicity of simultaneous drivers and contingencies that constrain both community dynamics and the importance of individual drivers. Perhaps we have also been too quick to dismiss succession as an important phenomenon of study just because we have been working on it for well over a century. Researchers have been examining competition for nearly as long and certainly no one would regard that avenue as passé or having outlived its conceptual usefulness (Trinder *et al.*, 2013). In fact, several authors have specifically touted the relevance of succession to modern ecology and to our contemporary environmental problems (Davis *et al.*, 2005; Walker *et al.*, 2007; Cramer *et al.*, 2008; Prach and Walker, 2011). We certainly believe that succession has retained its importance in ecology, not just for the reasons cited in those critiques, but because it offers real insight into the dynamics and structure of all plant communities. You may not care specifically about successional communities, but the lessons learned from them may be applied to any ecological system. It is with this overarching vision that we have undertaken this book.

In this chapter, we will not only outline our motivations and define our terms; we will attempt to be very clear about what we are trying to do and what we are not trying to do. We will also build a context for succession and the data that we will be presenting. This is an important first step, because if there are any misunderstandings in how we are using terms or what our perspective is, our message may be lost. This is the "getting to know you" chapter of our book. As with any conversation, the first moments are not always the most exciting, but they lay the groundwork for all that follows.

The goals of this book

Our goal in writing this book is simple – to develop a greater understanding of plant community dynamics. We came at this challenge from having worked to further our conceptual understanding of community organization and from conducting empirical studies on community dynamics. The resulting book is part monograph and part conceptual treatise. This structure was not based on our inability to agree on how the book should be developed, but rather from a strong conviction that the two approaches are necessary to form a full understanding of plant communities. Empirical studies are wonderful in that they isolate drivers of community structure and dynamics, and lead to clear and compelling, publishable papers. However, at the end of the day/career, it can be difficult to see exactly how the individual pieces that result from this work all relate to each other. On the other hand, conceptual approaches to ecology can stimulate ideas that may generate new experimental studies or research linkages that were previously unexplored. The drawback to this approach is that the broader the framework, the less specific it necessarily becomes. At broad levels it can be difficult to see how a particular conceptual organization applies to an individual system and so we often default back to more specific theories. Our approach has been to develop both the conceptual and empirical aspects of succession simultaneously. We will present a conceptual framework that organizes our thoughts on community dynamics and place it into the historical context of other successional ideas. We will then use the conceptual framework to guide our analysis and discussion of a long-term study of successional dynamics. We do this to specifically test the utility of our conceptual approach and to ensure that the data that are presented (and there are a lot) build a broader, integrated understanding of the system and of community dynamics in general.

The major sub-theme of this book is integration, the opposite of reductionism. Reductionism excels at finding *an* answer and forms the foundational basis for much of what is presented here. However, it can be difficult to gain a broader understanding of ecological phenomena if all empirical studies do not find the same pattern/effect. Integration, in contrast, embraces contingencies by building a larger context in which a study may be understood. The lack of a conceptual framework and an appreciation of contingencies may lead to arguments and the development of simplistic dichotomies in science. Examples of such dichotomies include: Is diversity related to ecosystem function?; Does disturbance lead to invasion by non-native species?; Are communities individualistic? (as an aside, if your research question can be answered with yes or no, then you are way down on the reductionist scale in ecology). All of these questions have led to arguments in the literature and can suffer from the lack of an appropriate context. By moving from simple yes/no questions to understanding when a particular pattern/effect can be expected, one moves into the ecological realm of integration and contingencies. The more complex the system, the greater the number of contingencies, the greater the need will be for a conceptual framework to organize all drivers into a logical structure.

One analogy that can help to illustrate the utility of a conceptual framework is that of a jigsaw puzzle. When working a complex puzzle, people generally begin by sorting the pieces using some sort of rubric. Often, edge pieces are a key first step in that they delimit the boundaries of the system (puzzle). Pieces may then be sorted based on color or distinctive aspects of the image. This will help the researcher (puzzle worker) to assemble the puzzle more efficiently. Of course, this all assumes that you looked at the box top and know what the resulting image should be. In ecology, we rarely, if ever, know how the system works, which is why we are studying it in the first place. If you did not know what the target image was, the sorting structure (conceptual framework) may become even more important to assembling the puzzle. The pieces of the puzzle represent individual reductionist aspects of the system – competition between species A and B, availability of resource X, herbivore damage by a generalist, etc. It is difficult to know which piece(s) will be critical in forming an understanding of the image. Likewise, it is difficult to know a priori which individual interaction or driver will be more important than others in forming that understanding. The framework provides a way to organize the pieces so that you know how they relate to each other. If you have a very simple puzzle, organizing the pieces may not be necessary, though the puzzle would not be very intellectually stimulating. If you are dealing with a complex system, the contingencies (linkages among pieces) that generate the final image are multiplicatively greater and organization becomes crucial. Of course, in ecology, we rarely have all, or perhaps most, of the puzzle pieces. Our view of the final image is constantly changing as we find new ecological pieces on the conceptual carpet.

As integration is our goal, we have tried to form a more complete view of our study system. We of course acknowledge that there are many other ways to explore our data. While we have tried to be exhaustive in our conceptual development, we have not tried to be exhaustive in our literature citations. Our goal for each topic has been to represent the general ideas and move on. The literature that we have selected to include is primarily from successional systems, especially when it concerns succession in abandoned agricultural land – our model system. Whenever possible, we have focused on studies based in the same site as the data presented here. Much of this work has been done by our colleagues and students over the years. The remaining papers cited represent classics, personal favorites and other selections that have caught our eye for various reasons. We apologize if we offend in any papers omitted.

Why is succession relevant today?

Succession as a concept is foundational to the development of ecology (McIntosh, 1985). To a certain extent, the history of succession *is* the history of ecology. Setting that aside, there are many reasons why succession as a concept and successional systems as models are still of importance. As the economics and logistics of agriculture have changed, the rate of land being retired from agriculture has increased (Foster, 1993; Foster *et al.*, 2004; Flinn *et al.*, 2005; Hatna and Bakker, 2011). Historically, this has meant the abandonment of less financially lucrative lands, but now even fertile lands may be retired from agriculture. Much of this has likely been driven by the transformation of agriculture from small, family farms to a larger industrial mode of agriculture. Lands too steep for row crops would have once supported a family's livestock; but as animal production has shifted to feedlots and crop production to larger equipment, these areas have been allowed to undergo succession. Similarly, societal constraints on logging have pushed wood production to private landholdings leading to extended periods of forest regeneration in some areas. In tropical areas, shifting agricultural practices and human populations may at least temporarily generate secondary forests (Lambin *et al.*, 2003). These socioeconomic changes mean that there is, and will continue to be, much more land experiencing succession. Understanding successional processes means that we will understand the forces regulating a significant portion of our modern landscape. This also means that the opportunity to remediate and restore these communities through manipulating succession is great (Clements, 1935; Luken, 1990; Cramer and Hobbs, 2007; Walker *et al.*, 2007; Prach and Walker, 2011).

A completely different value of successional communities comes from their importance to theoretical studies. Successional systems, particularly old fields, have become model systems for testing theories of biodiversity, competition, ecosystem regulation and many other contemporary ideas (Tilman, 1985; Carson and Pickett, 1990; Wedin and Tilman, 1993; Clay and Holah, 1999; Stevens and Carson, 1999; Symstad, 2000; Klironomos, 2002; Schmitz, 2010). Aside from the ubiquity and lack of conservation status of successional areas, these systems are particularly amenable to manipulation because they are dynamic over relatively short time scales. We may be more inherently interested in other ecosystems, but we can get results within a season or two in a successional system. The hope is that by understanding the dynamics and structuring forces of successional systems, we can apply these lessons to systems that are of much greater conservation concern and that experience dynamics over a much greater time scale. Finally, and we must be honest here, no one is concerned with protecting old fields. From an experimental view, this means that these systems can be manipulated in very severe ways with no societal consequences. Take, for example, the usage of old fields as model systems to understand how fragmentation alters the persistence and movement of small mammals (Robinson et al., 1992; Barrett et al., 1995). Doing similar things in forested landscapes, and with larger animals, would be much more difficult (e.g. Brookshire and Shifley, 1997; Laurance et al. 1998) and would require a massive, coordinated research effort, instead of a mower.

Interestingly, successional systems are somewhat of an oddity in nature. One can argue that logging mimics natural disturbance regimes, though the scale, periodicity and severity may be much different. Old fields, however, are very anomalous. What is the historical analog of abandoned agricultural land? Old fields on one hand could be envisioned as similar to very large forest gaps. A critical difference, however, is the long history of soil disturbances and the legacies that are generated by those disturbances which would be absent in gaps (Foster, 1993; Walker et al., 2010; Kuhman et al., 2011). If we look at the native species that dominate the herbaceous stages of succession, we find disturbance-adapted species that were once relegated to large forest gaps or persistent forest openings such as glades (Marks, 1983). As succession proceeds, herbaceous species are replaced by opportunistic woody species that would also likely colonize large gaps, persisting until canopy closure limited their recruitment. On top of this odd assemblage of native species is a large suite of non-native species that are typically dominant early in succession (Inouye et al., 1987; Bastl et al., 1997; Meiners et al., 2002a). These species were largely introduced intentionally and accidentally through agricultural practices, though some were introduced for ornamental purposes. What results is an odd assemblage of native and non-native species that have come together to form a community that covers a large portion of the planet's surface under relatively novel physical conditions.

Finally, we must address whether there is really anything new to learn in succession. After all, ecologists have been studying succession for over 100 years – have they been lazy? The answer is of course, yes there is more to learn and no, ecologists have not been lazy. The view of understanding succession as largely a completed task comes from confusing description with understanding. We have described successional dynamics from many, many areas and know many of the general transitions that occur over time. In addition to this work, decades of reductionist ecology have provided us with a long list of potential regulators of community dynamics that may or may not be important in any individual system. What we are still lacking is a full appreciation of the contingencies that constrain succession and the interactions that may lead to dramatic shifts in the trajectory of individual species or the entire community.

As an example of the importance of description vs. understanding, let us examine gene expression. Cell biologists know that there are regulators of gene expression that either encourage transcription or inhibit it (promoters, transcription factors, methylation etc.). These are the basic drivers of gene expression. Mechanistically, the basic drivers are known, and introductory textbooks can outline a simple process of control. However, there may be multiple genes that are involved (epigenetics) and so contingencies may occur based on the gene's context in a cell. The net effect of all controllers is an up- or down-regulation of a gene product. Of course reality is even more complicated in that another gene may be turned on that represents the same cellular function. The one-gene–one-product rule has, as well, been thrown out, as there can be manipulation after transcription. All this is to say that while we know the parts that may be involved, no one would (or should) walk up to a cell biologist and claim that we understand gene expression. There are quite simply too many contingencies at the cellular level to make broad sweeping statements. Similarly, we would argue that succession, or more broadly community dynamics, is still a fertile area for research.

Finally, successional biology contains many aspects of population and community ecology. For this reason it is a fabulous context to integrate often disparate fields of study. Succession touches on issues of allocation, tradeoffs and reproduction at the population scale; plant strategies, diversity and competition at the community scale; and nutrient cycling, productivity and consumer interactions at the ecosystem scale. One of us (STAP) has even announced at a symposium on ecological theory that "nothing in community ecology makes sense except in the light of succession" mirroring the famous comment by Dobzhansky on evolution. While this drew the hoped-for laugh from the audience, much can be made of the ability of succession to integrate ecology.

Definitions and controversies

For such a foundational concept of ecology, succession has more than its share of controversies. To avoid these, we need to define our terms and place them into an appropriate historical context. This would seem a simple task, but as you will see, it can be tricky.

General ecology texts often contain simple definitions such as succession is the directional and predictable change in species composition in response to disturbance. This definition is perhaps suitable for the pre-med student who will never think too deeply on the subject, but it rapidly fails for those who pursue the ecological sciences. This is where the issues begin. We could add "more or less" and "often" in a few places, but that really just muddies the definitional waters. Interestingly, the restrictive conceptualization of succession in textbooks is a modern construct. Many early ecologists treated all vegetation change as succession (e.g. Warming, Watt etc.). This definition would certainly simplify things conceptually, but would be unsatisfactory for a great many people who view succession as a sort of special case. As a compromise, we pose that succession is a special type of vegetation change, bounded by certain constraints. We will define these constraints shortly, but it is important to realize that the general mechanisms and processes involved in successional and non-successional vegetation changes are the same. It is in this vein that succession can be used as a unifying concept in ecology (Davis *et al.*, 2005).

Succession and disturbance are often conceptually linked, so we will start with defining these terms:

Succession – The temporal change in species composition or three-dimensional structure of plant cover in an area (Pickett and Cadenasso, 2005). This definition does not imply directionality, predictability or an endpoint, nor is it restrictive to only following a disturbance. Successional systems can be divided into primary and secondary, based on the lack or presence of established soil. Though the underlying processes should be the same in primary and secondary succession, relative importance of individual drivers will likely be quite different.

Disturbance – An event that alters the structure of vegetation or the substrate the vegetation is growing on, often resulting in a change in resource availability (Pickett and White, 1985; White and Jentsch, 2001). Missing from this definition is any mention of the time scale of a disturbance, though it does suggest a potential mode of effect via alteration of resources. This definition differs from that of Grime (2001) who defined disturbance solely as the destruction or removal of biomass.

These definitions seek to avoid the challenges posed by other conceptualizations of succession, largely espoused in the views of Clements (1916) as they advocated a clear starting point, directionality, predictable trajectories analogous to the development of an organism and, finally, a clear and stable endpoint (Pickett and Cadenasso, 2005; Pickett et al., 2009). Our definitions are broad enough to encompass the range of vegetation changes traditionally incorporated under the successional umbrella – cyclical community dynamics, progressive and retrogressive change, autogenic and allogenic drivers, the filling in of depressional wetlands, retreating glaciers and volcanic activity. They also allow for the contingencies and variability in dynamics that have come to characterize succession. If a more restrictive definition of succession that includes some form of directionality and a discrete disturbance is preferred, that will limit succession to exclude cyclical community dynamics, retrogressive changes and hydroseres. Though this is not our approach, this definition can work well and simply shifts more of the range of potential successional dynamics to the more inclusive vegetation dynamics (Figure 1.1). However, the delimitation between succession in a strict sense and vegetation dynamics in the broader is purely artificial. As the system that forms the foundation of this book represents succession on abandoned agricultural land, it luckily fits even the most restrictive definitions of succession.

If, as most contemporary ecologists, we reject the idea of a climax community as it is too entrenched with the idea of stability, how do we describe the end of succession? One candidate would appear to be the concept of dynamic equilibrium. In this concept, composition and structure at a landscape scale are maintained, despite constant turnover and compositional changes occurring within individual patches. While this addresses the idea that systems are constantly changing, the notion of stability at broader geographic scales seems unlikely in an age of global climate change, nutrient deposition, biological invasion and habitat fragmentation. Also, it does little to inform what is going on within an individual successional site. We might alternatively describe the end of succession as the point at which the dynamics of the site are no longer being driven by the event that initiated the succession. For hydric successions, this would be when soil development and the drying of the site no longer influence the composition of the site. Similarly, for old field succession, this would be when vegetation dynamics shift from being driven by the disturbance of the plow to being driven by other processes. In many instances this will also represent a shift to internal processes such as gap formation and regeneration. Of course this endpoint is difficult in application as land clearing often generates even-aged stands that more or less reach maturity and senescence at the same time, leading to

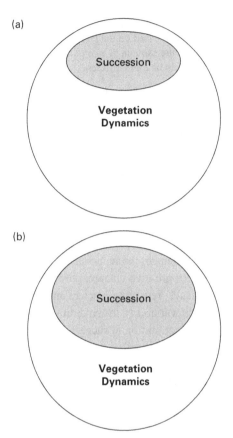

Figure 1.1 Conceptual representation of the relationship between succession and vegetation dynamics using a restrictive (A) and more broad (B) definition of succession. In all cases the dynamics represented by succession are a subset of vegetation dynamics as a whole. At the most extreme, succession and vegetation dynamics may be considered synonymous, with complete overlap. This illustrates the commonalities of mechanisms in succession and vegetation dynamics.

pulses in the mortality and regeneration of canopy trees. Furthermore, soil disturbances such as plowing may leave persistent legacies for up to centuries (Dupouey *et al.*, 2002; Flinn and Marks, 2007; Walker *et al.*, 2010). Operationally, it may be simplest to envision a gradual shift from successional to non-successional dynamics and abandon the notion of a definable endpoint because of its inherent artificiality (Figure 1.2).

The data

The core of this project is a unique data set from a series of permanent plots in abandoned agricultural land (Chapter 2). Following the initial abandonment treatments, these plots and fields have been left unmanipulated as the vegetation changes have been documented. The observational nature of the data have from time to time drawn criticism from

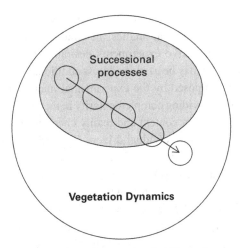

Figure 1.2 Representation of successional shifts in the processes that regulate dynamics within plant communities. Early in succession, dynamics are driven by the processes that initiated succession (e.g. disturbance). As succession proceeds, driving forces shift towards those not related to the initiation of the system, signaling the end of succession.

Figure 1.3 Diagrammatic representation of the parallel gradients of experimental to observational study and of mechanistic specificity in ecological studies. While the increased ability of manipulative studies to isolate mechanisms over observational studies is commonly noted, there is also a tradeoff in the broader applicability of such data. Highly manipulative studies provide direct and specific mechanistic tests, but sacrifice the ability to test the importance of other hypotheses, particularly the further those hypotheses lie from the original motivating question.
Observational studies lack the ability to clearly isolate mechanisms, but data from these may be applied to a wide diversity of questions with little constraint.

reviewers and (unfairly!) from granting agencies. This criticism stems from the inability of such data sets to definitively link community dynamics to specific mechanisms. We believe that while reductionistic approaches have their merit, there is also real value in data derived from observational approaches (Austin, 1981).

Our logic behind this claim follows that of Levins' (1966) critique of population models. There is an inherent tradeoff between the generality of a model to a variety of systems and its accuracy in forming predictions within individual systems. We posit that there is a similar tradeoff between the level of manipulation and control imposed on a research system and the ability of a study's data to address questions beyond the original scope (Figure 1.3). The more controlled the experiment, the greater the ability to isolate the individual mechanisms that operate. However, this mechanistic specificity comes at a cost – the more limiting the controls on the system, the less utility the resulting data have to address other questions. Observational studies, in contrast, lack the strong mechanistic linkage that makes manipulative experiments desirable. However, there are also few limitations on how that data may be used to address additional hypotheses as there are no constraints on the data imposed by the experimental manipulation. Both approaches are important in ecology, providing complementarity in the results achieved. Observational studies provide an important context and reality check for manipulative studies that may overemphasize the role of individual drivers. Similarly, observational studies may suggest avenues for reductionist experimentation and may further the development of ecological theory (Austin, 1981).

For this book we exploit observational data following experimental variation in abandonment conditions. The original goal of the project was to focus on successional dynamics. We are not constrained by the original framing goal, allowing us to explore issues of assembly rules, biological invasions, heterogeneity and functional ecology. The long-term data pre-date these ideas in ecology, but can still be successfully used to explore them.

The organization of this book

This book is organized into four main sections, each building off the previous in a (hopefully) logical and clear manner. Each section has specific goals that constrain the material and approach within each. We have strived to make the progression of themes logical so that the reader will have all the pertinent information by the time a particular idea arises. This approach will be useful for someone reading the text in its entirety, but means that someone reading an individual chapter may need to skim back through previous sections to fully understand what is being presented. There would quite simply be too much repetition to fully develop each chapter as a stand-alone contribution. Indeed, had this been possible, there wouldn't really be a need to write an entire book. Our goals for each section are as follows:

Context – In this section we will provide the background for both the data that we will be presenting and the conceptual premise of the book. We will start with a history of the data and the site that forms the basis for this book. We will also provide a somewhat abbreviated history of successional thought, particularly focusing on linking ideas to show conceptual development and key advances over the years. From there we will build the conceptual framework that forms the guiding principle for this text – that although community dynamics are complex, there is a strong value in simply organizing drivers of community change hierarchically. Finally, we will briefly review the primary drivers of community dynamics and place them into our hierarchical conceptual framework.

Pattern – These chapters primarily deal with the long-term data and the dynamics seen within our system. They focus largely on ecological pattern rather than on developing a mechanistic understanding of temporal and spatial dynamics in plant communities. The specific purpose of these chapters is to set the stage for the hierarchical and mechanistic approach in the chapters that follow. This is not done to reduce the importance of these ideas, but rather to more quickly and efficiently move us on to the areas that we would like to address in more depth. This section will also address some of the basic successional hypotheses and observations that have been proposed over the long research history of succession. We will start with a treatment of the community and population dynamics of succession – the two ecological scales at which most of the descriptive work in succession has been conducted. These chapters will be followed with a discussion of the role of continued disturbance in succession, particularly on the influence of rainfall, and the lack thereof, on community dynamics. The section will conclude with an exploration of the many ways that diversity can be related to succession.

Integrated themes – In this, perhaps the most important section of the book, we will specifically link the long-term data with our conceptual framework. The goal here is to both explore the utility of the framework and generate a mechanistic understanding of community dynamics in the system. While we have attempted to be more comprehensive in the pattern portion of the text, we have been more selective in the themes explored in detail here. The choices made are partially a natural extension of our individual research interests and partially an effect of the specific nature of the site and data. This section will start by addressing community convergence and ecological assembly rules as these have been of interest for a long time period. We will then address the ecology of non-native species, as successional systems are characteristically heavily invaded by a diversity of species. From there, our approach will become somewhat more abstract as we address heterogeneity as a cause and consequence of succession. Finally we will place functional ecology into our conceptual framework and highlight the perspectives that approach can bring to plant community dynamics.

Synthesis – In the final section, we will take a step back and evaluate the broader implications of what has been developed throughout this book. We will highlight the lessons learned, the perspectives gained and the surprises uncovered. The first chapter in this section will specifically focus on the relationship between succession and restoration and how our conceptual framework can be used to more fully merge the two. Finally, we will revisit our call for integration in ecology and the utility of our conceptual framework as a mechanism for that integration.

Part 1

The conceptual background and development of succession

The chimeric nature of this project – part monograph, part conceptual framework – necessitates that the foundational book chapters form the basis for both. In this section we will provide the history and ecological context for the Buell–Small Succession Study (BSS). This goal is relatively straightforward to achieve and will include the motivation for initiating the study and the environmental conditions of the site. We will also present the methodology of the BSS to provide a context for the data presented in the following sections.

Providing the background for the conceptual framework is a much more daunting task. Much has been written about succession over the last century – too much, some would argue. Our goal of providing an integrative view of succession requires that we address much of this history. While there are many ways that this task could be divided, we will first address the historical development of successional thought. Though early successional work was primarily descriptive, it was often accompanied by a relatively mature conceptualization of the complexities of the processes involved. We will trace the development of successional thought from these early studies through the more mechanistic, but often unduly reductionist, approaches that have characterized more modern work. While modern successional work is often grounded in theory, the theory frequently lacks the broader view necessary to place the mechanism in an appropriate ecological context.

We will use the diversity of successional approaches to argue for the necessity of a conceptual framework to guide research in dynamic communities. It is only in developing such an approach that true integration is possible. We will present the conceptual framework that is utilized throughout this text and discuss the importance of such frameworks to the development of a mature and complete view of nature. Only after the framework is outlined will we review a sampling of the myriad successional mechanisms that have been presented over the years. We have done this to specifically place focus on a broader view and not the individual drivers that characterize most successional work. In doing so, we present our guiding structure as generally applicable to dynamic communities, and not bound to the importance of one or more successional mechanisms and empirical theories.

2 History and context of the Buell–Small Succession Study

> The saving of this forest and the opportunities it creates for research, now and in the future, will almost certainly be applauded even more in the future than it is today. Those who have made it possible have the appreciation of ecologists everywhere, plus a reasonably high assurance that what is learned here will have significance for all who are involved in managing biological phenomena.
>
> Henry John Oosting, Plant ecology and natural areas, presented at the dedication of the Hutcheson Memorial Forest (1957)

The Buell–Small Succession Study (BSS) has a remarkable history. It is one of the longest continual studies of old-field succession in the world and it has been managed over its 55 year span by only three people! The BSS tracks the change in plant communities undergoing successional transitions from released agricultural fields to now young forests. The initial theoretical motivation of the study was to test competing ideas about the patterns and mechanisms of succession. In 1958 the study was begun by establishing an array of permanent plots in old fields that would be sampled yearly for plant composition and abundance by Rutgers' graduate students, faculty, and alumni for years to come.

No one anticipated the duration of this study at the onset, but the sampling design and commitment to consistency has allowed the data to become invaluable in addressing a myriad contemporary theoretical and conceptual questions about vegetation dynamics. The project remains active and productive, contributing to the primary literature and to the training of graduate students in plant ecology. While the old fields have been undergoing a gradual change, the theoretical concerns of the field of ecology have dramatically shifted and the landscape surrounding the study has experienced increasing pressure from suburban development. Through the changes, the data collected and the understanding gained about how systems change and function remains highly relevant. In this chapter, we provide a brief history of the BSS study and its founders and outline the controversies within plant ecology at the time that motivated the establishment of the BSS. We will characterize the site of the BSS, the changes in the landscape surrounding the site, and describe the study design and sampling protocols. The BSS is a living study, and the goal of this chapter is to set a firm foundation for the data and discussion to come in the chapters following.

History of the Hutcheson Memorial Forest Center, location of the BSS

The Buell–Small Succession Study is located at the Hutcheson Memorial Forest Center (HMFC) in Somerset County, New Jersey, USA (40°30' N, 74°34') (Figure 2.1). The HMFC is composed of a 26-hectare old-growth forest, formerly known as Mettler's Woods, and 30 hectares of surrounding fields. The Mettlers acquired the land from the local Lanape tribe in 1701. They maintained the old-growth forest and continuously farmed the surrounding fields until a descendant, Thomas Mettler, considered selling the land in the 1950s.

When Thomas Mettler wanted to sell the property, the Citizens' Scientific and Historical Committee for the Preservation of Mettler's Woods began raising funds to purchase it. In 1951, a donation from the United Brotherhood of Carpenters and Joiners of America enabled the forest and adjoining agricultural land to be purchased in 1955,

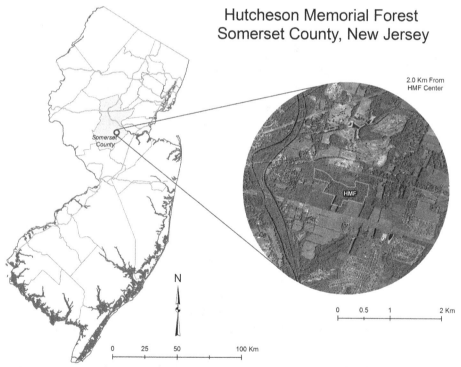

Figure 2.1 Map of New Jersey including counties, major highways and the coastline. The Hutcheson Memorial Forest Center is in Somerset County in central New Jersey. The 2 km radius area around the HMFC is expanded and an air photo from 2010 is the base layer with the forest outlined in white. Sources: (1) counties published by NJ Office of Information Technology, Office of Geographic Information Systems 2012; (2) highways published by US Department of Commerce, US Census Bureau, Geography Division 2013; (3) coastline published by the NJ Department of Environmental Protection 2009; (4) 2010 air photo from the National Agricultural Imagery Program, US Geological Survey.

saving it from development. The United Brotherhood provided the donation as a memorial to honor William L. Hutcheson, their president from 1915 to 1952. Hutcheson was born in 1874 in Michigan and began his career by cutting and framing timber for ships. In 1902 he helped to organize a local union. As president of the United Brotherhood, he built the organization by dramatically increasing its membership, improving pay and hours, initiating a pension program, and establishing a home for aged members (Stevenson, 1957). Hutcheson's leadership in labor issues was recognized in his public service under three US presidents – Wilson, Roosevelt and Truman. The General Executive Board of the United Brotherhood thought the old Woods a fitting memorial for their former leader and Mettler's Woods was renamed Hutcheson Memorial Forest (HMF).

On October 15, 1955, Hutcheson's son, Maurice A. Hutcheson, who succeeded him as president of the United Brotherhood, presented trusteeship of HMF to Rutgers, the State University of New Jersey and The Nature Conservancy. In his comments at the dedication ceremony in 1957, Hutcheson recognized the important role of scientific research in improving and maintaining forests which would affect future generations of United Brotherhood members and the land was dedicated to the preservation of the old-growth woods, and to research and education (Hutcheson, 1957). When the land was deeded to Rutgers, farming in the fields surrounding the forest ceased. Though the land was to be used for ecological research, the deed stipulated that the forest could not be disturbed or manipulated in any way and if it was, the property would be turned over to The Nature Conservancy for management.

At the time of the dedication, the forest was described as consisting of a tall tree canopy of mixed oak species with a continuous dogwood layer below. The structure and composition of the mature oak forest was a result of past climate and human influence, and the goal of preventing disturbance in the forest was to remove the human influence, allowing the physical environment alone to influence the future forest composition and structure (Buell, 1957). It was thought that no tree had been cut and the ground never tilled in the forest, making HMF one of the last uncut forests of this type in the Mid-Atlantic States if not the United States. It is apparently the only uncut upland forest in New Jersey and was designated a Natural Landmark by the National Park Service in 1976. This doesn't mean, however, that the forest was immune to disturbance. In November 1950 a storm-related blow down occurred and some salvage logging was done to remove the broken and downed stems (Buell et al., 1954). This disturbance perhaps played a role in Mettler's decision to sell the land.

The story presented in this volume is the story of the surrounding fields that have generated more than 50 years of plant succession research as part of the BSS, not the HMF per se. The fields abutting HMF were slated by Professor Murray Buell of Rutgers University, the first director of the Center at HMF, to be used for research on succession and other ecological, botanical and zoological phenomena. The fields and forest are linked because the released fields were intended to serve as a buffer protecting the ecologically sensitive old-growth forest from the influences of the surrounding landscape. Therefore, eight of the ten fields of the BSS are adjacent to the old-growth forest (Figure 2.2). Other fields on the original property were to be used for demonstration or

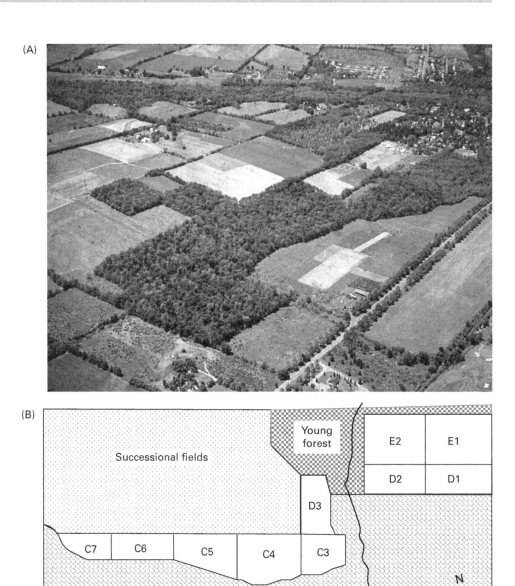

Figure 2.2 Panel A: Air photo of the Hutcheson Memorial Forest and surrounding lands. This photo was taken by Jonathan D. Moulding, Ph.D. in 1973. Panel B: Map of the BSS fields and the Hutcheson Memorial Forest. Study fields indicated by an alpha-numeric code and located along the northern perimeter of the old growth.

manipulative research, but experimental manipulations were not to be permitted in the old woods, and only sparingly in the BSS fields.

Landscape context of the BSS

The BSS is located in the Piedmont physiographic province of New Jersey which is about 4143 km^2 and occupies approximately one-fifth of the state. It is located between the Highlands and Coastal Plain provinces and the elevation of the Piedmont ranges from 91–122 m above sea level (New Jersey Department of Environmental Protection – New Jersey Geological Survey 2006). This central region of New Jersey was intensively used for agriculture, but the farms were characterized by high erodability and droughtiness. Many farms and fields in the Piedmont were abandoned in the 1940s for several reasons, including the opening of more easily farmed soils of the Midwest, changing economic and social situations for agriculture in the east, and pressure from spreading urbanization (Foster, 1993; Foster et al., 2004).

The soils at HMF belong to the Penn soil series, derived from the Triassic red shale of the Brunswick Formation (Ugolini, 1964). There are only slight variations in soil texture, drainage, and depth between the fields used in the BSS (Ugolini, 1964). Robertson and Vitousek (1981) compared three fields of various ages representing different life-form dominance – annual, perennial and shrub. They found no significant differences among the fields in soil structure, soil texture, pH, percentage organic C, concentrations of NH_4-N or percentage total N. They did, however, find significant differences in concentrations of NO_3-N, Ca, K, Mg, and acid-soluble PO_4-P. These differences are difficult to interpret as no consistent trends existed (Roberston and Vitousek, 1981) and the specific field used in the analysis was not reported. Consequently, it is assumed that initial soil differences were small between the old fields in the BSS.

The Piedmont region is in the temperate zone and precipitation is evenly distributed throughout the year. The mean annual precipitation is 116.1 cm and since the BSS began, year-to-year variation has ranged from less than 70 cm in 1965 to 167 cm in 1975 (Figure 2.3). As is typical of the region, HMF has warm temperatures in the summer and cool temperatures in the winter, and the mean monthly temperatures range from –1.6 °C in January to 22.4 °C in August (New Jersey State Climatologist; National Climate Data Center); year-to-year variation in temperature is minimal (Figure 2.3).

At the time the BSS was initiated, the landscape immediately adjacent to the HMFC was primarily agricultural fields with minimal development. Individual or small clusters of houses were scattered within a kilometer radius of HMF and the town of East Millstone is located between 1 and 1.5 km west of HMF. There was little change in the landscape surrounding HMF between 1930, our earliest images, and 1956. Over these 26 years, some development occurred to the northeast of the HMF and more than 18 ha of land was developed within 1 km of HMF, and 38.65 ha within 1.5 km (Figure 2.4). During the 55-year run of the BSS, additional changes have occurred in the surrounding landscape and these changes will be described in more detail below.

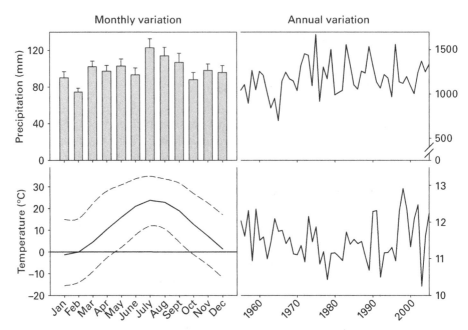

Figure 2.3 Mean monthly and year-to-year variation in precipitation at HMF, and mean monthly temperature +/− max and min and year-to-year variation in temperature (note y-axis values). All figure panels include the years 1955–2005 and the average precipitation and temperature is calculated over the 30-year span.

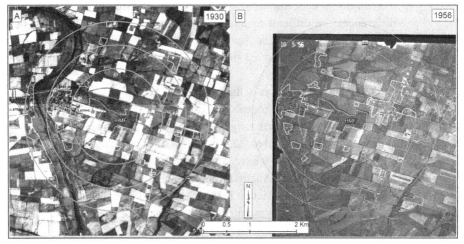

Figure 2.4 Air photo of the HMFC and surrounding landscape from 1930 and 1956. Concentric zones of 1, 1.5 and 2 km radii are shown calculated from the centroid of the old-growth forest. The old-growth forest is in the center and patches of development are outlined in white. (Source: 1930 photo downloaded from the New Jersey Geographic Information Network and published in 2009 by the NJ Office of Information Technology, Office of Geographic Information Systems. 1950 photo from the USDA FSA Aerial Photography Field Office and the digital scanning and processing from Grant F. Walton Center for Remote Sensing and Spatial Analysis (CRSSA), Rutgers University. Buffers and land-cover polygons added by the Cadenasso lab for this chapter.)

Motivation for establishing the BSS

Understanding change in plant community composition through time and space has been a central concern in ecology since the dawn of the science. One of the first studies of succession was conducted by Cowles (1899), who investigated the change in plant communities along the dunes of Lake Michigan. Dunes further inland were presumed to be older, and therefore the plant community on these dunes would be of a later successional state than plant communities on dunes closer to the water. The study of succession quickly spurred theoretical controversy in the nascent field.

Two ecologists, Frederic Clements (1916) and Henry Gleason (1917), proposed different views of plant succession and the nature of the plant community. Clements suggested that successional change in plant communities was orderly and direct such that plants present at one time facilitated and eventually gave way to plants that dominated later stages. He compared the succession of different plant communities in a particular place to the development of organisms and, similar to organisms, proposed that plant communities had specific stages. The ultimate stage for a plant community was referred to as the climax state and this state was the plant community predicted to exist in an area of a given set of environmental conditions. Clements did recognize that this climax state may never be achieved if the plant community experienced disturbances, such as fire, that prevented its progression through the various stages. In contrast to this view of community development, Gleason suggested that plant communities consist of a collection of species all individually responding to environmental conditions and the constraints of their own physiology, as well as interacting with each other. Therefore, Gleason proposed that plant community composition is a result of physiological constraints imposed on individual species at a particular location, and that species' response.

These two contrasting views of plant succession raged for years in the field of ecology. With time, nuanced understandings and mechanisms of plant interaction were proposed to refine the debates. One particular conceptual advancement proposed by Egler in 1954 provided a crucial motivation for the initiation of the BSS. Egler put forth the initial floristic composition hypothesis, which suggested that all species present in the plant community during succession are present from the initiation of the successional processes following the initial disturbance. As the succession proceeds, different species dominate the community, based on their individual life-history traits. Therefore, faster-growing species will dominate immediately following the disturbance, and as time passes and the community changes the slower-growing species, which have been present since the beginning, start to dominate the community as the faster-growing and short-lived species die out. This hypothesis was in opposition to the prevailing assumption that species arrived in succession in order of their dominance.

Egler's hypothesis seemed unreasonable to the founders of the BSS based on their experience. The only sure way to tell, however, was to sample specific permanent plots through time and determine whether communities came and went as wholes, or whether populations rose and fell through time. Egler had a reputation for harshly critiquing the work of other ecologists (see "A commentary on American plant ecology," 1951), so it is perhaps not surprising that his hypothesis was viewed as a challenge and, as such, served to motivate the study. The Buells were in graduate school together with Egler in Minnesota and, despite their spirited rivalry, maintained friendly correspondence with him for years (http://www.atonforest.org/Four2.htm). An additional motivation for the BSS was provided by a study conducted by Bard in 1952 characterizing plant communities in fields near HMF that had been released from agriculture. Bard used a space-for-time substitution method where she considered the old woods of HMF to be the climax state and found fields within 5 miles of HMF that were at various stages of succession; she sampled 27 fields at 9 different stages.

In the early days of ecology, the only method available to discover the patterns of community change through time was to compare sites of different ages since disturbance or abandonment. This method, called either space-for-time substitution or chronosequence, assumes that the different sites are subject to the same conditions and have the same species available to them. If these crucial assumptions are not met, the patterns may reflect permanent differences between the sites or other ecological processes rather than successional change (Pickett, 1989). This is the approach used by Cowles (1899) in one of the first studies of succession and by Bard (1952) in the fields near HMF. Therefore, establishment of the BSS was motivated by: (1) testing Egler's initial floristic composition hypothesis and (2) testing the assumptions embedded in the space-for-time substitution approach to studying plant succession. These motivations required that the same plots be sampled through time, and changes to plant community composition be directly quantified.

People of the BSS

The Buell–Small Succession Study (BSS) is named for its originators (Figure 2.5A). Murray Fife Buell was Professor of Botany at Rutgers, The State University of New Jersey. He earned his Doctor of Philosophy degree under the great plant ecologist, William S. Cooper, at the University of Minnesota. Helen Foote Buell earned her Ph.D. in phycology at the University of Minnesota. Although Dr. Helen Buell was not a member of the Rutgers faculty, she was an important member of the intellectual community in botany and ecology, and contributed significantly to the training of students and to research. Dr. John Alvin Small earned his Ph.D. at Rutgers University studying disease in tomatoes. He remained at Rutgers teaching botany and ecology and played a major role in the development of the Ecology Graduate Program at Rutgers that was started in 1946 with Murray Buell joining the faculty (Gunckel, 1978).

Murray Buell and John Small both worked on the BSS until their deaths in 1975 and 1977, respectively. Helen Buell worked on the project until the mid-1980s, but enthusiastically continued to share her energy and knowledge until she died in 1995. Dr. Steward T. A. Pickett, who joined the faculty of Rutgers in 1977, began to work on the project in the summer of 1978. Pickett was joined by plant ecologists Drs. Scott J. Meiners and Mary L. Cadenasso, and a community ecologist Dr. Peter J. Morin

(A)

(B)

Figure 2.5 Panel A: Photo of the BSS originators Murray Buell, Helen Buell and John Small (l to r) taken in December 1963. Photo courtesy of Norma Reiners. Panel B: Photo of the BSS leaders today, Mary Cadenasso, Steward Pickett, Scott Meiners and Peter Morin (l to r) taken in July 2007. Note the same HMF entrance arch in the background of both photos.

(Figure 2.5B). Pickett continued to lead the project until 2002 when leadership was passed to Scott Meiners, who continues to provide leadership today.

Although in the early days, the fields were sampled by the Drs. Buell alone, later joined by Dr. Small, the study soon grew too large for them to conduct by themselves. Over the years large numbers of graduate students in botany, zoology and ecology, several undergraduates and post-doctoral researchers at Rutgers, and even visiting scientists, have assisted in the sampling (Appendix 1). The continuation of the BSS would not have been possible without the expert and careful assistance of this array of dedicated scientists.

Methods: site design and sampling protocols

The Buell–Small Succession Study was initiated during the first growing season, 1958, after the dedication of the Hutcheson Memorial Forest Center (HMFC). During this first year, two fields were established with two additional fields being established every other year (1960, 1962, 1964 and 1966). The complete set of 10 old fields that make up the BSS exhibit four different initial conditions: the year of abandonment, season of abandonment, the last crop and mode of abandonment (Table 2.1). Fields were abandoned in either the spring or the fall, were last in row crop or hay, and were abandoned with intact litter or bare soil. The sampled portion of the old fields also vary in size between 0.35 and 0.87 ha (Myster and Pickett, 1988).

Because the research was motivated by testing Egler's initial floristics composition hypothesis and overcoming the limitations of space-for-time substitution approaches to studying succession, the BSS was designed to describe the pattern of succession in specific fields through time by sampling the very same plots year after year. Based on consultations with a statistician, 48 plots, measuring 2.0×0.5 m, were permanently marked in each of the 10 fields. The size and shape of each field determined the placement of the grid. These plots were rectangular to capture heterogeneity in the herbaceous and shrubby communities expected to dominate in the first decades of succession. The permanent plots were sampled every year in late July from 1958 to 1979, after which they were sampled every other year. The sampling is ongoing.

Table 2.1 Year of release and treatment at time of release for the 10 old fields used in the BSS. Treatment includes a three-letter code that represents season of abandonment (F = fall and S = spring), last crop type before abandonment (R = row crop and H = hay), and mode of abandonment (L = leaf litter and B = bare soil).

Field	Year abandoned	Treatment
C3	1958	FRL
D1	1958	FRL
D2	1960	FRL
D3	1960	SRB
E1	1962	FHL
E2	1962	SHB
C6	1964	FHL
C7	1964	SHB
C4	1966	SRB
C5	1966	FRB

The BSS fields are typically sampled by teams of two people each. Each team samples one permanent plot at a time with one person calling out species presence and cover and the other recording the data. When a team is finished with the plot, they leap-frog over the other teams to the next plot to be sampled. This process insures that all plots are sampled and that sampling teams are passing each other in the field providing opportunities for consultation with plant identification. Each permanent plot is marked by two posts on the long edge of the plot farthest from the HMF. When a team arrives at a plot to sample, a wooden sampling frame is put into place. The frame consists of four pieces – two 2 m pieces and two 0.5 m pieces. One of the 2 m pieces is placed along the two posts marking the plot and the piece is placed as close to the ground as possible, but above the leaf litter. This ensures that the frame is placed with as little disturbance to the vegetation as possible. The other three pieces of the wooden frame are then placed to create a rectangular plot. The plots are sampled by estimating the percentage cover of each species in the plot. If a species overhangs the plot, the percentage cover overhanging is recorded. The percentage cover of bare ground, litter, lichen and moss is also estimated. Samplers start with the topmost layer, usually the trees, and work their way down to the species closest to the ground. In addition to estimating the percentage cover for trees, the number of stems of each tree species rooted in the plot is also noted. This includes all sizes from seedlings to mature trees. Before the sampling teams leave the field, a check is done to verify that all plots have been sampled by calling out the plot numbers. These plot roll calls are filled with anticipation and, if all plots are called, are a cause for celebration, especially on the hottest and most humid days.

In 2000, significant tree canopy had formed over all or much of each field. Counting stems in the permanent plots and the percentage canopy cover hanging over the plot seemed an incomplete way to capture the changing vegetation structure of the fields. Therefore, all trees that rooted in or were hanging over the plot were identified to species, assessed for placement in canopy (*a la* Smith, 1986), tagged and DBH measured so that they could be tracked through time.

Changes since the BSS began: scientific and landscape context

The BSS was motivated by several debates within ecology about how plant communities changed through time. One controversy was between the different views of Clements and Gleason about the nature of plant communities. The second controversy was Egler's initial floristics composition hypothesis. These two controversies have in some ways been solved as a result of the BSS, other permanent plot studies in forests and fields, and judicious use of certain chronosequences and experiments. Consequently, contemporary succession theory incorporates aspects of the extremes of the controversies by recognizing when each of the patterns or processes occurs (Pickett *et al.*, 2011, 2013b; Chapters 4 and 14).

Despite these particular controversies being settled, the need to continue long-term, permanent plot studies of succession is required by new motivating questions that have emerged. These questions are appropriately examined by the BSS and other similar studies. Questions that now rise to the top of the list of motivations for the study include those concerning: (1) patterns of species assembly and assortment in time and space; (2) the role of functional groups in succession; (3) the place and significance of invasive exotic species in mid- and late-successional communities; (4) how species life histories and morphologies relate to their invasion and persistence and (5) the role of episodic events in succession. Many of these topics form the basis of chapters in this volume and data from the BSS is used to address these questions. Succession remains a highly relevant area of ecological research today because succession is driven by a suite of processes that occur to different degrees in all plant and animal communities. Because successional trajectories are so obviously dynamic, they provide a powerful stage for disentangling the web of interactions that characterizes communities and ecosystems.

Just as the field of ecology has changed over the intervening 55 years of the BSS, so too have the fields of the BSS. The first years of sampling required much time in the field as the plots were filled with layers upon layers of agricultural weeds. As perennial species and shrubby species became more prevalent in the plots, the fields were more difficult to navigate through. For several years *Rosa multiflora* was a dominant species and the plots became treacherous to sample and had to be located by crawling under the prickly bushes on hands and knees. As *R. multiflora* phased out of dominance, the thorn-laden canes left behind continued to pose a challenge. Additional challenges were provided by poison ivy taking on the size and structure of trees and the more recent increase of deer ticks carrying Lyme disease. Many of the fields now contain substantial tree canopy, particularly along the boundary with the old forest. Because of the lower light, much of the understory has opened up and new invasives are moving into the fields. These plant community dynamics are more thoroughly depicted in the data chapters to follow.

The landscape surrounding the HMFC and the fields of the BSS has also changed over the past 55 years. In a 1 km radius from the old growth forest, developed land increased from 25 ha in 1956 to more than 43 ha in 2010. Land use change has been more extensive farther away from the HMFC, however, and the rate of that change has increased since the late 1980s (Figures 2.6 and 2.7). The primary change has been from agricultural fields and woods to housing, but a golf course (~65 ha) was added before 1970 to the north of HMFC (Figure 2.8). Within 2 km of HMF, developed land increased from 227 ha in 1988 to 280 ha in 1995 and 365 ha in 2010. By 2010, the 2 km buffer area surrounding the HMFC had 29% of the land area developed and more than 34% of the golf course is included.

While the BSS is primarily an observational study, it is in the context of a much larger body of research. In the fields and forests of HMFC that surround the BSS fields, decades of manipulative research have focused on successional processes and other interactions, resulting in 80 MS and PhD theses and over 190 publications. The combination of the observational BSS data with the experimental work throughout HMFC make this site an extremely well-studied system that can yield theoretical and practical information on plant population and community dynamics. This research context makes the BSS particularly well positioned to generate a unified view

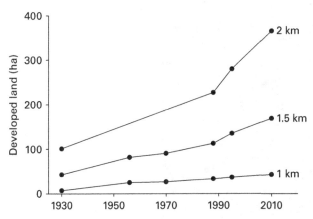

Figure 2.6 Land use change surrounding the HMFC. Air photos from the years on the x-axis were analyzed and all built areas were digitized within concentric rings around the HMFC at 1, 1.5 and 2 km radii from the centroid of the old forest. Imagery from 1956 and 1970 did not include a complete 2 km zone and, therefore, was not included in the data.

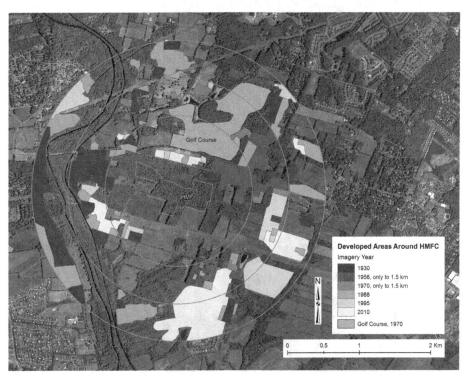

Figure 2.7 Successive change in developed land in the concentric zones around the HMFC from 1930–2010. Base image: 2010 air photo from the National Agricultural Imagery Program, US Geological Survey. All photo interpretation and digitizing in the concentric buffer zones completed in the Cadenasso lab for this chapter.

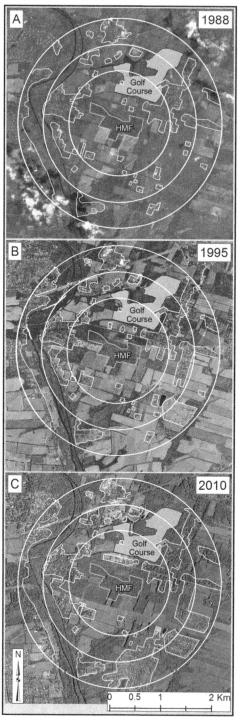

Figure 2.8 Comparison of developed land around the HMFC from 1988, 1995 and 2010. Concentric zones of 1, 1.5 and 2 km radii are shown, calculated from the centroid of the old-growth forest. The old growth forest is outlined in gray and patches of development are outlined in white. Air photos from 1988 (source: National Aeronautics and Space Administration (NASA)) and 1995 (source: US Geological Survey) were digitally scanned and processed by the Grant F. Walton Center for Remote Sensing and Spatial Analysis (CRSSA), Rutgers University. The 2010 air photo is from the National Agricultural Imagery Program, US Geological Survey. Buffers and land cover polygons added by the Cadenasso lab for this chapter.

of community dynamics that integrates processes operating at multiple ecological scales. In addition, while the initiation of the BSS preceded many contemporary theories and hypothesized mechanisms of plant community dynamics, the detail, continuity and temporal scale of the data make it an ideal opportunity to develop an integrative and synthetic view of community change.

3   Succession theory

> "Hardly has the soil been left to itself before it is invaded by wild plants. The first to settle are numerous annual and other herbs, also shrubs, which form a commonplace vegetation of weeds, whose seeds and fruits blow hither from all sides or are brought by birds. Thus arises a community which is gradually converted into weed bushland (a 'secondary formation'). But soon the forest-plants start to grow afresh; shoots sprout forth from stumps and roots, and perhaps from seeds that lay hidden in the soil: after a number of years forest once more occupies the site."
>
> Johannes Eugenius Bülow Warming. *Oecology of Plants* (1909)

The long history of succession has resulted in a long and diverse history of how researchers have viewed succession. Some ideas have been more philosophical in their application, others more predictive theories. All have attempted to describe and understand the dynamics of vegetation. The quote by Eugenius Warming that starts this chapter at first appears to be a simple descriptive statement about succession. However, mechanisms are suggested in his text: the role of seed dispersal, for example. His quotation also embodies predictions, such as the life forms that will first dominate. As one of the primary goals of this book is integration, it is first necessary to revisit the foundations of successional thought. With this goal in mind, we can then build towards a perspective that will incorporate these ideas into a single conceptual whole.

This chapter provides an overview of the theory of succession. We begin with general, phenomenological – that is, pattern based – theories, and move to more specific theories or contributing models that explore mechanisms behind the patterns. The narrative moves from simple, general theoretical perspectives to more complex and nuanced conceptual understanding of succession. Along the way, we address the classical community theories, theories based on ecosystem functioning, plant-by-plant transition models, mechanistic models that incorporate life history strategies and individual plant performance. We broaden out using the cyclic model of resilience that emerged substantially from succession theory, but which reminds us that social-ecological systems also undergo succession and are susceptible to disturbance and disruption. We end by pointing to the limitations of phenomenological approaches to succession, as well as those that embed confining assumptions into the basic definition. This sets the stage for exploration of the process approach to succession later in the book.

Phenomenological theories

Recognizing and explaining successional patterns – Succession theory was one of the first contributions of the young science of ecology as it emerged in the late nineteenth century. Although Darwin's theory of evolution by natural selection had alerted botanists, zoologists and the other biological disciplines of the day to the dynamism of species, much of mainstream biology still approached its objects of study as a static collection of things that could be placed in drawers. Plant ecology was thus mainly a biogeographical pursuit, matching collections of species to climate, substrate and elevation. These investigations provided a firm foundation for the science of ecology.

Still, dynamism was in the air. In geomorphology, the dynamic mantel was taken on by William Morris Davis, who proposed an erosional cycle for landforms. Low-relief, stable land forms with slow, meandering streams near the base level of the regional watertable, were susceptible to geologic uplift and subsequent downcutting of the streams. The newly uplifted landscapes were labelled as immature, and were characterized by swift streams in deep, V-shaped valleys. During the long periods between uplift events, landscapes gradually became rounded rather than rough as erosion extended back from the main streams. In developmental biology, dynamism was illustrated by the presumed recapitulation of long-term evolutionary trends in the ontogeny of individual embryos. The idea of life cycles and predictable progress characterized much of this dynamic thinking.

The dynamic fever took hold in ecology through the fieldwork and theorizing of such leaders as Henry Chandler Cowles (pronounced "coals") at the University of Chicago (Cowles, 1899). His outdoor laboratory comprised the dunes on the eastern shores of Lake Michigan. This landscape had been glaciated some 12 000 years previously, and was still gradually rebounding from the weight of the melted ice sheet. Consequently, new beach was being continually created, becoming available to invasion by terrestrial plants. Cowles made extensive observations in the complex mosaic of plant communities and generated a temporal model of how vegetation changed on older and older dune surfaces. This was a radical approach, because it recognized that the physical environment, to which plants were assumed by all to respond, was not static. Nor then would the plant communities be static. Bare sand was invaded by beach grass and a few vines. Cowles' model proposed further steps: Upon the dunes stabilized by the colonizing grasses, shrubs invaded. These were replaced by longer-lived, but still light-demanding trees, which ultimately gave way to long-lived, shade tolerant trees such as American beech and basswood. The soil beneath these changing communities gradually accumulated organic matter from the leaf litter and woody debris of generations of plants, which enhanced the nutrient- and water-holding capacity of the soils. More deeply rooted perennials prevented the loss of nutrients to deeper soil layers, and conserved limiting resources within the ecosystem. Clearly the plants and the physical environment were interacting. This was perhaps the second most radical idea that Cowles contributed from his observations on the Lake Michigan dunes.

Soon after Cowles published his huge monograph on plant succession in the highly respected *Botanical Review*, Frederic Edward Clements, initially skeptical, adopted succession as his principal theoretical vehicle. Indeed, by 1916, Clements had not only adopted succession, he had made it his own. In the more than 650 pages of his *Plant Succession: An Analysis of the Development of Vegetation*, he reviewed the knowledge of successional patterns from many regions, presented a theoretical statement of the expectations of succession, and erected a comprehensive classification of the ideal trajectories of vegetation change along with classes recognizing exceptions to the ideal trends. Like the usual interpretations of the Darwinian tree of life, or of Davis' landform change, or of organismal development, Clementsian succession was progressive and deterministic. It proceeded in one direction and reached an inevitable, climatically controlled composition and architecture – the climatic climax. Exceptions at relatively local scales might be found based on substrate differences or on frequent disturbances that prevented the ideal climax from emerging. Notably, Clements' theory emphasized the replacement of life-form groups, from herbs, through shrubs, through trees as the sequence of expected plant types. Furthermore, his explanations focused on the regional scales where climate zones are readily recognized and contrasted. His theory was firmly founded on the generalization that the plants in an area would tend to match the environmental resources and physical features of the environment (Clements, 1905). The "area" he emphasized tended to be relatively coarse-scaled, large enough to either summarize climatic regularities or organize large exceptions to those regularities.

Shortcomings of early succession theory – Not all ecologists were entirely satisfied with the Clementsian theoretical system, however. Within months of the publication of Clements' *magnum opus* in 1916, Henry Allen Gleason (1917) published a pointed critique. Gleason emphasized the idea that the dispersal of plants could be a predominant determinant of community composition. He further emphasized that the boundaries between communities were most often indistinct. Hence, plant species could be modeled as having characteristic and idiosyncratic distributions across environmental gradients. This was in contrast to Clements' assumption that communities were discrete and obligate units, essentially equivalent to individual organisms. Gleason emphasized the population, whereas Clements had emphasized the assemblage. Both began and ended with the match between plant and environment (Eliot, 2007). However, each focused on different scales of observation, and hence emphasized different predominant mechanisms. Clements, conceptualizing at the coarse spatial scale, assumed all species as equally available in his conceptual model. Gleason, paying attention to quite local scales, necessarily put greater emphasis on dispersal and the vicissitudes of arrival at particular sites. Gleason's finer scale of theoretical structure might have been considered a walking scale – local and gradual transitions emerged more clearly at that scale, while Clements took a synoptic view of vegetation, viewed from horseback on the Great Plains, or traveling with his wife and collaborator, Edith Schwartz Clements, in a Model T Ford (Bonta, 1991). Such a relatively elevated and powered trajectory would match a map of average weather front locations. At such a scale, boundaries would appear more distinct and abrupt.

Other aspects of Clementsian hegemony were threatened by the realities observed in various ecological systems. For example, William Skinner Cooper (1913) at the University of Minnesota observed that gaps in forest canopy were crucial to the dynamics of the plant community. In fact, old forests were mosaics of canopies of different successional stages, including recently disturbed open gaps, gaps that hosted dominants that required high levels of light and other resources, and areas of continuous old canopy, comprising dominants that were conservative resource accumulators. This kind of mosaic and dynamic canopy was not restricted to forests. Alexander Stuart Watt (1947), in Great Britain, noted that grasslands, heathlands and bogs also exhibited what would later be called patch dynamics: there was a pioneer phase, a "mature" closed-canopy phase, a degenerate phase and a gap phase in the many communities he examined. Watt's phrase "pattern and process," introduced in describing gap phase dynamics, has been widely adopted by plant ecologists for various models explaining the structure and functioning of communities and ecosystems (van der Maarel, 1996). Similar gap phase dynamics were observed early on in Swedish boreal forest by Rutger Sernander (1936, in Hytteborn and Verwijst, 2011), and in tropical moist forest as well (Aubréville, 1938). These insights were echoed by a growing number of observations in vegetation regularly or periodically affected by fires, floods and various kinds of windstorms, including hurricanes, downdrafts and tornadoes (Peterson, 2000).

While the early theories of succession may seem from a historical distance to focus mainly on the phenomenon of succession, and the pathways to be expected, early theorists were equally concerned with the causes and mechanisms of succession. Clements claimed that six events jointly describe the mechanisms of succession: (1) creation of an open site; (2) dispersal or migration of plants to the site; (3) establishment; (4) competition; (5) reaction of plants and environment and (6) stabilization. Creation of an open site, called "nudation" by Clements, involves factors that physically affect environment – disturbances like abandonment of agriculture, or fire, or windstorm – generating a substrate, or opening up or virtually destroying a previously existing plant community. The remaining five factors he considered to be actions of the plant community itself. With the exception of stabilization, which is better conceived as the *result* of the other factors, this list of causes is helpful today (MacMahon, 1981; Pickett *et al.*, 2009; Pickett *et al.*, 2011). However, its components have been expanded to include actors and interactions that were neglected by the first generation of successional theorists (Cadenasso *et al.*, 2009). For example, one reality abundantly demonstrated over the last 100 years of succession research is the role of conditions and organisms that survive during and persist through many events that begin succession. Other post-Clementsian insights include interactions with consumers and disease-causing agents, for example.

Alternative models – One of the ways that mechanisms of succession have been described is as three alternative processes: facilitation, tolerance and inhibition (Connell and Slatyer, 1977). Facilitation is the promotion of the success of new species at a site due to changes resulting from the establishment and growth of earlier invaders. If this process causes succession, then later dominants would only be successful after a period of dominance by early successional species. In addition, the order of dominance would be expected to be rigorously fixed, based on increasing requirement for altered resources, greater vulnerability to stress or enhanced protection from disturbance. Facilitation would mean that early successional species would have a positive effect on later successional species.

Tolerance, the second category of successional cause, is interpreted in one of two ways. One is as a meshing of life cycles, which determine such things as longevity, season of establishment, length of juvenile period, mode of dispersal and persistence (Pickett *et al.*, 1987). Under this "active" tolerance model, species would be arrayed through succession simply based on their longevity and patterns of growth and reproduction. The second interpretation of tolerance is as a statistically neutral description of patterns through time. Under either interpretation of tolerance, "early successional" species would have no effect on later successional species.

Inhibition is the third category of successional cause. Observations by Katherine Keever as early as the 1940s suggested that this was a real and common cause of succession (Keever, 1950). Under this hypothesis, the presence of established early- or mid-successional species in a successional community would inhibit the invasion or performance of later-successional species. Empirical research associated with the Buell–Small Succession Study shows inhibition to exist in the community of plants that constitute this succession, as summarized elsewhere in this book.

All three of the "causes" of succession identified by Connell and Slatyer are in fact net effects or outcomes of other, more specific interactions. Because they represent a complex of causes, it may be that any one succession actually incorporates several of these possible outcomes. Entire successions cannot be characterized as belonging to one of these alternatives. This realization emerged from experiments that attempted to test the alternatives as discrete hypotheses, but found the interactions to be mixtures of outcomes that could not be cleanly separated into strict logical types (Hils and Vankat, 1982; Armesto and Pickett, 1986). Thus, the separation of net effects from active interactions as causes characterizes the contemporary understanding of succession (Pickett *et al.*, 1987). More will be said about the complete roster of successional causes as recognized by contemporary ecology in Chapter 4, on successional drivers.

Ecosystem patterns – The overview of succession theory has to this point emphasized the perspective of organisms and assemblages of organisms. There is a second major approach to theorizing succession that emerges from the ecosystem paradigm of ecology. Ecosystems are defined as units comprising the biological organisms and the environmental factors with which those organisms interact within a bounded volume of the Earth (Coleman, 2010). They may be aquatic, terrestrial or mixed, of any size, and can be delimited to match boundaries that appear in the natural world or boundaries that match the scale of a research question or some practical application (Allen and Holling, 2002). For example, an ecosystem study may be focused on a clearly delimited catchment under the assumption that the integrative power of water movement will mainly operate within the watershed boundaries (Likens, 1984). Of course, some important fluxes, say of limiting nutrients, may be accomplished by atmospheric deposition, below-ground flow, or movement by organisms. These fluxes across watershed boundaries may be significant (Cadenasso *et al.*, 2003a). An alternative way to select ecosystem limits is to employ jurisdictional boundaries, as when the ecosystem of a managed forest or of a city is to be understood (Landres *et al.*, 1998). In such cases the number of important fluxes and vectors can be very large indeed.

The issue of boundaries is critical because ecosystem ecology usually focuses not only on the energy and material exchanges and transformations within ecosystems, but also on the transfers across their defined boundaries. Ecosystem ecology therefore deals primarily with quantities of energy, energy embodied in matter, or quantities of nutrient or contaminant materials (Chapin et al., 2002). Energetics are viewed through the lens of thermodynamic theory, and matter transformations and exchanges are governed by conservation of mass, and by the stoichiometry of the chemicals in ecosystems, the fact that elements appear in biologically important molecules in certain proportions (Elser, 2003).

Eugene P. Odum (1969) provided the classic statement of the patterns of ecosystem processes expected to change through succession. This paper has fueled a great deal of the research into ecosystem succession. Its hypotheses are clear and testable; however, the underlying assumptions behind some of them are suspect, and share a great deal with the discredited theory of Clements. Odum, for example, stated that ecosystem succession occurred because the system followed a strategy of increasing homeostasis of control of the physical environment, associated with maximum biomass or information content. Like Clements' "stabilization" mechanism, ecosystem homeostasis may be a result of succession, but not its cause. Nevertheless, the hypotheses about ecosystem functioning and structure erected by Odum are important.

Most relevant to a biogeochemical perspective are his energetic and nutrient-oriented hypotheses. A principal shift in succession is the increase in allocation to maintenance, compared with allocation to growth. This leads to hypotheses of: (1) a reduction of the ratio of gross production to community respiration toward 1; (2) a shift from a high ratio of gross production to standing crop to a low ratio; (3) a shift from low amounts of biomass supported per unit of energy flow to high amounts of biomass per unit energy flow. In addition, energetics are expected to shift from high yield to low community production, based on a shift from linear, grazing-based food chains to complex food webs dependent mostly on detritus. These energetic trends are hypothesized to be associated with low amounts of organic matter in early succession moving to large amounts later in succession. Furthermore, Odum expects inorganic nutrients to be mainly outside of organisms early in succession but contained within biota later. Nutrient cycling is thus expected to be open early, but closed or "tight" later in succession, with rapid exchange between organisms early contrasting with slow nutrient exchanges between organisms and environment later. Hence, homeostasis is generally expected to be supported by increasing internal symbioses, conservation of nutrients and enhanced resistance to external disturbances.

These hypotheses have provided significant stimulus for subsequent research and refinement. The adaptive cycle of resilience, described later in this chapter, is one refinement. Resilience theory acknowledges that an old, complex, conservative ecosystem may be more susceptible to some disturbances than younger, less-complex ones (Gunderson, 2000). Rather than seeing the climax or mature ecosystem as an unmitigated improvement on younger ecosystems, the resilience cycle explains how the interaction of organisms and physical environments reflects different kinds of adaptation throughout the history of systems (Holling and Gunderson, 2002; Folke et al., 2012). Different degrees of resource conservation and different levels of complexity reflect

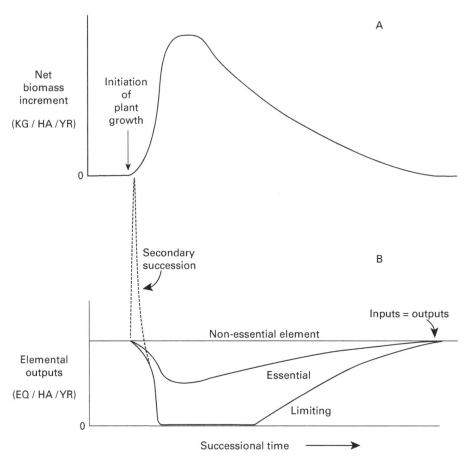

Figure 3.1 Expectations of nutrient limitation during succession. From Vitousek and Reiners (1975). Used by permission of Oxford University Press.

different responses to constraint and opportunity, and these may be patchily distributed across changing mosaics.

A further refinement to the ecosystem thinking of Odum's nutrient cycling ideas was introduced by Vitousek and Reiners (1975). They proposed, like Odum, that young ecosystems would have leaky cycles of limiting nutrients. In contrast, ecosystems in which growth and maintenance were balanced – the presumed "mature" state in Odum's framework, would necessarily not absorb additional limiting nutrients. If an ecosystem were not continuing to accumulate biomass, additional conservation of nutrients would not be possible. Consequently, limiting nutrient retention would have a hump-shaped pattern, with lower retention during early succession where there were few organisms present to take up nutrients, and then again when production came to equal respiration. Vitousek and Reiners present different expected patterns for: (1) nutrients that are not essential for organisms; (2) those which are essential and (3) those which are strictly limiting to organism performance (Figure 3.1). Further work by Vitousek traces ecosystem nutrient dynamics over the very long term, reaching into geological time. Such a trajectory is illustrated by different aged volcanic soils in Hawaii (Vitousek, 2004). Young volcanic substrates exhibit limitation by nitrogen, a biotically controlled element. On ancient, weathered and vegetated substrates, limitation shifts to phosphorous, a sedimentary nutrient cycle.

Because our focus here is on plant populations and communities, we will not further describe the theories of mineral nutrient and carbon dynamics through succession. A good summary of these expectations and their mechanistic justification is provided by Chapin *et al.* (2002).

Statistical models of transition

As more and more data accumulated on succession, and as theoreticians sought generalizations to replace the problematic propositions from the era of Clements through the mid-twentieth century, quantitative models of succession have become important tools.

Markov models – One way to model succession is based on the probability that one community type will transform into another one. Henry Horn (1975) of Princeton University was a pioneer of this approach. In it, a row vector of the abundances of tree species in a forest plot is multiplied by a matrix of probabilities that a species will be replaced by itself or by another species. The probability matrix was based on empirical observation of forest plots. Assuming that the transition probabilities are stable over time, the composition of the forest can be projected into the future. Additional assumptions about the lack of coarse-scale disturbance, the equal availability of species across the forest, or the absence of invasion by some new species, for example, must also be met. If those assumptions are not "seriously violated" Markov models successfully represent steady state forest composition (Culver, 1981). Jose Facelli in a critical review, suggested that the assumptions of the Markov matrix approach would rarely hold well enough to provide reliable forest projections (Facelli and Pickett, 1990). However, the failure of the assumptions provides insight into the historical contingencies that affect particular successions, rather than just nullifying the approach.

Gap phase models – Markov models focus on the group of potential successors beneath the canopy of an existing dominant tree in a forest. Herman "Hank" Shugart exemplified this approach to predicting turnover of composition and structure in forests. Following the insights of pioneers such as Watt and Cooper, cited above, Shugart constructed gap phase replacement models (Shugart and West, 1980). In early versions of these models, a forest is conceptualized as a mosaic of small, independent patches, each dominated by a tree with a simple canopy layer (Bugmann, 2001). These models have been applied to a remarkable variety of forests, ranging from low elevations to treeline, and from boreal to tropical latitudes (Holm *et al.*, 2012). Over the decades, gap models have been extended by incorporating tree growth and patch interactions, among other factors. The current generation of these models is better considered to be landscape models, and they have had great success at representing coarse-scale forest succession (Elkin *et al.*, 2012). The use of aggregate forest properties as output, such as biomass and architecture, is particularly successful. Current generations of such landscape models are commonly supported by remote sensing data and can address climate change (Shuman et al., 2011).

Landscape and gap models have become important generators of theoretical insight in succession. For example, the empirical generalization that tradeoffs in shade tolerance – that is the tradeoff between survival in low light and growth in high light by different tree species – explain forest dynamics has been explored via modeling (Gravel et al., 2010). Coexistence via light-based gap partitioning is in fact a special case of coexistence in variable environments. There are three mechanisms of this kind of coexistence. One is the storage effect (Chesson and Huntly, 1989), in which populations of suppressed individuals depend upon spatially heterogeneous disturbance to persist within the system. The second mechanism is the relative non-linearity of light requirement across species. Relative non-linearity generates coexistence under variable light conditions compared to stable light environment. The third is heterogeneity of light in allowing coexistence of species, expressed in the successional niche. The successional niche concept refers to the capacity of fast growing species to pre-empt resource pulses after disturbance, and the storage effect.

Mechanistic models

Models have increasingly focused on mechanisms, and following the wisdom of Clements and Gleason, have built on the idea that plant traits and performance are the foundation of plant succession.

r-K – One of the earliest plant strategy models is that of r-selection versus K-selection. This conceptual contrast was introduced by Robert MacArthur and E.O. Wilson (1967) in their invention of modern island biogeography. It was used to explain the differential colonization of oceanic islands by different species. r-selected species were those that were characterized by rapid growth, relatively short lifespans and prolific reproduction. Parental investment in individual offspring would be low, and dispersability of the offspring would be favored. r-selected species were expected to predominate in environments in which resources were readily available. In contrast, environments where most resources were locked up in other organisms would select for species that drew lightly on those resources at any one time, but which conserved resources once assimilated. These conservative species were described as K-selected, investing heavily in each individual offspring. The costly offspring would have limited or specialized dispersal modes, and would individually have higher probabilities of successful establishment than the individual offspring of r-selected species. The ideas behind the r-K continuum have evolved or been replaced in the larger field of life-history ecology. In particular, age-specific mortality, and models explicitly examining density-dependent and density-independent mortality factors (Stearns, 1992) have become more important. The fundamental idea that contrasting suites of assimilation and allocation provide the raw material for successional change survives.

R* – This conceptual tool, pronounced R-star in conversation, is formally labeled the resource ratio hypothesis as it applies to succession (Tilman, 1987). It posits that in resource-limited environments, the species capable of reducing the amount of resource in the environment to the lowest level will dominate. An important assumption is that plants can be limited by either above-ground resources, that is light, or by below-ground resources, such as soluble forms of nitrogen. There will be an R* for each resource dimension, and the mix of light and nitrogen available in a site at any given time will determine the identity of the dominant species. Succession can be conceived as a trajectory tracing shifting dominance through time as the mix of species affects the availability of ambient light and nitrogen. This logic follows Tilman's proposal that soil resources are most limiting in unproductive environments, while light is the main limiting resource in productive environments. Succession in relatively well-watered habitats is seen as a shift from soil limitation to light limitation. Notably, Tilman's models do not address spatial heterogeneity of resources. Of course, in some situations, such as very old terrestrial systems, phosphorus may be the main limiting soil nutrient, and under arid conditions, light may rarely or never effectively be limiting (Vitousek, 2004).

Grime's c-s-r – Philip Grime has emphasized that plants differ in their competitive abilities, tolerance of physical stress, and dependence on a colonizing life history. In essence, plants divide an environmental space defined by disturbance and stress (Grime, 1979). Where disturbance and stress are both low, competition predominates. Ruderals are plants that exploit periodic or patchy disturbed sites, reaching them before others and pre-empting the resources available there. Grime expects ruderals to be annuals, and to have little persistent litter, but high reproductive output of small seeds produced at the end of the period of favorable resource acquisition. Competitive species are expected to be tall statured, long-lived and capable of rapid growth and resource assimilation associated with the time of maximum leaf production. Leaf longevity is short, resulting in copious leaf litter that can persist for a long time. Reproductive effort of competitors is expected to be devoted to a few, large seeds relative to the plant's biomass. The third class of stress tolerators is expected to be slow growing, long-lived, often evergreen and unpredictable reproducers. Such species should generate little leaf litter, but this litter may persist for long times. The three species groups should differ in defensiveness against consumers, litter decomposition and the extent of their canopy spread. Succession can be conceived of as an excursion of the community through the three-dimensional space defined by control of resources by competition, degree of stress or occurrence of disturbance. During a long succession, each of the three strategy types should find situations suitable for dominance. Because the environment is determined in part by the plants, as well as the base resource level, succession again appears as a complex interaction of species traits, physical environment and effects of neighboring plants. The contrasting leaf characteristics and reproductive allocation of the different classes also open the door for assessing the impact of consumers, such as disease organisms, herbivores and browsers, along with the effect of animal vs. physical dispersal vectors.

Some scholars have noted that the c-s-r model of Grime and the R* model of Tilman are complementary, but in the aggregate, still incomplete (Craine, 2005). Joseph

Craine's critique includes a useful summary of the fundamental differences between the two theories. The differences highlight contrasts in the locus of allocation within plants, the degree of selectivity of disturbance events, and the role of pre-emption in temporally and spatially patchy environments. He schematically contrasts six potential ways in which light vs. nutrient competition can play out in plant communities with different patterns of supply: (1) uniform, (2) pulsed or (3) patchy. Not surprisingly, the spatio-temporal distribution of either soil or light resources has dramatic impacts on the traits necessary for competitive dominance. Additionally, Fernando Maestre and colleagues (Maestre *et al.*, 2009) have proposed that the concepts of stress need to be refined to separate stressors that involve resource limitation from those that are environmental regulators, such as temperature.

Individual-based models – Individual-based models differ from the gap- or patch-based models described under the heading of statistical models, in being focused on the kinds of traits and performance features identified as plant-based mechanisms of succession rather than on phenomenological changes. SORTIE is one of the most widely used of individual-based models (Pacala *et al.*, 1993, Canham and Pacala, 1995). Individual-based models have been especially useful in forest where following trees and locations is easy. Individual-based models provide several advantages over the more aggregated statistical and patch-based models (Gratzer *et al.*, 2004). These models are based on shade tolerance of seedlings and saplings, survival of saplings under contrasting light regimes and nutrient requirements (Pacala *et al.*, 1996; Dube *et al.*, 2001). In addition, dispersal distance can be dealt with effectively in SORTIE and similar models (Ribbens *et al.*, 1994). It is important that this kind of model is spatially explicit, and based on individual performance throughout the life span of the tree. Forest dynamics over very long time periods can be effectively simulated using this kind of model. For example, the role of small-scale disturbances was assessed by Menard *et al.* (2002). Their analysis revealed that disturbances of 500, 800 or 1100 m^2 could be realistically modeled, and that initial conditions of aggregation and density were important only over the first 300 years of a 1000 year sequence. Furthermore, occurrence of small-scale disturbance at 400 or 600 years did not alter the later results. This robust family of models has proven useful in forest management projections, as well as in discovering the spatially explicit array of mechanisms in forest succession.

The individual-based models embody one of the key tenets of contemporary succession theory: even though succession is defined as a change in communities or assemblages, those changes emerge from the characteristics and interactions of individual plants with each other and with the physical and environmental factors that constrain growth.

Adaptive cycle of resilience

Succession theory is one of the roots of a theoretical structure that explains system behavior as a cycle of resilience (Holling, 1973). Resilience in this case is defined, not as the capacity of a system to recover to an equilibrium state after some perturbation, but as the capacity to adapt and adjust in the face of environmental disruptions while still maintaining its basic structure and functioning (Peterson et al., 1998). Indeed, this definition of resilience, labeled "ecological resilience" to distinguish it from the equilibrium or engineering view of recovery, suggests that adaptive systems will in fact change through time (Brock, 2000).

Succession in its long-term and broad-scale extent can exemplify a cycle of resilience. Take a boreal forest system of large size. Any one spot in the forest may over the long term trace out the kind of changes identified by Watt or Cooper. An intact forest canopy may be disrupted by a very intense fire that kills the canopy trees (Heinselman, 1973; Lavoie and Mack, 2012). Resources are released by the consumption of forest community and soil biomass, and some of these resources are now available at the site. Of course, some resources may be lost by volatilization, as is the case for nitrogen, or by wafting away of particulates in smoke, or via soil leaching by rainfall. The resources that remain are no longer controlled by the former community, which was likely dominated by slow-growing, resource-conserving species and associated microbes, as suggested by r-K selection theory or by Grime's c-s-r model. Resource-demanding species invade or are stimulated to germinate after the fire, and these come to dominate, relying on the pool of released resources. The community established at that time is likely to be structurally relatively simple, and to have fewer species than later in succession. However, new species that demand less resource and live longer come to dominate by a combination of interactive mechanisms. With the passage of sufficient time for all species from the regional pool to reach the site, a community structurally and compositionally much like the one that burned is expected to develop. All phases of this process are together conceived as a system, and the various species that play different roles over time are also distributed across space in patches that suit their life histories, resource demands and abilities to interact with other species, whether plants, animals or microbes. The system as a whole retains its fundamental character as a boreal forest because different spots experience different dynamics offset in time, and individuals of different species come to occupy or predominate in different spatial patches. The adaptive behavior of the whole system requires that components shift in time and space. One conceptualization of this sort of system is patch dynamics, and the adaptive cycle of resilience expresses the temporal aspects of those dynamics.

The adaptive cycle of resilience combines two conceptual axes derived from aspects of succession theory (Holling, 1994; Wu and Wu, 2013). One axis is the shift from resource acquisition and ruderal behavior at one extreme, to resource conservation and investment in a few well-stocked propagules on the other (Figure 3.2). The table from Craine (2005), discussed earlier, points to some other important features of this contrast. The other axis summarizes the expected shifts in complexity as systems accumulate structure and experience undisrupted interactions with increasing time since a major disturbance (Figure 3.2). Ecologists have observed that mesic terrestrial systems often accumulate canopy layers through time, develop increasing spatial heterogeneity, and support more species that rely on specialized biotic interactions. To be complete, it is important to recognize that as mesic systems reach great age, there can be simplification in terms of species diversity, although structural complexity may remain high. However,

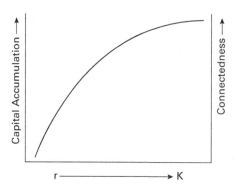

Figure 3.2 Major drivers of shifts in resilience, portrayed along an axis of contrasting allocation strategies from r extreme which favors responsiveness to uncontested resources, high mobility and great reproductive output, to the K extreme of competitiveness and resource use efficiency. The left-hand axis describes the accumulation of capital within the system components, while the right-hand axis describes the increasing connectedness as systems change from resource exploitation to resource conservation. Following concepts in Holling and Gunderson (2002) and Scheffer et al. (2002).

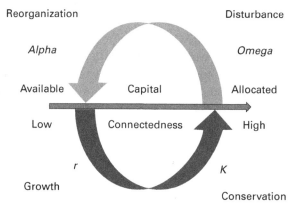

Figure 3.3 The adaptive cycle of resilience. The conceptual space is defined by the drivers of capital accumulation and connectedness. The bottom part of the loop, called the "front loop," represents the relationships of capital accumulation, while the top part of the loop represents the response of a system to disturbance and its potential for reorganization. Resilience is an axis in the third dimension that represents the capacity of a system to respond to disturbance by reorganizing and initiating a new phase of capital accumulation. See citations for Figure 3.2.

in water-limited systems, structural complexity and diversity may continue to increase over the entire successional sequence (Auclair and Goff, 1971).

The adaptive cycle framework requires one more axis. This third axis is that of resilience, or the capacity of the system to adjust through changes along the other two axes while maintaining its aggregate identity (Scheffer et al., 2002; Redman and Kinzig, 2003). The model of the adaptive cycle consequently traces through four contrasting quadrants of potential system behavior (Figure 3.3). Although the cycle represents a temporal sequence, there is no set duration for any part of the cycle, and the system need not spend the same amount of time in all phases. Furthermore, the phases can either grade into each other or suddenly shift, depending on the nature of disturbance events, the environment in general and the traits of the contributing organisms (Gunderson and Holling, 2002).

The cycle generalizes the shift from resource exploitation, labeled r in the model, to resource conservation, labeled K. In parallel, the system shifts toward increasing complexity. High levels of complexity make a system susceptible to many disturbances, and a severe event may disrupt the system and put it in a phase in which resources and the network of interactions are released, labeled omega. If resources are not lost, and species that can exploit those freed resources are available, then the system can reorganize during a phase labeled alpha.

Not all systems will succeed in the adaptive cycle, since collapse is a possibility (Carpenter and Gunderson, 2001). Collapse can occur for a number of reasons, including the loss of resources during or after disturbance, and the inability of the conservative component of the system to release (Biggs et al., 2010). The resources available for reorganization can be lost due to an especially severe disturbance, or due to a sequence of different disturbances that follow one another very quickly. Fire followed by torrential rain is often a sequence of events that sets an area on a new trajectory of dynamics due to loss of resource-holding capacity or persistent propagules. The absence of species that can exploit released resources is also a potential cause of collapse of a system into a different state. For example, resource-demanding species of trees may be lacking in very old mesic forest due to decay of the seed bank over time. This may imperil the retention of released soluble forms of nitrogen after disturbance, which may in turn limit rate or composition of recovery.

Causes of collapse can reside at the other extreme of the adaptive cycle as well. Here, physical firmness of the system may be associated with increased damage from disturbance events. Similarly, systems in which relationships are "flat" and very broadly linked throughout may allow the impact of disturbance to spread rapidly throughout the whole assemblage (Turner et al., 1989). Compartmentalization of interactions and modularity of networks are two adaptive responses that reduce the chance that a disturbed system will not be able to respond to disturbance. A useful image for the factors that reduce capacity to control release is the concept of rigidity. Physical and interactive aspects of a system that prevent it from releasing resources and interactions are said to enter a rigidity trap, and may consequently shift a K-phase system into an entirely different state and identity. The loss of resources or reduction of the roster of species that can exploit reduced resources and engage in opportunistic interactions is labeled a poverty trap affecting the *alpha* phase of the cycle.

Humans in the adaptive cycle – The adaptive cycle has been presented as though it were a strictly biological concept so far. However, the two drivers of resources and complexity can both have human dimensions. For hybrid social-ecological systems, which are now the majority of systems on Earth (Ogden et al., 2013; Williams and Crutzen, 2013), social resources can be conceived of as various kinds of capital (Ostrom, 2005), and complexity can refer to the social, political and economic apparatus of problem solving in a society or culture (Tainter, 2006). Anthropologist Charles

Redman and ecologist Ann Kinzig (2003) have described how the rise of civilization in Mesopotamia fits the model template provided by the adaptive cycle over the long term from 300 to 2500 BC, for example. They summarize alternative periods of spatial consolidation and fragmentation of settlements and the agricultural support network in conformity with expectations of the adaptive cycle. They also summarize a similar, but shorter, example of the Hohokam collapse in the American Southwest.

Joseph Tainter (1988) has provided mechanistic substance to the social axis of increasing complexification that corresponds to the r-K and capital axes. As social systems seek to solve problems, they add complexity, at an economic cost. Tainter presents a number of well-worked examples, including the growth and collapse of Rome and of the Southern Maya. He indicates that human societies do not voluntarily simplify when faced with new problems, and if a political-economic settlement system is isolated from others, ultimately the investment in maintaining the system does not repay returns in terms of risk spreading. Proximate events, such as military invasion or drought, may appear to be the causal event, but ultimately it is the literal cost of complexity as capital, production or administration that is the limit. The examples of collapse Tainter presents can be called rigidity traps (Biggs *et al.*, 2010).

Humans also enter into the picture explicitly in considering the role of succession as a tool in landscape conservation. The potential of landscape memory to affect the adaptive cycle is an important recognition of a mechanism of conservation (Bengtsson *et al.*, 2003). Landscape memory is a metaphor for the various capacities that allow coarse-scale spatial systems to be adaptive in the face of sudden or pervasive change. These capacities are the ingredients for resilience embodied in a spatial mosaic. Focused on conservation, the idea of landscape memory acknowledges that success is not likely to be achieved by establishment and management of protected areas, because such sites are too small in almost all cases to accommodate the dynamic processes of unmanaged patch dynamics. Consequently, resources, reservoirs of biodiversity at various levels and aspects of disturbance regimes must reside outside preserved areas. Hence, Bengtsson *et al.* (2003) call for the maintenance of successional fallows, dynamic successional reserves and the practice of mimicking natural disturbance regimes throughout complex landscapes. The conservation potential of the cultural, rural and even urban landscape components thus must be considered part of effective adaptive design of landscapes that provide ecosystem services and preserve ecological structure and function (Pickett *et al.*, 2013b).

Summary of the adaptive cycle – While the adaptive cycle suggests that systems can persist because of adaptive response to disturbance, collapse is a real possibility for both natural and human systems. Disturbance is a component of all such systems, and its spatial and temporal role is important. The interaction of disturbance with system structure and capacities to respond determine whether a system will adapt, persist or radically alter its form and function to become something different. The adaptive cycle thus reminds us that there is not just one direction and that stasis may be illusory. Indeed, there can be multiple states and trajectories in succession across landscapes. The phases of the adaptive cycle likely meld, and should not be thought of as discrete. However, over the long term, disturbance and release may appear as relatively short events. The role of landscape memory is a practical application of the adaptive cycle of resilience to conservation. Importantly, the adaptive cycle of resilience is appropriate to human components of systems as well as to wild systems.

Need for integration

The variety of theoretical and conceptual approaches to succession is large. Since the birth of the specialty, there have been many examples of fragmentation, yet there remains a common core. All the conceptual approaches that have survived rest on the understanding of what individual organisms require, how they perform in different habitats and how they interact with each other and with the physical environment. This is as true of animals as it is of plants. Contemporary disciplines or specialties that are closely related to succession, though sometimes tacitly, are these: landscape ecology, invasion ecology, restoration ecology, community assembly, competition theory, island biogeography theory and life history theory. There is much opportunity for re-integration and mutual invigoration (Davis *et al.*, 2005).

Succession grew out of the melding of two venerable ecological, or strictly, pre-ecological, perspectives: biogeography and physiology or autecology. The first investigates coarse-scale distributions of organisms and the second pair investigate the adaptation, structure, life history and performance of organisms. As the length of data sets has increased (Weatherhead, 1986), and the scale of observation has extensified (Johnson, 2000), succession has become a dynamic landscape discipline (Pickett *et al.*, 2011).

An especially important feature of contemporary succession theory is that it goes well beyond the strictures of biology. While the organismal perspective remains central, the role of the physical environment at various scales, and its interaction with organisms and assemblages occupies an equally important position. Succession can be reduced to the processes of arrival of species at a site interacting with the various processes that govern the sorting of species through time. Physical and biological factors are involved in both of these major successional processes.

Contexts

Two aspects of context are important in understanding any succession. The first is scale. Succession as a part of very large landscapes or regions is explained by the relationships of patches of different successional ages, with each patch responding to disturbance events of various extents, origins and severity. The patterns of succession at specific locations have increasingly called for understanding their relationship to inputs and influences from adjacent or even distant patches. Succession theory has shifted, therefore, from a within-system focus, to a regional or spatially contextualized focus.

The second aspect of context is temporal legacy. Early theories of succession started with a blank slate, requiring in the most extreme cases, input of both resources and organisms. Clements' term, "nudation" implies such bare naked conditions. However, as data on successional patterns have been extended across the event horizon of time zero, a role for legacies has become apparent. Rare is the disturbance that in fact leaves a completely blank slate. Even such extreme cases as the eruption of Mount St. Helens in 1980 left in some environments buried seeds, rhizomes and even surviving gophers in their burrows (del Moral, 1993). The subsequent succession was shaped both temporally and spatially by these patchily distributed legacies. Legacies are all the more common in secondary successions following agricultural abandonment, as will be illustrated later in this book, or in forest re-growth following windthrow or fire.

Although all successional theoreticians who have proposed or generalized patterns through time have also been interested in mechanisms, the shortcomings of many hypothesized patterns have been clear. Some of the early hypotheses embedded final or teleological causes in their statements. Such statements as Clements' mechanism of stabilization thus confound cause and effect. So too did the productive hypotheses of Connell and Slatyer that identified three logically contrasting models of successional interactions: facilitation, tolerance and inhibition.

Many other expectations and hypotheses have been limited in utility by the fact that they hide important assumptions. The common assumptions embedded in most discussions of succession are often problematical. The review of models in this chapter has richly illustrated this fact. For example, beyond the limits of Clementsian "stabilization" as a cause, or of the illusory logical distinctness of facilitation, tolerance and inhibition as models, we have discussed the limiting assumptions embedded in Markov models, gap models, resource limitation/stress models and so on. The limits of individual models arise because the causes of vegetation dynamics in specific places are remarkably diverse and contingent upon the local and historical contexts. Indeed, models are a key mechanism for viewing assumptions about succession as hypotheses. This suggests that there may be no single universal model of succession. Rather, it is appropriate for models to be based on narrow assumptions so that different and perhaps complementary dynamics can be explored in isolation.

If the assumptions of individual sorts of models face limits when tested against the plethora of patterns and causes presented by successional reality, how much more problematic are definitions of succession that have limiting assumptions attached. Take, for example, the definition of succession provided by Clements in 1916. The first few pages of his book lay out the concept in a leisurely fashion, and unfortunately do not present a simple statement of the concept. However, the structure of his argument emphasizes several formative assumptions: (1) succession occurs within an organic entity, indeed is an organism; (2) succession is development of the life cycle of a vegetation formation and (3) succession consists of six causes, that may act in series or together. A similar take on successional assumptions appears in the definition Odum (1969) provides in his classic paper: (1) succession is orderly, directional and predictable; (2) succession results from the modification of a physical environment by a biotic community and (3) succession "culminates in a stabilized ecosystem in which maximum biomass ... and symbiotic function between organisms are maintained per unit of available energy flow" (Odum, 1969: 262). He summarizes this set of conditions as leading to increased control of the ecosystem by the biotic community. He moves on from this theoretical framework to propose or summarize the temporal trends in ecosystem succession, some of which are presented earlier in this chapter. Each of these assumptions has been found wanting in a number of empirical studies. Yet the concept of succession remains viable and compelling. How can this be?

Transition to the process approach

There are two reasons that successional theory has survived what may seem to be insurmountable attacks on its basic assumptions. One is that the core definition can be divorced from limiting assumptions or those that do not match the spatially and historical contingency of most successional trajectories. The second is the shift from theoretical emphasis on the form of trajectory or pattern to an emphasis on process.

If many of the assumptions that tag along with some definitions of succession are problematic, or outright wrong, is it possible to avoid the problem by stripping the definition of succession down to its core nugget? That is, succession, or more inclusively vegetation dynamics, is simply the change in the composition or architecture of a plant assemblage at a specified location over time.

This definition is purposefully agnostic concerning some things that have bedeviled the study of succession ever since Clements: (1) it does not specify directionality; (2) it does not require stabilization, or indeed termination; (3) it does not restrict cause to features or interactions that occur within the community; (4) it does not privilege either dispersal from outside or internal sorting within the community; (5) it does not specify a time scale. Consequently, a theory emerging from the fundamental, relatively assumption-free definition of succession can apply to any scale in time and space. Perhaps most importantly, it invites the construction and comparison of specific models that intentionally examine the definitionally excluded assumptions. Indeed, as new empirical observations accumulate, or the concepts within succession theory are refined, new models can be erected to examine any new assumptions that come to exist. The assumptions are in fact important and interesting, but they can cause problems if embedded in the general definition of succession.

The result of cutting free the confining assumptions from the definition of succession is the emergence of a process approach (Vitousek and White, 1981) to the subject. Rather than being bound by expected idealized patterns, real patterns extracted from permanent plot studies, chronosequences and experiments can be used to expose and evaluate the assumptions that have often been made sacrosanct by inclusion in some definitions of succession (Pickett, 1989; Johnson and Miyanishi, 2008). The processes of succession, many of which go back to Clements, along with new kinds of processes or resultant refinements to Clements' roster of causes, become the focus of succession research (Cadenasso *et al.*, 2009). Cause is sought, not in an analogy with organisms, or in the assumptions of what is good for the system, but in what plants do, how they are adapted, how they interact with each other and with microbes and animals, and how these interactions are conditioned or interrupted by both sudden and gradual changes in the physical environment. What does a process approach to succession look like? That is the job of the next chapter.

4 Conceptual frameworks and integration: drivers and theory

"The ecologist employs the methods of physiography, regarding the flora of a pond or swamp or hillside not as a changeless landscape feature, but rather as a panorama, never twice alike. The ecologist, then, must study the order of succession of the plant societies in the development of a region, and…must endeavor to discover the laws which govern the panoramic changes. Ecology, therefore is a study in dynamics. For its most ready application, plants should be found whose tissues and organs are actually changing at the present time in response to changing conditions."

Henry Chandler Cowles. The ecological relations of the vegetation on the sand dunes of Lake Michigan (1899)

The overview of successional theories and models in Chapter 3 has shown how ecologists have attempted to deal with successional patterns and associated causes. Because there are so many patterns of compositional turnover and architectural change in plant communities, the variety of theories and models has likewise been rich. Often, different theoretical perspectives or types of models have, by definition, relied on one or a few causes of change. We approached that complexity by starting from a very precise yet general definition of succession as *the temporal change in species composition or three-dimensional structure of plant cover* in an area (Chapter 1). Pointedly, the definition is stripped down compared to most familiar statements defining succession. It is important that the definition omits mention of the operative mechanisms or of expected trends through time. In particular, it is silent about end point, progressive change, the nature of transitions and many other things that have exercised succession researchers (Odum, 1969). In Chapter 3, it became clear that the specific mechanisms and expectations could be dealt with by using contrasting or complementary models, each of which might assume a different mix of mechanisms, constraints and opportunities for species interaction with particular kinds of sites and with each other. This chapter complements our overview of successional theory by sorting out the mechanistic complexity of the process using a comprehensive framework. In a sense, this chapter completes the work begun with our overview of successional theory and models in Chapter 3.

The literature on successional mechanisms is immense. Reviews of successional mechanisms appear in several key publications (Miles, 1979; West *et al.*, 1981; Burrows, 1990; Glenn-Lewin and van der Maarel, 1992; Walker, 1999; Cramer and Hobbs, 2007; Myster, 2008; van der Maarel and Franklin, 2013), among others. Citations to both classic and more contemporary work can be found in two of our recent publications, for example: Pickett *et al.* (2011) and Pickett *et al.* (2013b). Rather than repeating summaries that can readily be found elsewhere, we will focus here on articulating a framework for addressing the complex array of causes that have motivated the great diversity of approaches to successional theory and models.

We begin our construction of a comprehensive causal framework of succession by acknowledging the complexity of causes embedded in the classic distinction between primary and secondary succession. However, these two categories have several causal dimensions that the contemporary theory of succession separates conceptually. An additional classical conceptual tool in ecology is the distinction between initial floristic composition and relay floristics. This distinction, like that between primary and secondary succession, does two things. It reminds us of the interaction among different kinds of successional causes, and provides fodder for the more general, comprehensive framework that we aim to construct. In making sense of successional causes, therefore, we move from the acknowledgement of complexity in classical modes of succession and in classical hypotheses, to a statement of the kinds of differentials in both environment and species that can drive succession and end with a general dynamic model that shows how successional causes can act as sequential or simultaneous filters on the arrival and sorting of species at disturbed sites. This dynamic model template helps organize the analyses and interpretations that appear in the remainder of the book.

The nature of the site and its history

A crucial contextual feature of successional change is the nature of the environmental arena in which it takes place. This has been recognized since the establishment of succession as a field of study. Its most familiar manifestation is in the contrast between primary and secondary succession. These terms at their most fundamental reflect the kind and level of resources available at a successional site, along with the nature of any biological legacies that survive the events that create an open site. The distinction between primary and secondary succession combines two processes – site conditions and biological legacies – that we separate for the sake of clarity and comparability across successions. Here we focus on the nature of the site itself, but recognize that the biological legacies are determined in part by the kind and intensity of disturbance that initiates succession. We discuss the specifically biological legacies later in our construction of the causal framework of succession. The site conditions of resource base and disturbance effects provide the template in which the interactions of organisms play out in succession (Box 4.1).

Sites are made available by disturbances that: (1) disrupt the previously dominant community, while leaving some or all of the substrate in place (White and Pickett, 1985; White and Jentsch, 2001) or (2) by events that create new surfaces that by definition were not previously occupied by plants. Creation of new surfaces can be abrupt or gradual (Peters et al., 2011). Primary successions are those that occur on new sites (Walker, 1999), while secondary successions occur on sites that had been occupied by vegetation before. The process of primary succession reflects that the soil conditions of a disturbed site are a critically important feature of vegetation development. The existence

> **BOX 4.1** Differentials in available sites
>
> Catastrophic disturbance
>   Size
>   Intensity
>   Timing
>   Spatial pattern
> Site resource base
>   Nutrients
>   Moisture

of soil at all, compared to parent material, is one key determinant of successional pattern and rate (Chapin et al., 2002). Primary successions are initiated on sites that lack a well-developed soil profile. Nutrient and water-holding capacity must be built up via soil formation on substrates initially consisting only of parent material. Succession on such primary sites will necessarily be slow due to the limited resources available. Sites at which disturbance leaves in place much or some soil will have a greater resource-holding capacity from the start.

A refinement to the simplistic contrast between any vs. no available resources is that the degree of resource availability can depend on the intensity of a given disturbance (Cramer and Hobbs, 2007; Johnson and Miyanishi, 2007; Myster, 2008). A crown fire that consumes the above-ground canopy of trees and much of the soil organic matter will leave a more depauperate resource legacy than a windstorm that destroys the canopy, but leaves the litter and forest understory layers intact. Large canopy gaps created by windstorms may be subject to greater environmental stress, hence constraining resource use, than small gaps created by the fall of only a few trees. Intensity of disturbance also extends to the below-ground realm. Disturbances, such as landslides that remove the entire soil profile (Myster and Fernández, 1995), will likewise leave less of the resource pool than a superficial erosion event. Animal burrowing also disturbs the soil profile, but typically does so in relatively small focal areas (Boeken et al., 1995).

The variety of conditions left by a disturbance overlay the basic, slowly changing template of resource levels in an area. Thus, to understand how site conditions constrain or promote successional change, the underlying resource pools of a site and the degree to which these are freed from the control of existing dominants or are reduced or increased by disturbance must be known. The initial conditions of abandonment in the Buell–Small Succession Study (BSS) were intentionally diverse, including abandonment in: (1) different years; (2) in spring vs. fall; (3) after hay vs. row crops and (4) with crop residue intact vs. bare soil. As we will document later, all of these conditions affect succession during the first few years, while the contrast between abandonment after a perennial hay crop vs. after row crops maintains a legacy lasting more than a decade (Myster and Pickett, 1990). The diversity of abandonment conditions contrasts with the high degree of similarity of the soils across the site (Ugolini, 1964).

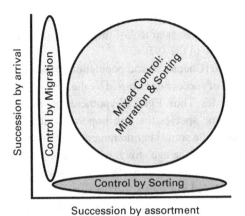

Figure 4.1 Succession controlled by the dynamics of species arrival – the "Gleasonian" dimension – of migration of species to the site of interest, or by the "Clementsian" dimension of assortment at a successional site, and the potential for complex, mixed reality of joint control.

The importance of contrasts in site conditions expressed in the succession literature and in the patterns of the BSS highlight that differences in site conditions are a crucial explanatory factor for succession. We label this set of causes based on site conditions as differential site availability. The differences among successions can be set up by the nature of the large disturbances that are commonly thought to initiate succession. The size, severity, seasonal and super-annual timing, and the spatial pattern of large disturbances are the important characters to understand. Such large disturbances interact with the resource pools of the site, which are thus also key to site contrasts affecting succession (Box 4.1).

From site conditions to species availability and sorting – The nature of the open site plays a role in a particularly persistent controversy in the study of succession. The most famous controversy in succession theory suggests two important dimensions for understanding the drivers of succession. Although Gleason and Clements have much more in common than typically realized in discussions of succession (Eliot, 2007), the perceived controversy suggests a contrast between emphasis on succession as a process of *assortment* among migrants to a site – the Clementsian dimension – vs. emphasis on the process of *arrival* or migration to a site – the Gleasonian dimension. In fact succession or vegetation dynamics can occur as a result of either process acting on its own, or of both processes acting together (Figure 4.1), as will be shown in the long-term patterns presented in the book.

A specific conceptual framing of this contrast between migration and sorting was an important motivation for the Buell–Small Succession Study. This contrast is fundamental to Frank Egler's (1954) "initial floristics hypothesis," which posits that the species that will dominate at different periods during successional time are in fact all present at the beginning of the process of sorting. The contrasting hypothesis was identified as "relay floristics," in which species arrive in the order in which they dominated the community. The Buell–Small Succession Study was established at least in part to test Egler's initial floristics hypothesis (Chapter 2). Murray and Helen Buell, and John Small, the study's founders, were skeptical of the validity of Egler's hypothesis, which was viewed as iconoclastic in the 1950s.

The patterns of community change (Chapter 5) and population dynamics (Chapter 6) that have emerged over the 50 years of succession exposed by the BSS exhibit aspects of both initial floristics and relay floristics. Thus, Egler's hypotheses are not characteristic of an entire succession. Rather, some species that are important later in succession invade quite early in certain fields, while some later dominants do in fact arrive later in time. For instance, a complete census of the age structure of tree populations in a 14-year-old field adjacent to the BSS fields revealed that 90% of surviving individuals of the dominant, *Acer rubrum*, had invaded during the first seven years of abandonment after cropping, whereas individuals of the second most important species, *Fraxinus americana*, invaded primarily in the latter half of the 14-year period (Rankin and Pickett, 1989). The remaining tree species, which were less abundant, all showed delayed invasion.

We can re-frame the two processes of arrival and assortment as differentials among species and environments that affect species availability and species performance, respectively. Thus, succession can be caused by: (1) differentials in species availability at a site and (2) differentials in species performance at a site, or a combination of the two (Figure 4.1).

Differential species availability

Differential species availability can be caused by a number of specific mechanisms. In the BSS, long-term seed dormancy in the soil is an important mechanism for species such as the annual herb *Ambrosia artemisiifolia*, which in the 1950s and 1960s was the typical dominant in plowed fields (Willemsen, 1975). Currently at HMFC, the usual annual herbaceous dominant after spring disturbance is the introduced *Setaria faberii*, which has only short-term dormancy, but does maintain a large soil seed pool for a short time. Dormancy and seed banking are important mechanisms of species availability in succession in general (Leck *et al.*, 1989). Seed banking may be considered to be dispersal through time (Baskin and Baskin, 1988), especially when the germination of dormant seeds is cued by conditions associated with disturbance of the soil. Bringing buried seeds to the surface where they are exposed to light, to low levels of CO_2, to greater temperature fluctuations and, in some cases, higher levels of nitrate is a signal that an open site is available for colonization by these resource-demanding species (Bazzaz, 1996). In some successions, including some fields in BSS, vegetative propagules such as stolons or rootstocks can persist through a disturbance event, and sprout to contribute to the active plant community.

Arrival from a distance is the other major mechanism of species availability. Most relevant to the BSS is dispersal by birds and mammals that can carry seeds from adjacent or nearby forests, hedgerows, domestic landscapes or older fields. In the case of mammals, seeds may be carried after ingestion, such as the fleshy fruited *Rubus* spp.,

> **BOX 4.2** Differential species availability
>
> 1. Dispersal
>   Vectors
>   Landscape connectivity
>
> 2. Propagule pool
>   Mortality rate of propagules
>   Land use/land cover

or on the fur of the animal, as in the case of barbed seeds of *Galium aparine*, for example. Birds most often transport seeds after ingesting them. The seeds of *Cornus florida* and *Juniperus virginiana* often arrive via this animal-mediated pathway (Holthuijzen and Sharik, 1984). In the BSS, as in many other secondary successions, arrival of seeds via animal vectors usually occurs later than the first few years of succession. Both birds and mammals are more likely to visit plant communities that offer perches and cover. Food is also an attractant – oil-rich fruit in the case of birds, and sweet fruit or the availability of browse in the case of mammals. Habitat structure that provides cover or other protective functions is important to both birds and mammals. At HMFC, the presence of experimental perches in low-statured oldfields facilitated the arrival of bird-dispersed species (McDonnell and Stiles, 1983), while shrub cover enhanced foraging by small mammals for seeds (Meiners and LoGiudice, 2003).

The most significant aspect of arrival for the process of succession is that species vary in their dispersal vectors, and that the landscape in which a field is embedded determines how effective each vector will be in bringing a species to a site that matches its requirements. The ability of a landscape to support reproductive adults of a species, and the extent to which the matrix situated between a seed source and a successional site promotes or inhibits the movement of seed vectors also contribute to differentials in species availability. The landscape context is also dynamic, changing with human use, the successional status of adjacent patches and species introductions in the surrounding mosaic.

We summarize the components of differential species availability in Box 4.2.

Differential species performance and assortment

Once species become available at a successional site, an additional filter determines whether and how they participate in succession. Performance refers to the physiological, life history and interactive actions that plants engage in. These activities permit growth, survival and reproduction (Bazzaz, 1986). Performance depends upon the genetic makeup of individuals, and the plastic responses to the conditions they experience. In other words, performance is plant behavior and architecture (Figure 4.2).

Such a broad arena of life processes necessarily includes many detailed components and processes. These components can act as differentials among species that drive

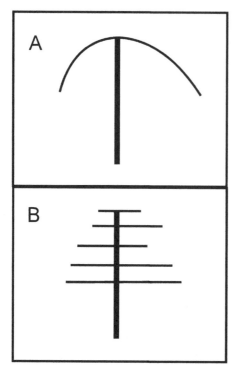

Figure 4.2 Architectural contrasts in succession. (A) Idealized late successional "monolayer" canopy adapted to shady conditions, and (B) idealized early successional "multilayer" adapted to maximizing light capture in sunny environments. Monolayers allocate more biomass to leaves compared to multilayers, which allocate more biomass to non-photosynthetic support structure.

succession based on the capacities and adaptations of the plants themselves. We divide performance into six major categories: physiology; life history; stress tolerance; competition; chemical interaction and consumers. Because plants are integrated but modular organisms, their capacities to respond in these different ways are not independent, of course, but contingent on other processes simultaneously occurring as plants interact. Differential performance encompasses the majority of the work that has been done in succession, and so this driver contains much more diversity in the types of processes involved (Box 4.3). Examples of some specific aspects of performance show how these differences among species can drive succession.

Physiology – Whole-plant physiology generates differential capacities among plants in three main processes: germination, assimilation and growth. We have already noted some of the ways in which germination biology differs among species in discussing the role of site conditions. Germination requirements of species contrast in many dimensions. Whether the seeds are dormant when dispersed, or enter dormancy if favorable conditions are not met during their first possible growing season, and how long seeds may remain dormant in the substrate are important contrasts (Baskin and Baskin, 1998). The kinds of conditions required to break any dormancy also differ among species. These range from simple degradation of thick seed coats, through the scarification by

> **BOX 4.3** Differential species performance
>
> 1. Physiology
>   Germination
>   Assimilation
>   Growth
>
> 2. Life history
>   Allocation
>   Reproductive strategy
>
> 3. Physical stress
>   Climate
>   Biotic legacy
>
> 4. Competition
>   Identity
>   Disturbance (fine-scale)
>   Consumers
>   Resource base
>
> 5. Chemical interference
>   Microbial
>   Plants
>
> 6. Consumers
>   Identity
>   Cycles
>   Distribution

animal digestive tracts, to complex patterns of soil freezing and thawing, for example. Once dormancy is broken, or in some cases, as seeds simply mature over time, the need for high levels of light, the requirement of certain daylengths, the need for high levels of some nutrient ions or the need for the heat of fire can signal some seeds to germinate. In secondary successional systems like the BSS, light, daylength and thermoperiod are the most common environmental signals for germination (Baskin and Baskin, 1998).

Assimilation requirements of plants refer to their needs for high levels of resource supply, vs. tolerance for low rates of supply (Bazzaz, 1986). In the BSS, photosynthetic light requirements are a good example of differential assimilatory demands. Species that typically dominate early in old-field succession have high photosynthetic rates, coupled with high compensation points based on their rapid respiration rates (Bazzaz, 1979; Bazzaz and Carlson, 1982). They are adapted to capture resources that are freely available in post-disturbance conditions. Similar trends occur in forest succession as well. Along with high photosynthetic rates, early successional species in post-agricultural fields have high rates of transpiration. The high stomatal conductances that permit the fluxes of CO_2 that support high photosynthesis are associated with profligate water use. High assimilation rates are the foundation of high rates of growth that are a common feature of early successional species (Bazzaz, 1986).

In contrast, later successional species tend to have more conservative assimilation and growth rates. This is because the resources that were available at the start of succession will have been captured by other individuals in the absence of subsequent disturbances. Hence, high rates of assimilation would not be satisfied by the resources available later in succession. Again, photosynthetic adaptations are a good example. Late successional species, as seedlings and saplings, have low maximum rates of photosynthesis and respiration. They can thus persist on the low levels of light available beneath closed and complex canopies in later successional communities (Marks and Gardescu, 1998).

Nutrients are likewise locked up in detritus or in the bodies of dominant plants as succession proceeds. Symbioses with mycorrhizae and low rates of nutrient demand characterize later successional plants in deciduous forest biome (Johnson *et al.*, 1991). Shifts in the predominant form of nutrients, especially N, can also appear in succession.

Resource availability may change through time, and not just act as a boundary condition at the beginning of succession, as it does in the case of differential site availability. For example, water availability can fluctuate through successional time with climate cycles, or even extreme spatial heterogeneity of thunderstorms or location relative to major cyclonic storm tracks. Years of high rainfall would be expected to set up different competitive relationships in a successional community than years during which drought might constrain canopy closure. Chapter 7 presents data on contrasts in water availability in the BSS succession.

Variation in resource availability can have profound effects on the dynamics of species in succession. When nutrient availability is maintained at artificially high levels, early successional populations may persist, restraining successional changes (Carson and Barrett, 1988). Similarly, reduction of soil fertility can increase the performance of late successional species with more conservative patterns of resource utilization (McLendon and Redente, 1991; Blumenthal *et al.*, 2003). These experiments suggest that variation among species in resource utilization can interact with local conditions in ways that will cascade through many aspects of differential performance.

Life history – Life history refers to the timing of events throughout the life of individuals, and the allocation of assimilated resources to different structures and processes in plants. Plant species and individuals differ in the proportion of assimilate they allocate to growth, maintenance (respiration) and reproduction. Species that typically dominate earlier in secondary succession allocate proportionately less to growth and more to reproductive functions. This strategy favors their arrival in patches that have relatively uncontested resources. Growth functions, such as increase in size, storage of carbon reserves and retention of limiting nutrients within the plant body, are contrasting allocation strategies of later successional species. In particular, the reproductive strategy, that is the number, size, timing and provisioning of seeds or vegetative propagules, is a key component of life history. Colonization success is favored by large numbers of mobile seeds or vegetative propagules. Persistence in closed communities, even when ultimately growth to adulthood takes place in localized canopy gaps, is favored by high investment in relatively few offspring (Leishman *et al.*, 2000). The large seeds of late successional species are an example of such provisioning.

Life history characteristics clearly differentiate the role of summer annuals, winter annuals and biennials in succession, both compared to each other and to perennials. In our North Temperate climate, summer annuals germinate in the spring, and become reproductively mature within months, before dying in late summer or autumn. In contrast, winter annuals germinate during or late in the growing season, and persist as ground-hugging rosettes over the winter, flowering and reproducing early in the next growing season. Biennials germinate in one year, but remain vegetative for their first year, often as rosettes, to finally reproduce in their second year. The order of dominance in BSS in spring plowed fields follows this order of life history contrasts. Fall plowed fields have many more winter annuals in the mix, and biennials are later dominants in all fields disturbed at abandonment. The role of life history in this sense was early on recognized by Catherine Keever (1950) in her studies of Tennessee old fields. Life history research was relatively neglected during the early years of succession research (Keever, 1983), but now has exposed a significant driver of succession. A refinement on the life history approach, which also combines understanding of physiology and allocation, among others, is the trait based study of succession (Garnier *et al.*, 2004; Raevel *et al.*, 2012; Chapter 12).

Stress tolerance – Tolerance of physical stresses vs. sensitivity to them is another dimension along which species differentials appear. This insight is also captured in Grime's c-s-r concept, discussed in Chapter 3. If stress is seen as the reduction of physiological functioning as a result of some physical condition, such as temperature or high atmospheric moisture demand, adaptation to stress may be seen as a mechanism to reduce the loss of assimilated limiting resources. Physically rigid, sclerophyllous leaves that maintain structure without exclusive dependence on turgor pressure is one example. Evergreen leaves can also retain limiting nutrients in plants for a long time, not exposing them to risk of loss during annual leaf turnover and subsequent re-absorption of the nutrients from soil or from symbionts.

Response to stress can also be affected by the specific microclimate experienced by an individual. In small fields, such as those in the BSS, which are all less than 1 ha in size, the light, humidity and temperature conditions near field edges can govern the resource availability and stressors experienced by different species. Edges with older fields, forests and hedgerows can produce such influences (Chapter 11) and are an important source of heterogeneity in species sorting.

Competition – There are many dimensions of the competition process that play out in succession. Competitive outcomes are determined by the particular mix of competing species present in the site and their spatial arrangement. Competitive ability is determined by the capacity to pre-empt resources, for example by beginning growth early in the season. A different aspect of competitive ability is the capacity to draw down limiting resources to levels lower than competitors can tolerate (Tilman, 1991). Both of these dimensions of competition are shaped by the resource base in the site. We expect that sites having higher levels of limiting resources would allow competitive exclusion or displacement to proceed more rapidly than sites with low levels of resources. This is simply because of the more rapid rates of growth under more fertile conditions.

Competition among plants usually takes time to play out. Competition in plant communities is spatially constrained to those individuals within close proximity that will be sharing a largely immobile resource pool. Thus, the competitively superior species or individuals are only interacting with a sub-set of the community, generating heterogeneity in competitive interactions. Within a successional site, there are also fine-scale processes that disrupt the temporal process of competitive displacement. For example, fine-scale disturbances, such as those that thin canopies or root mats within an essentially intact community, can slow competitive exclusion. This kind of mechanism seems to be the cause of periodic, but diffuse presence of individuals of annuals like *Ambrosia artemisiifolia* within later successional fields in the BSS. Browsers and herbivores can act in this way as well, and are especially effective if they prefer dominant or canopy-forming species as forage.

Chemical interference – The chemicals that organisms release into the environment can have substantial effects in succession. Plants and microbes are remarkable sources of chemicals that make their way into the environment. This porosity to chemical products may be adaptive or simply incidental. Adaptive chemicals are often defensive against herbivores and disease organisms. Incidental chemical release may involve metabolic byproducts that are harmful to other organisms or fairly innocuous carbon exudates. All such chemical effects in the environment are grouped under the heading of allelopathy. However, not all allelochemical effects documented in the laboratory, especially those using purified or concentrated extracts, have clear effects on succession in the field.

A strong suggestion of allelopathy is found in the case of *Ambrosia artemisiifolia* and *Raphanus raphanistrum* (Jackson and Willemsen, 1976), which dominated spring- or fall-abandoned fields, respectively. Neither species maintained dominance for a second year even when the dominant that usually replaced them was removed. Jackson and Willemsen took this as compelling evidence of a chemical inhibition mediated by the soil each species had dominated for a year.

Chemical interactions may also be context dependent. Liana species at the HMFC were examined for allelopathic potential in laboratory assays (Ladwig *et al.*, 2012). While all liana species had the potential for allelopathic inhibition of other species, only the toxicity of early successional species was induced by shade, suggesting a successional role for chemical interactions. Finally, chemical interference need not be direct (Hale and Kalisz, 2012). One of the predominant non-native invasive species in the young forests of the BSS is *Alliaria petiolata*, which inhibits the formation of mycorrhizae on associated plants (Roberts and Anderson, 2005).

Consumers – Although we have mentioned consumers as potential regulators of the outcome of plant–plant competition, their effects are important enough to merit distinct focus on their own. Indeed, such focus is also useful to counteract the long neglect of the effects of animals and disease organisms on plant succession. Consumers affect seed mortality, canopy and branch architecture, leaf biomass, or root biomass and structure. Through this variety of effects, consumers can exclude or greatly reduce some species from a successional sequence, and they can retard the invasion of some species or speed the demise of earlier dominants. Their indirect effects though habitat modification, such as trampling or burrowing, are fine-scale agents of disturbance. A mammalian exclusion experiment adjacent to the BSS fields documented that the order of colonization of *Juniperus virginiana*, *Acer rubrum* and *Cornus florida* into old fields was the reverse of the order of their sensitivity to mammalian browsing (Cadenasso *et al.*, 2002).

The dimensions of the consumer environment can be described as identity, cycles and spatial distribution. Insect consumers vs. large mammalian browsers, will have different effects on succession (Shure, 1971; Brown, 1985; Davidson, 1993). Flower or seed predators will act on a different collection of successional processes than will burrowers or grazers.

The temporal cycles or episodes of consumer population growth and decline will also affect how they act in succession. Outbreaks or peaks of predators or herbivores are well known. For example, both the introduced gypsy moth (*Lymantria dispar*), and the native cherry scallop shell moth (*Hydria prunivorata*) produce episodic population explosions that open tree canopies and can affect the dynamics of plant populations. Similarly, outbreaks of *Microrhopala vittata*, a beetle that specializes on *Solidago*, result in dramatically reduced stem density and inceased colonization by trees in old fields (Carson and Root, 2000). Cycles of small mammal grazers such as voles or seed predators such as mice can also affect succession and will be explored later (Chapter 11).

The spatial distribution of consumer populations may also generate patterning within succession. For example, the clumping of grazing rabbits beneath shrub patches in old fields, or the association of deer with forest/field edges add spatial complexity to successional dynamics. In combination with the degree of plant susceptibility, the spatio-temporal pattern of consumer activity is a critical regulator of successional dynamics.

Interaction among the drivers – The overview of successional drivers in the last few sections is intended to emphasize completeness rather than distinctness. In other words, it is less important that the different categories of drivers not overlap than that they summarize the whole range of interactions that can affect the tempo and mode of succession. In fact, in describing the operation of several of the drivers, it has become clear that different kinds of drivers interact. For example, the role of consumers depends on the ranking of competitive abilities of plants in the community, a resource base that permits competitive differentiation and the selectivity of the consumer. Similarly, life history allocation patterns affect the structure of plant canopies, which in turn affect the rates of above-ground resource capture and, in some cases, defensiveness against some consumers. Many other kinds of interaction could be mentioned, but we will leave those details to later chapters where we will explore the specific roles of drivers in the BSS in detail. Suffice it to say here that connections among all the successional drivers are possible, and are often operative in causing the patterns of species composition and community architecture through time. Such controlling contingencies highlight the need for conceptual frameworks to organize our thoughts on community dynamics. This is the topic of the next section.

A causal framework of succession

Organizing the variety of causes is an important tool in understanding the complexity and variety of successional trajectories that appear so prominently in the BSS and elsewhere. A causal framework performs several functions. First, it collects all the potential causes of a complex phenomenon in a unified conceptual structure. Second, it shows how the many dimensions of causes, including mechanisms, interactions, constraints and enabling conditions relate to one another. Third, it indicates how general causes can be disaggregated into more specific causes (Figure 4.3). Causal frameworks thus show the parts of a system of interest. Succession as a causal system consists of differentials in the sites available, differentials in how species become available at those sites and differentials in how species perform and interact. The ability of causal frameworks to specifiy the linkages between individual mechanisms and place them into an appropriate ecological context makes them a necessary tool for integration.

There is another way to show the linkages among the causes enumerated in a conceptual framework. The causes can be assembled into process models. Such models

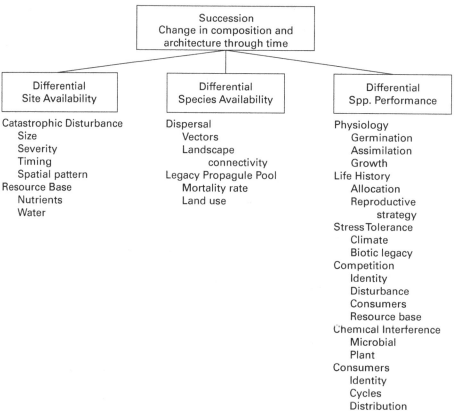

Figure 4.3 A hierarchical framework of successional causes or drivers, indicating the more general processes and phenomena at higher levels, and more specific or detailed mechanisms nested at lower levels. Based on Pickett *et al.* (2013a) with modifications.

show the temporal or dynamic linkages that are possible among the mechanisms, constraints, interactions and enabling conditions that govern system dynamics. Like all models, they also call for the specification of the spatial and temporal boundaries, and the spatial grain of system components.

A dynamic model template of succession

We use the three causal differentials in succession or vegetation dynamics as ingredients for an ideal flow of succession as a process. This successional process can be considered to start with an open site that filters the species that can take part in the dynamics. There are two other filters beyond site conditions and history that determine the order and turnover of dominance, and indeed the composition and architecture of the entire plant community through time. Thus, succession is the cumulative result of three potential filters, each of which corresponds to one of the three major causes of succession (Figure 4.4). The first filter is site conditions and history, the second filter is species availability to the successional site and the third filter is the performance of species at the site. Each of the three filters involves interactions among organisms or between organisms and physical conditions within a specified spatial boundary. Our presentation of filters here emphasizes causes we know or expect to operate specifically in the fields of the Hutcheson Memorial Forest Center where the BSS is located.

Site conditions and history filter – The site on which succession will be studied possesses a set of conditions, some of which reflect its deep history. For old-field succession, the nature of abandonment is a prominent aspect of the filter. Historical agricultural practices and their legacies are another major feature of the site. The resource availability of the site reflects the agricultural legacy, that is, the kinds of crops, the nature of cultivation and weed control and the manuring or fertilization that took place over the agricultural life of the field. Agricultural practices and the nature of abandonment determine the spatial patchiness within the field, although these are also affected by the adjacent plant communities and their animal and plant residents. Coarse filter components of site conditions include the climate and the native soils and parent materials underlying the field. Site conditions can also be affected by inputs from adjacent or distant systems, including such things as nitrogen pollution and acidic precipitation.

Species availability filter – Species availability is the next potential filter. Within a field, the presence or absence of a persistent seed bank, and the location of any surviving mature, reproductive individuals are major determinants of the initial species complement in the field. The timing of reproduction or maturation of propagules and the emergence from any dormancy relative to the timing of disturbance of a field will also determine what species can be part of the succession. The nature and behavior of dispersal vectors, including wind, gravity or various species of animals, is part of the availability filter. The action of dispersal vectors is affected by the configuration of the adjoining landscape, and hence the connectivity of pathways over which vectors can travel. A coarse-scale component of species availability is the invasion of species from other regions or continents, often with human assistance.

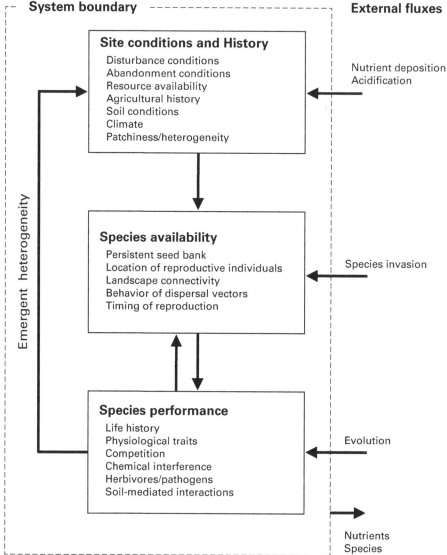

Figure 4.4 A process model template for succession showing how specific causes identified by the hierarchy of Figure 4.3 can be arranged as a sequence of filters. In addition to interactions with other ecosystems or landscape patches, the internal dynamics of a community are shown. In an ideal sequence, the first filter to act would be the nature of the site, based on its inherent physical and chemical conditions and legacies of prior occupancy or use. The next filter to act would be constraints and opportunities determining what species were available at the site in question. The third filter acts upon species that are actually present at the site, and is based on differentials in species performance. The action of the three filters together may involve two important feedbacks. Species performance can generate new, spatially heterogeneous site conditions that in turn alter successional dynamics. In addition, species performance at a given time can influence the availability of species at later times.

Species performance filter – Once species arrive at the site, and have been filtered by the conditions at the site, their role in the succession is determined by their performance. Performance depends on the timing and duration of different life history events and stages, the physiological traits each species possesses, how it does in competition with other plants, whether it is susceptible to the chemical interference from conspecifics or other species and the sensitivity to herbivores or pathogen effects. Notably, many aspects of plant performance are mediated by soil conditions that result from the actions of other organisms. This is especially the case with mycorrhizae. Species performance can be affected by the coarse filter of evolution, which may be influenced by gene flow from elsewhere, or simply to temporal adaptation. Finally, just as Clements noted, the conditions of a site can be modified by the species that occupy it. Therefore, there is a feedback between the heterogeneity of conditions within the successional site as time unfolds and the performance of species that have occupied or currently dominate the site.

We combine the three filters into a process model template of succession (Figure 4.4). The template employs insights from the causal hierarchy presented earlier in this chapter (Figure 4.3), it specifies the boundary of the system – a field or system of fields and associated older-successional patches and it acknowledges the potential role of external inputs, either those from some distance or those reflecting temporal, evolutionary trends. Certainly, climate change is an external influence of increasing importance in understanding all successions.

This model template (Figure 4.4) helps frame the remainder of this book. It identifies the causes of the patterns observed over 50 years in the Buell–Small Succession Study, and alerts us to the potential for influences that come from beyond the individual fields themselves. It reminds us that the external conditions and inputs may have changed over the period of observation, including the modification of the habitat template by a sequence of dominants. Finally it codifies the interaction of causes, and of scales over which causes operate. With this model template in mind, we turn first to presenting the community and population patterns in our succession in Part 2, followed by a section in which we explore causes ranging from plant–plant interactions and the role of invasive exotic species to spatial heterogeneity and trait-based interactions.

Part 2

Successional patterns in the BSS data

This first collection of data-intensive chapters focuses on describing the patterns and dynamics within the BSS data. These chapters are not specifically structured by our conceptual framework, though they certainly still reflect the influence of the framework on our thoughts. Instead, these chapters aim to present a more complete view of the general pattern and process of succession. Many of the generalizations and ideas about succession are based in individual constituent theories that are contained within our broader framework. These patterns represent some of the major theories that have been posed over the years. It made sense to revisit these ideas with the BSS data early in the text as these ideas form our foundational knowledge of succession and community dynamics. The pattern data have been grouped into three chapters – one focusing on population dynamics, one on community dynamics and one on the role of disturbances.

The goals of this section are two-fold. As these chapters deal with the most fundamental views of succession, they are necessary to build an appropriate context for the chapters that follow. Contained within these patterns are the dynamics that constrain the processes that will be explored in more depth in the next part of the book. Secondly, this was also our opportunity to present and test many of the successional generalizations that have been discussed for decades. There have been so many hypotheses about succession that developing a full treatment of each would make an inappropriately long book.

This section represents an opportunity to quickly capture the dynamics of the BSS and apply the data to many different questions. Many of the analyses contained here started out as a simple exploration of the question "Can we find this pattern in our data?" That may not have been the most theoretically focused way to address the data, but it has been quite fruitful. The problem with many of the logical and obvious successional ideas that are in the ecological literature is that they just don't describe what has happened in the BSS. These surprises are important to address early on because they represent potential misconceptions that can mislead our understanding of the system.

5 Community patterns and dynamics

> "The developmental study of vegetation necessarily rests upon the assumption that the unit or climax formation is an organic entity. As an organism the formation arises, grows, matures and dies. Its response to the habitat is shown in processes or functions and in structures which are the record as well as the result of these functions. Furthermore, each climax formation is able to reproduce itself, repeating with essential fidelity the stages of its development. The life-history of a formation is a complex but definite process, comparable in its chief features with the life-history of an individual plant."
>
> Frederic Edward Clements. *Plant Succession*: *An Analysis of the Development of Vegetation* (1916)

While contemporary ecologists certainly reject the tone and implications of the "super-organism" metaphor of Clements' work (however literal his usage may have been), succession is still largely viewed as a community-level phenomenon. Researchers vary greatly in the location of their focus along the gradient from emphasis on emergent community properties (i.e. a Clementsian approach) to emphasis on the ecology of constituent populations (a Gleasonian approach). Despite variation in ecological perspectives on the nature of communities, succession is first and foremost a community phenomenon. When one thinks of the phrase "successional community," a variety of habitats, such as abandoned crop fields full of goldenrods, forests recovering from clear-cut logging or plants emerging from volcanic ash, may all come to mind. It is the changes in composition over time that engage our interest. Even in those cases where the roster of species is largely set, the relative importance of species shifts as we observe the site through time. The community provided the first entrée into succession for the profession of ecology and for most field ecologists. Therefore, we will begin to delve into the successional dynamics of the Buell–Small Study by examining the community-level patterns and processes that occur over time. We will begin by describing the general changes in community structure, diversity and composition that occur during succession and then compare data from the BSS with a previous chronosequence analysis of successional patterns for the area. At the end of the chapter we will finish with a discussion of the relationship of these data to historical predictions made by ecologists about succession. We will return to many of the issues discussed here in the next chapter, where we will specifically deal with the successional ecology of the constituent populations.

Structural changes during succession

One of the most fundamental changes that occur during nearly all successions is that of the transition among life forms. This is perhaps the most predictable aspect of successional dynamics and was the major perspective utilized by Clements in much of his work. This perspective may have been one of the reasons he developed the deterministic and predictable view of succession that is so widely rejected today. The BSS data provide a remarkably clear and classic series of life-form transitions common to all communities with forests as their endpoints.

Starting at the beginning, annual species dominated the community for only the first year after abandonment (Figure 5.1). Within four years these species were greatly reduced in abundance and after 10 years were relegated to a small minority of the community. The rapid collapse of annuals in early successional communities has been attributed to interference by leaf litter, which prevents re-establishment (Jackson and Willemsen, 1976; Facelli and Facelli, 1993), though nutrient addition may prevent this transition (Carson and Barrett, 1988). Although they declined rapidly, annuals persisted at very low abundance throughout succession as colonizers of local soil disturbances and other establishment opportunities. They made particularly large resurgences following droughts (see Chapter 7). After decades of decline, annuals increased again late in succession as opportunities for establishment re-appeared in the much less dense herbaceous layer of the forest understory. These late successional annuals are a combination of invasive exotic species such as *Microstegium vimineum* and native woodland herbs such as *Pilea pumila*.

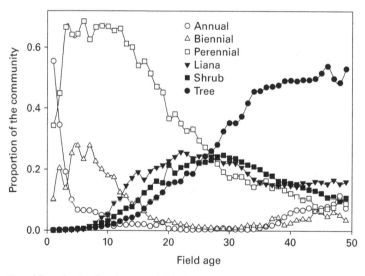

Figure 5.1 Transitions in the dominance of life forms over succession. Herbaceous species are indicated in open symbols, woody species in closed. Data are averaged across all BSS fields. Data from Meiners and Pickett (2011). Reprinted with permission from University of California Press.

Both biennial and perennial herbaceous species increased dramatically following the collapse of annuals. Biennials peaked between years 5 and 7, comprising around 30% of the community during their peak. The decline of biennial abundance was much slower than the decline of the annuals, with the biennials persisting in greater abundance until the community reached 20 years post-abandonment. As seen with the annuals, biennials again increased late in succession, once canopy closure occurred. Longer-lived herbaceous perennials dominated the successional community for over 20 years and were ultimately replaced as the dominant life form by trees. Though the initial dominance of herbaceous perennials very early in succession may partly have been a consequence of the perennial grasses planted in the hay fields, they clearly dominated the community well beyond the influence of these species. Perennial cover reached a sustained peak of around 65% of the community from years 3–12, before beginning a slow decline as woody species began to increase. Despite this decline, perennial herbaceous plants were a substantial component of the community during all successional stages.

Shrubs and lianas are both very opportunistic life forms (Grime, 2001) and had very similar successional trajectories. Despite arriving early in succession, both life forms initially expanded much more slowly than any of the herbaceous groups (Meiners, 2007). After 10 years of succession, both lianas and shrubs began to increase dramatically, with each contributing 20% of the community at their peak. Lianas initially expanded more quickly than shrubs did, but shrubs eventually peaked once lianas had begun to decline. Both life forms exhibited marked decreases during the last 20 years of succession observed, probably a result of tree canopy closure (Robertson *et al.*, 1994; Hutchinson and Vankat, 1997; Flory and Clay, 2006; Ladwig and Meiners, 2010a, 2010b). As some lianas have the potential for access to the canopy, this life form has persisted later into succession, while shrubs as a life form have continued to decrease.

Tree dominance represents the final structural transition in this succession. Many trees became established very early following the cessation of agricultural disturbance in the BSS (Rankin and Pickett, 1989; Myster and Pickett, 1992b), but dominance of the community by trees was delayed until 28 years after abandonment. While all life forms have persisted within late successional communities, the abundance of herbs, vines and shrubs has been depressed by tree canopy closure. We will see in later chapters many other changes that occur as a consequence of the transition from herbaceous to woody dominance. After nearly 50 years of succession, the communities were the most structurally complex of all those generated through time, with all life forms represented. The species that compose some of those life forms may represent future challenges to the community as they are largely non-native species (Chapter 10).

One of the other major changes that occurs during succession is an increase in the total cover of the community (Inouye *et al.*, 1987; Bastl *et al.*, 1997; del Moral, 1998). This is thought to follow similar changes in the total biomass of the community (Odum, 1969) Within the BSS, total cover consistently exceeded 100% as the canopies of different plant species often overlapped. Temporally, total cover increased somewhat consistently during the first 25 years of succession (Figure 5.2). After 25 years, total cover remained relatively constant, though certainly total community biomass would be expected to increase as canopy trees continue to increase in size and produce permanent woody

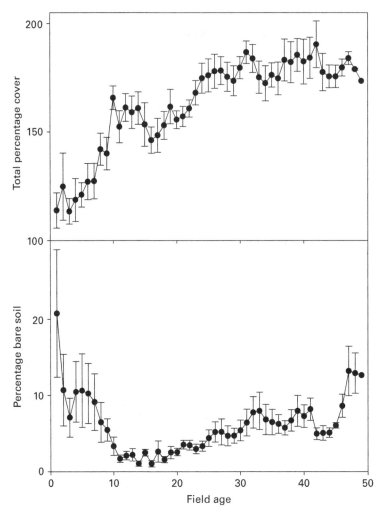

Figure 5.2 Change in total cover (summed across all species) and bare ground during succession. Bare ground is recorded as the abundance of bare mineral soil that is not covered by leaf litter. Data presented are the averages (± SE) of all BSS fields.

tissues. The lack of continued increase in cover suggests limitations to the usage of cover as a surrogate for biomass in late successional communities.

An important community characteristic closely related to changes in the total cover of the system is the availability of bare ground. Patches of bare soil are critical to the germination and establishment of many species, particularly annuals and other less-competitive species (Goldberg and Werner, 1983; Tilman, 1983; Goldberg, 1987; Goldberg and Gross, 1988; Carson and Pickett, 1990; Vankat and Carson, 1991; Wilby and Brown, 2001; Gibson et al., 2005). As is usually seen in systems that have been plowed, bare soil was relatively abundant in the first year after abandonment (Figure 5.2). However, variation among fields was very large due to differing starting points following abandonment. Some of the fields were abandoned following hay production, some were allowed to develop an early cover of winter annuals and some were plowed last in the spring of abandonment. Across the system, availability of bare soil reached a minimum 10–20 years after abandonment and then slowly increased over time. This increase, particularly in the oldest fields, may be related to the competitive suppression of the understory flora by canopy trees. Small-scale disturbances, such as that from deer or small mammal activity are also likely sources of exposed soil. Regardless of their origin, late successional communities appear to have many more microsites appropriate for establishment than mid-successional communities. The lack of openings in mid-successional communities may be related to the rapid recolonization of soil disturbances by clonal old-field species (Goldberg and Gross, 1988).

Mosses are not typically followed in studies of old-field succession. In the BSS data the coverage of mosses has been recorded, but individual species have not been identified. There were two distinct peaks in moss cover during succession, which likely represent two different suites of moss species as they occurred in very different environments (Figure 5.3). Mosses, as in vascular plants, vary dramatically in their environmental tolerances, which generates turnover during succession (Bliss and Linn, 1955). Early moss abundance closely followed the availability of bare soil in the system. After low coverage in the first year after abandonment, moss coverage increased to an average of 8% and then slowly declined until year 10. These early successional mosses would have been exposed to full sun and to long periods of dry soils as well as freeze–thaw cycles at the soil surface (Buell *et al.*, 1971). As the plant canopy filled in and litter covered the soil surface, cover of shade-intolerant mosses would have declined (Whitford and Whitford, 1978; Wilson and Tilman, 2002). In systems where soils are poor and retard the development of a closed herbaceous canopy, mosses may be persistent (Whitford and Whitford, 1978). After a 10-year period, moss cover again began to increase. As noted earlier, this period of moss expansion coincided with the timing of tree expansion at the site. Mosses increased greatly over the next 15 years, reaching a peak cover of 14%, then precipitously dropped. Mosses typically thrive in forest understory environments in microsites where they do not become covered by leaf litter (Bliss and Linn, 1955; Kimmerer, 2005), so this decline was unexpected. As bare soil was available for colonization, suitable habitat would not appear to have been limiting throughout this period of moss decline. The timing of moss decline also coincided with the decline of the invasive species *Rosa multiflora* at the site. At the plot scale, moss and *R. multiflora* cover were not correlated, suggesting that microclimate associations were not important in generating the changes in moss cover. Tree cover increased much earlier in succession, suggesting that neither shading nor increased deposition of leaf litter was a driver of moss decline. A more likely mechanism appears to be indirectly mediated by increasing deer activity (*Odocoileus virginianus*). With the loss of large areas of this thorny shrub, physical barriers to deer disappeared and would have allowed more foraging by deer in the successional forests. Though deer probably do not directly consume mosses, the more important impact may result from trampling (Studlar, 1980, 1983). The potential linkage between forest mosses and deer overabundance has not been examined (Côté *et al.*, 2004).

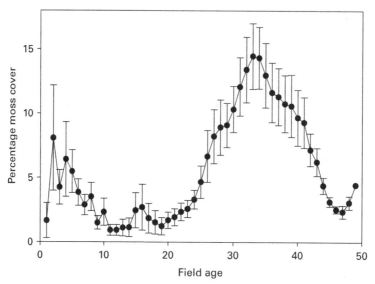

Figure 5.3 Change in the abundance of mosses over succession. Data presented are the averages (± SE) of all BSS fields.

Non-native species

Non-native species are major components of successional communities worldwide (Inouye *et al.*, 1987; Bastl *et al.*, 1997; D'Antonio and Meyerson, 2002; Tognetti *et al.*, 2010; McLane *et al.*, 2012). The linkage between succession and invasion has been noted in many studies and may represent real opportunities to merge two related bodies of ecological research and to inform invasion biology (Davis *et al.*, 2001, 2005). Within the BSS data, the relative abundance of non-native species decreased with successional age (Figure 5.4). The fields started out with nearly two-thirds of their cover composed of non-native species. Over the first 20 years of succession, the relative abundance of non-native species decreased, leveling off when approximately one-third of the community cover was contributed by non-native species. Following the initial decline, non-native abundance remained at that level for the next 30 years of succession.

While the decrease in the relative abundance of non-native species appears to suggest the overall decline of non-native species, this is not the case. When native and non-native species are separated, two different abundance patterns emerged over time (Figure 5.4). Though there were some fluctuations, native species were less abundant than non-natives for the first few years after abandonment. By year 10, total cover of natives exceeded that of non-natives and continued to increase for the next 20 years. In contrast, non-native species remained at a relatively constant cover throughout succession. The continued increase of native cover resulted in a decreased relative abundance of non-native species, even though non-native cover remained unchanged. During the period of constant non-native cover, the composition of non-native species would have changed dramatically. These data suggest that non-native species have formed a

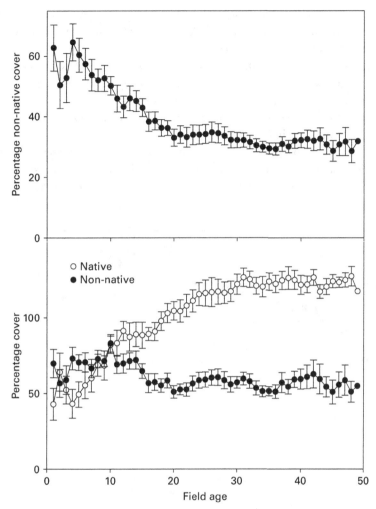

Figure 5.4 Change in the abundance of native and non-native species during succession. Data are plotted as proportion of cover (top) and as absolute cover (bottom). Data presented are the averages (± SE) of all BSS fields.

constant, and to this point persistent, component of the successional community. From a management perspective, the BSS data suggest that allowing succession to proceed is not a sufficient strategy to reduce the level of invasion within these communities.

Species diversity and abundance distributions

Despite Egler's suggestion that successional systems become largely closed to new immigrants as succession proceeds (Chapter 2; Egler, 1954), the BSS fields have accumulated species consistently throughout succession (Figure 5.5). Over the first 30 years, species richness at the field scale doubled from 43 species to an average of nearly

89 species. Most of the species that arrived late in succession were native species, some becoming more widespread over time. A few non-native species also arrived once the system had made the transition to a woody-dominated community. Of these late-arriving non-natives, *Alliaria petiolata*, *Microstegium vimineum* and *Rubus phoenicolasius* have become major components of the community and appear able to persist. Very late in succession, species richness declined somewhat from the mid-successional peak, consistent with the intermediate disturbance hypothesis (Loucks, 1970; Connell, 1978). We will return to this idea in Chapter 8.

In sharp contrast to the temporal dynamics of richness at the field scale, richness at the plot scale remained relatively constant (Figure 5.5). Throughout succession, the average number of species in each plot has remained within the narrow range of 11.5 to 14

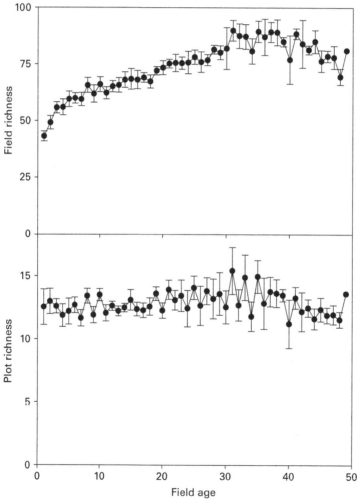

Figure 5.5 Change in species richness during succession measured at the field (top) and 1 m^2 plot (bottom) scales. Data presented are the averages (± SE) of all BSS fields. Plot richness is calculated as the mean plot richness within each field, averaged across fields.

species (Meiners *et al.*, 2002a). This range of species richness is within the range seen across a broad array of herbaceous communities (Gross *et al.*, 2000). The consistency of species richness at the 1 m^2 scale across herbaceous plant communities suggests that there is an inherent limit to the number of species able to coexist locally in these systems. In contrast, communities appear largely open to new species at coarser spatial scales (Stohlgren *et al.*, 1999, 2001; Bartomeus *et al.*, 2012), as seen when the data are aggregated at the field scale. The consistency of species richness through time in the BSS indicates that limits on local diversity do not change with succession. While the across-site consistency in richness was described for predominately herbaceous systems (Gross *et al.*, 2000), the pattern appears to have held through the young forests of the BSS. As these forests mature and transition to later successional trees and understory plants, we expect this pattern to break down, as forest understory abundance and diversity are typically lower at fine spatial scales (Small and McCarthy, 2003).

Evenness and diversity at the field scale varied in much more complex ways than did species richness. Though diversity incorporates both richness and the relative abundances of species within a community, the temporal patterns in diversity within the BSS appear largely driven by changes in evenness (Figure 5.6). Both evenness and diversity had the same general temporal changes, though the timing and magnitude of fluctuations varied somewhat. The first seven years after abandonment represented a period of rapid increase in evenness and diversity as both more than doubled during that period. The peak in both measures at year seven roughly corresponds with the dominance of biennials and herbaceous perennials within the community. Following this peak, both evenness and diversity declined, and then recovered to a similar peak in years 18 and 19, respectively. This period appears to correspond with the decline in herbaceous dominance and transition to woody-dominated communities. This secondary peak in evenness and diversity was followed by a 20+ year period of gradual decrease. Tree canopy closure would have slowly decreased the abundance of sub-canopy woody and herbaceous species and would lead to increased inequity in species abundances. The final few years of sampling appeared to capture an increase in the evenness and diversity of the community. Although this increase may have been partly generated by a severe drought in 1999 that increased canopy openness, it also corresponded with a widespread increase in understory herbaceous species within the site. Increasing understory species included both characteristic forest understory species (e.g. *Hackelia virginiana* and *Eupatorium rugosum*) and non-native forest invasive species (mostly *Alliaria petiolata* and *Microstegium vimineum*). The linkage between transitions in life form and increasing evenness is analogous to the idea presented by Armesto *et al.* (1991) that cycles of heterogeneity occur during succession and are driven by competitive dominance (Chapter 11). Increases in evenness early in succession have been noted in other systems (Wilson *et al.*, 1996) and may be more related to ecosystem function than diversity (Wilsey and Potvin, 2000; Wilsey and Polley, 2002).

Rank–abundance curves for the BSS fields (Figure 5.7) revealed abundance distribution patterns within each successional community that are common to most plant communities (Whittaker, 1975). Relatively few species dominated the community at any one time. From these dominant species, abundance decreased quickly, typically with

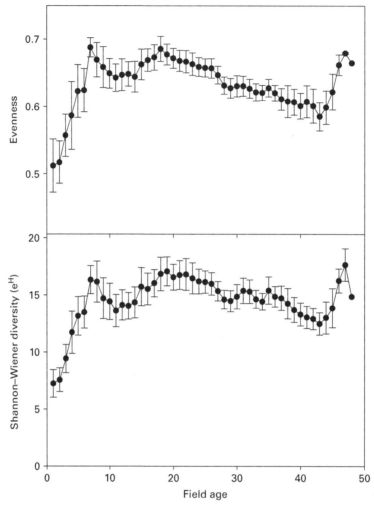

Figure 5.6 Changes in diversity and evenness during succession. Both measures are based on Shannon–Wiener indices calculated at the field scale. Data presented are the averages (± SE) of all BSS fields.

a tail of very low-abundance species. As succession proceeded and the richness of the communities increased, the number of dominants stayed relatively constant, but more low-abundance species were added to the community. The one exception to this general pattern is in the oldest age in the example. This field (C3) remained relatively constant in total species richness over the last 30 years of succession. In contrast, most fields increased in species richness during that period. This field became heavily invaded by the non-native species *Microstegium vimineum* between the last two years plotted. It appears that this invasion resulted in the loss of the rarer species, generating an attenuated rank abundance curve.

The shape of the rank–abundance curves also changed during succession. As succession proceeded, the overall steepness of the rank abundance line changed, indicating a

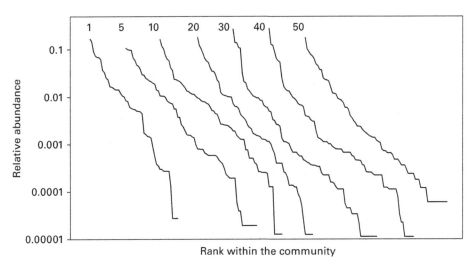

Figure 5.7 Dominance diversity curves for field C3 throughout succession. To separate field ages, ranks are separated along the x-axis by 20 species.

shift in dominance patterns. From years 1–10, the slope of the curves gradually decreased, indicating an overall lessening of dominance within the community. However, at year 20, the rank–abundance curve quickly increased in steepness. This period coincides with the heavy dominance of clonal perennial herbs, particularly goldenrods (*Solidago* spp.) and with the peak abundance of the non-native liana, *Lonicera japonica*. During years 30 through 50, the steepness of the rank–abundance curves gradually decreased again. These patterns are also visible in the diversity measures previously discussed. The general broadening of the rank–abundance curve has been noted in chronosequence successional studies of community structure (Bazzaz, 1975; Whittaker, 1975), and has been used to infer shifts in dominance.

Species composition

While we will spend a great deal of time examining the successional trajectories of individual species in the next chapter, it is worth examining the overall compositional changes that occur during succession. A non-metric multidimensional scaling (NMS) ordination of the BSS data revealed some characteristic dynamics of the system (Figure 5.8). The majority of changes in composition appear roughly directional and associated with the age of the system (rotated to correlate with axis I). While there was a great deal of compositional variation from year to year within a field, there were relatively few reversions along the X-axis, and those that did occur were relatively small. For this system at least, succession at the community level as a whole appears largely directional in nature (Clements, 1916; Odum, 1969).

A major feature of early successional communities, particularly immediately after abandonment, was the broad range in composition seen across the BSS fields. This

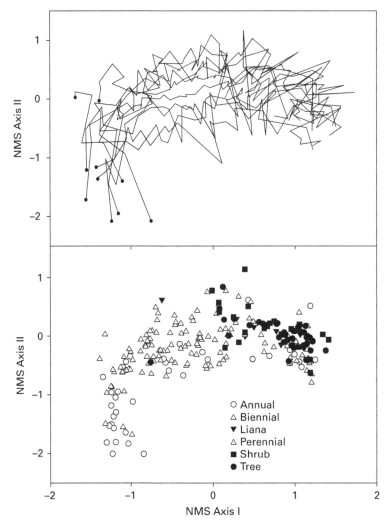

Figure 5.8 NMS ordination of the compositional changes occurring during succession. The ordination was conducted on species data relativized by their maxima to equally weigh all species. The ordination was rotated to maximize the correlation with successional age along the X-axis. The upper panel shows the trajectory of all 10 BSS fields over time with the initial composition indicated by a dot. The lower panel plots species, identified by life form, in the ordination space.

compositional variation, seen along the Y-axis in the ordination (Figure 5.8), was largely generated by the differences among fields in pre-abandonment agricultural treatment, though variation generated by differences in abandonment year and local dispersal limitation were probably also important (Chapter 9). Those initial plant communities show a remarkable amount of variation in the abundance of grasses and winter annuals, as would be expected based on contrasts in season of abandonment, degree of soil disturbance at abandonment and last crop. The composition of first-year communities was so different from each other and the rest of the data that most of them were flagged as potential outliers to the ordination analysis. However, within one year of succession the field's compositional differences decreased so that they converged on a composition more similar across communities. We will re-visit the idea of community convergence and assembly in Chapter 9.

Within the compositional changes, we also see changes in the species characteristic of the community. When the component species are plotted in the ordination space, we generate a visual summary of the role of each species in the ordination (Figure 5.8). The ordination shows similar transitions in life form to those described earlier, but provides better resolution on the successional positioning of each species. Annuals were strongly clustered with the youngest fields, though they were also represented throughout the entire successional sequence. There was great variability among fields in which annuals came to dominate, spreading these species along the Y-axis. As succession proceeded from left to right in the ordination, species life form shifted to those with longer life spans. Each life form dominated one region of the species plot, but was spread out in successional time. For example, trees were clustered on the extreme positive end of the X-axis, but some tree species, notably *Juniperus virginiana* and *Cornus florida*, appeared closer to the middle of the successional gradient. While life form is clearly a useful ecological abstraction to help understand successional transitions, species within life forms may vary dramatically in their successional role. Such differences in the successional patterning of species have been labeled as individualistic (Gleason, 1927).

Using the same analytical approach, we can also address the speed of successional change. Typically, the tempo of successional dynamics is thought to shift with the life spans of the constituent species (Prach *et al.*, 1993; Debussche *et al.*, 1996) leading to slower rates of change as succession proceeds (Anderson, 2007). Short-lived species will die and be replaced quickly, leading to large compositional changes. As succession generates communities dominated by longer-lived species, mortality rates will decline, reducing the speed of replacement. At geographic scales, the rate of succession to trees may be associated with soil fertility and latitude, which are related to the dominance of herbaceous species that compete with tree seedlings (Wright and Fridley, 2010; Fridley and Wright, 2012). This places the BSS towards the slower end of secondary succession in North America (Wright and Fridley, 2010), though transitions in herbaceous species are still likely rapid. We will re-visit the idea of herbaceous–woody interactions in Chapter 9.

Within the BSS data, there were clear changes in the rate of succession over time (Figure 5.9). As already discussed, first-year communities differed greatly from each other and from all later years in succession. The breakdown of these initial differences generated very large compositional changes within the first two years of succession. From those initial changes, the rate of compositional change decreased for most of the first 20 years of succession. This period of slowing succession corresponded closely with the transition to and dominance of herbaceous perennials. Following this period, the rate of succession increased substantially, corresponding with the shift in dominance towards woody species. While the growth and establishment of woody species was typically quite slow within the BSS, the collapse of herbaceous perennials, many of which were tall, clonally reproducing species, generated large compositional

Community patterns and dynamics

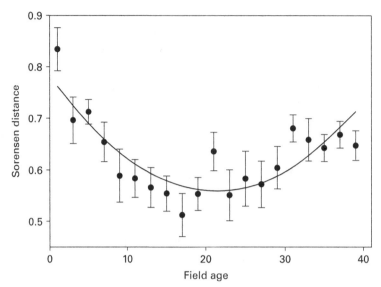

Figure 5.9 Temporal changes in the rate of succession as measured by Sorensen's dissimilarity index between years. Dissimilarities were calculated for two-year windows to keep the measures consistent during the alternate-year sampling of the later years. Data presented are the averages (± SE) of all BSS fields and the trend is a best fit line (quadratic).

changes in a relatively short period of time. As the BSS fields have aged, the colonization of forest understory species has also led to the maintenance of higher rates of compositional change. Looking to the future, as the last vestiges of the shade-intolerant species are displaced from the system and are replaced by shade-tolerant understory species, we would expect the successional rate to also slow down. Further into succession, the rate of canopy tree replacement by shade-tolerant species will depend on the lifespan of the current canopy trees and the rates of disturbance experienced by the system.

Relation of the BSS data to successional predictions

Permanent plots vs. the chronosequence approach

One of the motivations behind the initiation of the BSS project was the desire to assess the effectiveness of the chronosequence, or space for time substitution approach, to quantifying successional dynamics (Chapter 2). Under the direction of Murray Buell, Gily Bard had completed a chronosequence study of the successional processes in and around HMFC (Bard, 1952). While herbaceous species data were collected in plots like those of the BSS, she used different methodologies to assess shrub and tree abundance. These differences prevent direct quantitative comparisons, but the general similarities and differences can be easily seen.

The general transition of life forms that Bard described is repeated in the BSS data. However, the timing and composition of these transitions differ. Bard's study includes as one of the major stages of succession a grass phase dominated by *Schizachyrium scoparium*, a warm-season perennial grass. This phase persisted for nearly 30 years before yielding to dominance by woody species. Though this species was present in the BSS, and was sometimes locally abundant, it remained relatively unimportant within the BSS fields (Pickett, 1983). Despite the abundance of other grasses within the BSS, tree regeneration and dominance appears to have begun much earlier within the BSS than in Bard's study. As trees typically compete heavily with herbaceous plants (Kolb and Steiner, 1990; De Steven, 1991b; Gill and Marks, 1991; Debussche and Lepart, 1992; Inouye *et al.*, 1994; Davis *et al.*, 1998), particularly for water, the lack of a persistent grass phase may have sped this successional transition (Fridley and Wright, 2012). *Schizachyrium scorparum* is also commonly associated with poor soils (Billings, 1938; Niering and Dreyer, 1989), so it is also possible that the fields examined by Bard were on poorer, less agriculturally productive soils that may have limited tree growth. Fields on non-productive soils would also be more likely to be retired from agriculture than more fertile, and economically productive, sites.

A second difference between the BSS data and Bard's chronosequence is in the importance of shrubs and lianas, which Bard found to be in overall low abundance. In contrast, shrubs and lianas dominated the middle stages of succession within the BSS. The majority of shrub and liana cover within the mid-successional communities of the BSS consisted of two non-native, regionally problematic invaders, *Lonicera japonica* (liana) and *Rosa multiflora* (shrub). Both of these species were present, but at low abundance in Bard's chronosequence. *Lonicera japonica* was widely introduced as an ornamental, while *R. multiflora* was actively introduced for decades for use as a living fence (Patterson, 1976). The initiation of the BSS study coincided with the expansion of these species in the landscape and their presence would represent a shift in the local species pool. The older successional areas sampled by Bard would have had relatively little opportunity to become heavily invaded by either of these two species (Pickett, 1983). In contrast, the fields of the BSS would have initiated succession when these two species were already present in the surrounding landscape. Together, these two relatively new additions to the local flora have dramatically altered succession by changing the structure of mid-successional communities. However, they do not appear to have prevented the key successional transition to forests.

Though chronosequences have been widely used to address successional dynamics, they are clearly limited in their predictive ability. The critical assumptions of the chronosequence approach are that: (1) all sites will be, or have been, subjected to similar processes and that (2) the patterns among sites reflect temporal change (Pickett, 1989; Glenn-Lewin and van der Maarel, 1992; Johnson and Miyanishi, 2008). However, these assumptions may not be met and chronosequences may suggest quite different patterns from those revealed by long-term studies (Jackson *et al.*, 1988; Pickett, 1989; Bakker *et al.*, 1996; Johnson and Miyanishi, 2008; Walker *et al.*, 2010). Upon examining our conceptual model of vegetation dynamics, it appears unlikely that these assumptions will ever be sufficiently met. Over the decades of the BSS, deer (*Odocoileus virginianus*) densities have increased, nutrient deposition has increased and many new species have entered the landscape. In addition to this, we see clear evidence of the influence of initial abandonment conditions and localized dispersal limitation on vegetation dynamics (Chapter 9). Since the diverse pool of potential drivers of vegetation change is unlikely to remain static for long periods of time, it may be more appropriate to ask when the usage of chronosequences is justified. As our ability to make predictions about ecological systems is limited by the quality and temporal extent of available data (Clark *et al.*, 2001), chronosequences will remain a necessary tool to understand vegetation dynamics. However, this utility should be tempered by full acknowledgement of the changing spectrum of community drivers that generates variation among sites.

E. P. Odum's successional trends

In Odum's classic treatment of succession, "The strategy of ecosystem development" (1969) he makes some very clear and detailed predictions and generalizations about succession. Though the paper clearly espouses a more holistic and Clementsian view of communities, particularly in the language of community "development," it remains one of the foundational treatments of succession, and the table summarizing his views occurs in nearly all ecology textbooks in one form or another. Despite the wealth of conceptual and testable hypotheses in this paper, relatively few have been systematically addressed in experimental studies. This was true at the time the paper was published and remains so today. We have included Odum's table using his original phrasing (Table 5.1) and will address those trends to which the BSS data can be applied.

Under Odum's predictions for energy and nutrient fluxes, the BSS data have limited direct applicability. However, the accumulation of cover during succession is consistent with the shifts in production, production/respiration ratios and standing crop biomass that are predicted. When we address the functional ecology of succession (Chapter 12) we will see that the characteristics of the community also indicate a decrease in potential growth rates and a shift from low to high biomass investment in leaves, consistent with Odum's predictions.

Within Odum's discussion of community structure, we can address several of his hypotheses. We find mixed results with regard to the prediction that richness transitions from low to high during succession. At the field scale, species richness does increase over time, but early successional systems were not depauperate in species richness as is suggested. Also, we have begun to see the loss of species from the system once canopy closure has occurred, more in line with patterns described by Orie Loucks (1970). In contrast to patterns seen at the field scale, richness remained remarkably constant over time at the 1 m^2 scale. As Odum was largely employing an ecosystem perspective in his paper, the field-scale pattern would seem the most appropriate to his ideas and appears to follow the general prediction of increasing richness. The issue of scale is an important one and represents an opportunity to extend the scope of Odum's ideas on succession. For instance, if we link Odum's position that spatial heterogeneity should increase over time, becoming "well-organized," with the fixed level of richness observed at the plot

Table 5.1 A tabular model of ecological succession: trends to be expected in the development of ecosystems. (From Odum, 1969; reprinted with permission from AAAS.)

Ecosystem attributes	Developmental stages	Mature stages
Community energetics		
Gross production/community respiration (P/R ratio)	Greater or less than 1	Approaches 1
Gross production/standing crop biomass (P/B ratio)	High	Low
Biomass supported/unit energy flow (B/E ratio)	Low	High
Net community production (yield)	High	Low
Food chains	Linear, predominantly grazing	Web-like, predominantly detritus
Community structure		
Total organic matter	Small	Large
Inorganic nutrients	Extrabiotic	Intrabiotic
Species diversity-variety component	Low	High
Species diversity-equitability component	Low	High
Biochemical diversity	Low	High
Stratification and spatial heterogeneity (pattern diversity)	Poorly organized	Well-organized
Life history		
Niche specialization	Broad	Narrow
Size of organism	Small	Large
Life cycles	Short, simple	Long, complex
Nutrient cycling		
Mineral cycles	Open	Closed
Nutrient exchange rate, between organisms and environment	Rapid	Slow
Role of detritus in nutrient regeneration	Unimportant	Important
Selection pressure		
Growth form	For rapid growth ("r-selection")	For feedback control ("K-selection")
Production	Quantity	Quality
Overall homeostasis		
Internal symbiosis	Undeveloped	Developed
Nutrient conservation	Poor	Good
Stability (resistance to external perturbations)	Poor	Good
Entropy	High	Low
Information	Low	High

scale, we would generate increasing species richness at the field scale. This would largely be driven by increased turnover in species from plot to plot. As Odum's work largely preceded many of the discussions on scale in ecology, it is difficult to fully assess what his intended meaning was.

In dealing with diversity, Odum addressed species equitability separately from richness, predicting that it would increase through succession. An increase in evenness occurred only during the earliest stages of succession of the BSS and was much more complex than the general transition from low to high suggested by Odum. Evenness appears linked with transitions in dominant life form in the BSS. As future compositional changes will likely be more subtle and not involve changes in dominant life form, evenness may increase again as the forest matures, to follow Odum's language. Finally, later successional communities were more structurally diverse, as predicted, with much more vertical stratification of the community. However, a large component of the understory of the late successional communities consisted of non-native species. We will return to Odum's idea of pattern and spatial heterogeneity in Chapter 11, as this is a much more complex issue to address.

The BSS data clearly follow the general predictions of increasing life span and organism size at the community level as these are represented in the life-form transitions that occurred. These life-history transitions also followed the prediction of shifting from r-selection to K-selection during succession, as ruderal annuals and perennials were replaced by long-lived perennial and woody species. In a related idea, Odum predicted that niche specialization should transition from broad to narrow. The relationship of this prediction to the BSS data is a bit more complicated. Niche specialization could be interpreted as specialization along a successional gradient. In this case, early successional species were much more restricted temporally, and therefore exhibited greater niche specialization than later successional species. Early successional species have also been shown to have very specific responses to environmental conditions, which can determine the composition of annual communities (Pickett and Bazzaz, 1978; Parrish and Bazzaz, 1982; Myster and Pickett, 1988; Jensen and Gutekunst, 2003). These strong environmental associations also argue for overall specialization within early successional species. On the other hand, early successional species were typically widespread across fields and less patchy within fields (spatially homogeneous) within the BSS, indicating broad tolerances in addition to high vagility.

Stability, the resistance to external perturbation, was also hypothesized to increase with succession. The major disturbance that has influenced the vegetation of the BSS has been drought (Chapter 7). While we will deal with the complexities imposed by this disturbance later, it is difficult to directly assess Odum's proposition with the BSS data. Droughts have occurred throughout a range of successional ages, but have also varied in their timing and severity. As the impacts expressed by vegetation will be a combination of the characteristics of the disturbance and the community's response, it is not possible to isolate these factors with the available data. We will however, be able to examine the strength of the linkage between variability in rainfall and the structure of the community and use that to assess whether successional development (*sensu* Odum) is related to stability.

Several of Odum's predictions are, in general, supported by the BSS data, and many other predictions are yet to be addressed. It is a shame that we have not come further in our understanding over the last 40 years. The current attempt to link species traits with ecosystem properties and processes reflects a return to Odum's more holistic approach to communities in the context of ecosystems (Rees *et al.*, 2001; Lavorel and Garnier, 2002; Diaz *et al.*, 2004; Garnier *et al.*, 2004). Prediction, though, is not always simple even for the best of scientists. Ironically, some of the areas within the Savannah River Ecology Lab where Odum developed his views on succession (Golley, 1965; Golley *et al.*, 1994; Pinder *et al.*, 1995; Odum, 1960) persist as of this writing as open herbaceous communities. This is in stark contrast to Odum's prediction of rapid transition to pine forest.

F. E. Egler and initial floristics

Egler's contributions to succession, like those of many of the early ecologists, have largely been reduced to a caricature of his original ideas. Though he clearly argued for the primacy of species availability early in succession as a feature of his initial floristic composition concept, he did acknowledge that most successions would be jointly controlled by initial composition and relay or Clementsian floristics (Egler, 1954). The successional transitions of the BSS offer mixed results when it comes to Egler's hypothesis. The communities of the BSS were not closed to the arrival of new species once succession began. Even species that eventually came to dominate the system were able to establish relatively late in succession. Many of these successful late colonizers were non-native species that have become regionally problematic in forests and may not have been available to colonize earlier in succession. Other woodland species, such as the fern *Asplenium platyneuron*, or the understory herbs *Hackelia virginiana*, *Pilea pumila* and *Sanicula gregaria* have also appeared late in succession. Whether colonization by these species was initially limited because of the lack of suitable forest habitat (i.e. relay floristics) or because of limited dispersal is not clear. Following some of Egler's ideas on land management as it related to initial floristic composition (Pound and Egler, 1953; Wilson *et al.*, 1992), the establishment of perennial grass cover within the hay fields should have prevented, or at least greatly reduced, tree regeneration. However, there was no discernible delay in tree establishment in the hay fields (Meiners *et al.*, 2007), though the initial dynamics within hay fields were quite different from those following row crops.

In support of the initial floristic hypothesis, many of the species that were important during succession did arrive very early in the site. These species grew to abundance according to the temporal limitations of their life forms and allocation patterns. Therefore, short-lived species colonized and dominated very quickly. In contrast, woody species, with much greater allocation to stem biomass, grew more slowly and exhibited a temporal separation between their arrival and their ultimate dominance. Canopy tree composition has continued to vary across the BSS fields as the species composition of the fields largely reflects the canopy trees that were immediately adjacent to the fields. While variation in tree composition is mechanistically more related to dispersal limitation of the larger seeded trees, it generates similar patterns to those predicted by initial floristic composition. As Egler's ideas were formulated based on a small scale (15 × 15 m) patch of *Viburnum lentago* that had resisted colonization by trees (Niering and Egler, 1955; Niering *et al.*, 1986), it is difficult to determine whether the concept should be applied to broader spatial scales, though Egler certainly viewed this as a general phenomenon. Ecologically, this concept may be more closely linked with the idea of localized priority effects. As with Clements and Gleason, Egler also espoused a multitude of successional drivers and not only the one idea that he is best known for. Criticizing the initial floristic composition idea as not the only driver of succession places it in a role that it was never meant to occupy.

Summary

The broad strokes of the successional dynamics seen within the BSS largely fit within the range of community patterns documented in other systems. That being said, there is value in looking at the entire suite of temporal dynamics from a single system. One difficulty in relating the BSS data to other studies is that most have not examined community processes using the full palette of approaches done here. What has resulted is a very clear and complete view of succession that yields interesting dynamics and associations that would otherwise have been missed. This work forms the basis for the mechanistic approaches that follow and provides a richly textured context for the integrative topics addressed later in this text.

In contrast to the general patterns of succession, we find much more mixed support for many of the ideas and generalities upon which so much of modern ecology is based. It is always useful to re-visit the historical foundations of a discipline to continually re-evaluate and revise the ideas which may become entrenched to the point of being assumptions (Cooper, 1926). The diversity and interaction among successional drivers illustrated in our framework of vegetation change suggests a myriad contingencies that can generate novel successional dynamics. The value of re-visiting these ideas is not in verifying or refuting any of them, but in conceptualizing linkages among approaches to succession. The ultimate goal then becomes determining the conditions under which a particular pattern or process should be expected.

From the viewpoint of communities, we will now become more reductionist and address the dynamics of the individual species that have composed the BSS communities over the past 50 years. Again, we will be able to address long-held ideas in succession and determine some of the characteristics responsible for generating the dynamics of individual species.

6 Dynamics of populations through succession

> "In conclusion, it may be said that every species of plant is a law unto itself, the distribution of which in space depends upon its individual peculiarities of migration and environmental requirements"
> Henry Allan Gleason. The individualistic concept of the plant association (1926)

Gleason's approach to succession, and to the structure of communities in general, focused on the population as the primary determinant of structure and dynamics. This view is repeated by others that have dealt with succession as, first and foremost, a population process (e.g. Peet and Christensen, 1980). The diversity and structure of communities in this view is therefore an aggregation of the component population-level processes. The classic visual portrayal of community organization along a gradient is to plot the abundances of a suite of species along the dominant gradient (Figure 6.1, from Whittaker 1975). From this display comes the depiction of Clementsian vs. Gleasonian community organization seen in most textbooks. Clements' view is represented by discrete phases with rapid transitions in composition that are synchronous across species (Figure 6.1 A and C). In contrast, the Gleasonian depiction is a collection of broad and narrow species distributions with no synchronous transitions from one successional phase to another (Figure 6.1 B and D). Also included in Whittaker's depiction of species distribution is the role of competition, where shifts in the competitive superiority of species lead to rapid replacement of species along the gradient (Figure 6.1 A and B). Modern ecological thought, and certainly Whittaker's data, largely supports the gradual and individualistic transition of dominance from one species to the next (Figure 6.1 D).

Succession may be thought of as a complex gradient that encompasses both biotic and abiotic drivers of community composition (Tansley, 1935). Along with shifts in the identity and life form of competing species, there are shifts in resource availability and the identity of limiting resources (Tilman, 1985), shifts in the pools and fluxes of nutrient cycling (Bormann and Likens, 1979), and shifts in the activity and identity of animals that act as consumers and agents of dispersal (Elton, 1927; Brown, 1985; Myster and Pickett, 1992b; Edwards-Jones and Brown, 1993; Cadenasso et al., 2002). Time serves as a surrogate for this complex suite of changes that result from and influence succession. Unlike other ecological studies that focus on such factors as the distribution of species along moisture gradients, no mechanistic driver is implied in succession, as time itself does not lead to successional dynamics. Despite this limitation, we may certainly expect plant species to assort themselves along the temporal gradient of succession in analogous ways to how species are assorted along specific environmental gradients. A

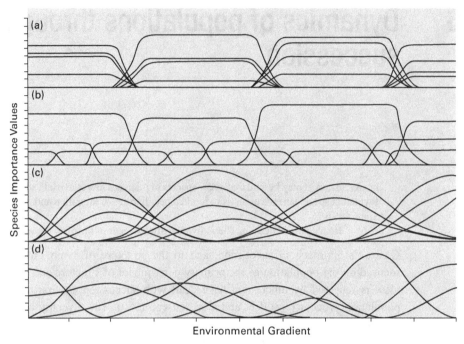

Figure 6.1 Theoretical distribution of species (each line) along environmental gradients from Whittaker (1975). Panels A and B represent species with strong competitive interactions that result in rapid shifts in dominance. Panels C and D represent communities where such tight associations do not occur. Panels A and C represent abrupt, Clementsian shifts in community composition, while B and D represent gradual, Gleasonian transitions. From Whittaker, Robert C., *Communities and Ecosystems*, 2nd Edition, © 1975. Reprinted by permission of Pearson Education, Inc., Upper Saddle River, NJ.

temporal gradient was used by both Gleason and Clements in predicting the nature of successional transitions.

In starting out the discussion of the population level dynamics of the BSS, we thought it would be illustrative to make a similar display of species distributions over time within the BSS data (Figure 6.2). The result, like any examination of the dichotomies that often characterize science, is something intermediate between the two extremes of Gleason and Clements. We can see some clear transitions among successional stages, particularly in the transition between annuals/biennials and herbaceous perennials, and the initial transition between perennial herbs and woody species. On the other hand, the data clearly reveal some species with broad temporal distributions that span a large period of successional time, in some cases overlapping different stages. In contrast to the stage-like transitions of early successional herbaceous communities, the compositional transitions of late successional communities are essentially continuous. Although the BSS data already encompass the major structural changes that are expected in succession to forest, further transitions are expected as the forest continues to change. Looking ahead to these changes, we would anticipate that these transitions would also occur gradually, as existing canopy trees are replaced by shade-tolerant species. As there is no reason to

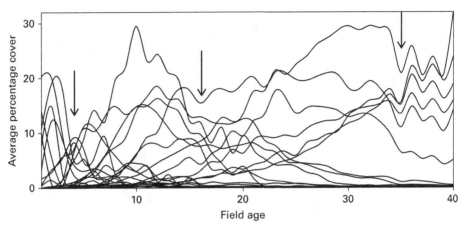

Figure 6.2 Successional trajectories of the 20 species with the highest peak cover in the BSS data. Arrows indicate periods of synchrony among species that indicate transitions in successional stages. These successional stages are further described in Figure 6.3.

expect synchronous mortality of the canopy trees that established early in succession, there should be no pulse of regeneration and therefore no abrupt transition to late successional forest.

No matter what your philosophical position along the Clements–Gleason gradient of community perspectives is, populations are clearly an important component of successional ecology. The many generalizations about the characteristics of early and late successional species are descriptions of the population ecology of succession. The prevalence of these ideas in the ecological literature strongly argues for the importance of population ecology to succession and to community ecology in general. Communities, by definition, are local collections of species. We may debate how strong and important the interactions among individual components are, but we must understand the limits on the building blocks of communities to understand the whole system.

Below we will explore the temporal patterns of species within the BSS, relate these trends to species characteristics, and explore the fundamental tradeoffs that result in much of the structure and dynamics that we see in communities. Along the way, we will discuss the limitations imposed by differences in resource allocation among species and begin our exploration of plant strategies (*sensu* Grime, 2001). These themes will be explored more fully in later chapters, particularly in those dealing with species invasion (Chapter 10) and functional ecology (Chapter 12).

The population trajectories of species during succession

If we use population dynamics to suggest the location of transitions from one successional stage to the next, we can generate four successional phases in the BSS: short-lived herbs, perennial herbs, mixed woody and forest (Figure 6.3). The boundaries of the first two phases are relatively clearly defined by synchronous population dynamics, and

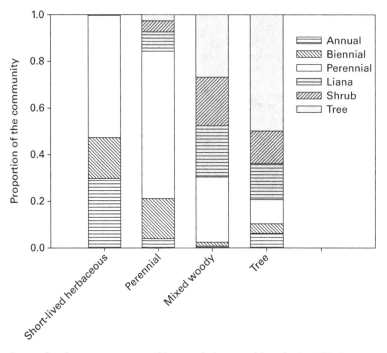

Figure 6.3 Successional stages as suggested by population transitions in the BSS data. Data plotted are the average abundance of each life form across all years within each successional stage.

generate communities dominated by short-lived and perennial herbaceous plants. While the beginning of the mixed woody phase is clearly defined by the collapse of herbaceous perennials and an increase in woody taxa, the boundary with the forest community is less clear. Woody species both increase and decrease more slowly than herbaceous species, leading to broad periods of transition. The distribution of life forms in each phase directly results from breaking up the continuous shift in community structure (Figure 5.1). While breaking up a continuous process, such as succession, into discreet phases may seem counter-productive, it does help us to categorize species and is a dramatic improvement over the classic dichotomy of early and late successional species. The population trajectories of most individual species actually place them fairly neatly into one of these four successional phases.

The dynamics of individual species largely followed the life-form dynamics seen in the previous chapter. However, aggregate changes in life forms masked the unique nature of individual species dynamics during succession. Within the 341 species encountered over the 50 years of the BSS, 84 have appeared in sufficient abundance to adequately follow and quantify their population trends through time. These species will form the foundation for much of the species-level work throughout this volume. To examine variation among species within a life form, we have plotted the temporal changes in cover for five key species within each life form. For most of the life forms, we plotted the five most abundant species within the BSS data. For these groups, the remaining species typically occurred in much lower abundance. Perennial herbs and

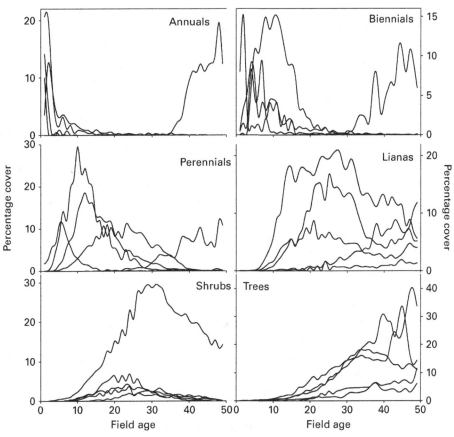

Figure 6.4 Population trajectories for selected species within each life form. Species were selected to represent the range of temporal dynamics within each life form. Annuals: *Ambrosia artemisiifolia*, *Digitaria sanguinalis*, *Erigeron annuus*, *Microstegium vimineum* and *Raphanus raphanistrum*; Biennials/short-lived perennials: *Alliaria petiolata*, *Aster pilosus*, *Barbarea vulgaris*, *Daucus carota* and *Trifolium pratense*; Perennials: *Elytrigia repens*, *Eupatorium rugosum*, *Fragaria virginiana*, *Hieracium caespitosum*, *Solidago juncea*; Lianas: *Celastrus orbiculatus*, *Lonicera japonica*, *Parthenocissus quinquefolia*, *Toxicodendron radicans* and *Vitis* spp.; Shrubs: *Cornus racemosa*, *Rhus glabra*, *Rosa multiflora*, *Rosa virginiana*, *Rubus flagellaris* and *Rubus allegheniensis*. Trees: *Acer rubrum*, *Cornus florida*, *Fraxinus americana*, *Juniperus virginiana* and *Quercus* spp.

trees were represented in the data by too many species with widely divergent successional trajectories to simply focus on the five most abundant. For these two life forms, we have selected representative species that span the successional range seen within each group.

Most of the annuals (Figure 6.4) were characterized by the expected peak in abundance immediately following abandonment followed by a rapid decline. After the first year or two, these species persisted at very low frequencies and abundance. There was remarkably little variation among species in this pattern. One annual, a forest understory species, exhibited the opposite temporal trend, with its peak at the latest stages of succession. As a forest annual, it persisted following its initial expansion across the site rather than declining. Though this species occurred much later than other annuals and persisted, it also exhibited the rapid rates of increase characteristic of the group.

Biennial and short-lived perennial species increased in abundance rapidly across the site, typically peaking between two and nine years after abandonment (Figure 6.4). These species tended to not become as abundant as did the annuals, but persisted for a much longer time. Biennial species also tended to persist at low abundances once the population had collapsed. Among biennial species, there was much greater variation in the timing of the population peak and duration during succession than was seen in the annuals. This may partly be an effect of the artificial separation between short- and long-lived perennials. However, there was also much greater interannual variation in species abundance within this life form. As in the annuals, the biennial/short-lived perennial group contained a single forest understory species that reached a sustained peak late in succession. This species still exhibited the rapid growth rate as well as the interannual variation in cover characteristic of the life form, however. Together with the annuals, the relatively synchronous population trends generated the first and the clearest of the separations of successional stages discussed above (Figure 6.3).

Perennial herbs exhibited a much broader range of successional trajectories than the other herbaceous groups. These species in general increased more slowly than the annual or biennial groups and also declined more slowly following their successional peaks (Figure 6.4). While herbaceous perennials as a whole were mid-successional dominants, they encompassed a broad range of early to late successional species. The predominance of mid-successional species within the group generated the second abrupt transition in community composition as woody species expanded. The broad range in trajectories of later successional perennials contributed to the gradual transitions in composition characteristic of mid-to-late successional communities in the BSS.

Woody species consistently peaked later in succession than nearly all herbaceous species, and also tended to be abundant for much longer periods of time (Figure 6.4). Shrubs and vines were dominant at mid-successional periods, though there was considerable variation among species. Shrub cover within the BSS was dominated by a single non-native species, *Rosa multiflora*, with the remaining native species having much less abundance. Shrubs as a whole peaked earlier in succession than lianas. Lianas as a life form also had a greater variety of successional trajectories than shrubs. While some species peaked in mid-succession, some lianas expanded later in succession and continued to do so (Ladwig and Meiners, 2010b). While mid-successional shrubs were largely displaced from the system as canopy closure occurred, late successional, understory shrub species may eventually colonize the young forests of the BSS. Native understory shrubs are typically not common in young forests with high tree density and dense canopies. The limited presence of late successional shrub species may also reflect limitation of the species pool, as many of these species are also in low abundance within the old-growth forest of HMFC.

Trees as a life form also exhibited a diversity of successional trajectories. Most species appeared relatively early in succession, but typically did not begin to expand in abundance until much later (Figure 6.4). The BSS data show two successional groupings of trees, those which peaked earlier in succession, around year 35, and those which have continued to increase. The earlier species represent shade intolerant and typically shorter trees that are soon overtopped by later successional species. These species, however, colonized the fields relatively quickly and became reproductively mature well before the late successional trees became abundant (*e.g. Cornus florida* and *Juniperus virginiana*). The later successional trees represent those species abundant in the adjacent old-growth forest such as *Acer rubrum*, *Carya* spp. and *Quercus* spp. *Cornus florida* was originally part of the subcanopy of the old-growth forest, but has failed to regenerate following an anthracnose (*Discula desctructiva*) outbreak and has become largely absent from the old-growth forest. Within the surrounding old fields, it spread rapidly and has clearly functioned as an early successional species.

Differences among herbaceous life forms represent positions along a gradient from short-lived to long-lived plants. Therefore it is not surprising that there is much overlap among herbaceous groups in their successional behavior. In contrast, woody life forms represent fundamental shifts in allocation patterns that are distinct from herbaceous plants. These constraints place even the fastest-growing woody species much later in succession. Woody life forms, however, do not represent positions along a continuum of woody plant strategies, but are rather discrete architectural types. Shrubs and lianas both represent opportunistic life forms that grow rapidly and reproduce when conditions allow. While lianas have the ability to enter the forest canopy and persist in high-light environments, shrubs are constrained to the understory. Trees allocate more resources to generating supportive stem structures than either shrubs or lianas and are therefore constrained to increase more slowly in succession. Gradients exist within each of the woody life forms with early and late successional forms present within each. For example, larger lianas, such as *Vitis* species, have the potential to reach and persist in a mature forest canopy, while the shorter *Lonicera japonica* dominated only during herbaceous and shrubby stages of succession when trees were shorter (Ladwig and Meiners, 2010a, 2010b).

Quantifying species population dynamics

In order to compare population dynamics of species within the BSS and to look for tradeoffs in population strategies, it was necessary to quantify the dynamics of each species. To do this, descriptions of population dynamics were calculated based on both abundance and frequency (Meiners, 2007; Meiners *et al.*, 2009). Plant abundance was measured as percentage cover and frequency as the number of plots occupied for species over time. Both frequency and abundance measures were used to account for the range of species sizes and abundances within the BSS. In other words, it was necessary to be able to differentiate between small, frequent species and larger, less frequent ones, which may have the same total cover. For each species, population trajectories were calculated by summing the frequency and cover across all 10 of the BSS fields. While pooling the data across fields would have obscured any variation among fields, it provided a conservative estimate of population dynamics across the entire site. This seemed the

Dynamics of populations through succession

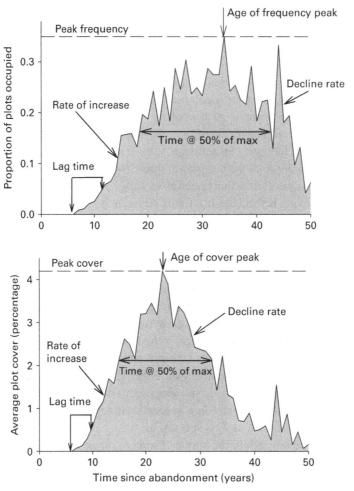

Figure 6.5 An example of the methodology used to generate the metrics of population dynamics for the 84 species that occurred in sufficient abundance to generate clear and quantifiable population dynamics. The species depicted is *Rubus allegheniensis*, a bird-dispersed shrub common in mid-successional communities. Not depicted are the two measures of plot dominance (total cover/plots occupied) – peak dominance and the year of peak dominance. Comparison of the panels shows the differences revealed by examining both cover and frequency.

best approach as we were interested in comparing species rather than examining variation within species. Based on the successional trajectory for each species (Figure 6.5) the following metrics were calculated for both cover and frequency data:

Rate of increase – the largest increase in abundance scaled to a single year. This metric was used as a functional estimate of each species population growth rate.

Peak abundance – the maximum frequency or average cover across all fields.

Age of peak abundance – the successional age when a species attained maximum abundance. This metric was used to determine a species successional position.

Span at 50% of peak – the length of time when a species maintained at least 50% of its peak abundance. This metric was used as a measure of temporal persistence during succession.

Rate of decline – the largest yearly decrease in abundance.

Lag time – the number of years between when a species first appeared and when it exceeded a critical abundance threshold – 5% for frequency or 10% for cover.

Two other metrics, peak dominance and the age of peak dominance, were calculated based on the total species cover divided by the number of plots occupied in each year. These two metrics were specifically included as a measure of a species' ability to control local space. Therefore, species that took over plots, even if the species occurred too infrequently to generate much cover, would be separated from frequent, but lower abundance, species. Together, these measures generated 14 metrics (6 frequency, 6 cover and 2 dominance) that adequately differentiated the range of population dynamics seen within the BSS data (Meiners, 2007; Meiners et al., 2009).

Variation among life forms

To explore the basic variation among life forms, we have selected six of the population dynamic metrics, half based on frequency, the other half based on abundance to explore in detail. These six metrics captured much of the fundamental variation among species in how they have performed during succession. We will return to the entire suite of population metrics when we address population dynamics using a multivariate approach.

The three population metrics based on frequency all measured aspects of the spread of the species across the site during succession. The rate of increase was much greater in annuals than in any of the other life forms and averaged a peak spread of 30% of plots in a single year (Figure 6.6). Perennials and shrubs were the slowest-spreading life forms, with an average rate of increase of 13% per year. In contrast, population lag times varied dramatically between woody and herbaceous species. Herbaceous species averaged a lag time of less than two years, while woody species had lag times nearly four times as long. Clearly, rapid rates of increase, once established, did not necessarily equate to rapid growth when species were first colonizing the fields. Rates of population decrease closely followed the pattern of population growth, as life forms with rapid rates of increase also had rapid rates of population decline. Similarly, life forms that were slow to colonize plots were also slow to decrease in frequency.

The population metrics based on plot coverage in general provided information about the performance of species during succession, though plot occupancy would have also contributed to overall abundance. The maximum cover attained by species did not vary dramatically among life forms. In general, woody species tended to have greater peak cover than herbaceous species, but the variance within woody life forms was very large. There was a clear break between herbaceous and woody species when the populations reached peak abundance; herbaceous life forms peaked between years 8 and 12, while woody life forms peaked after 23 years of succession. The comparatively late peaks for

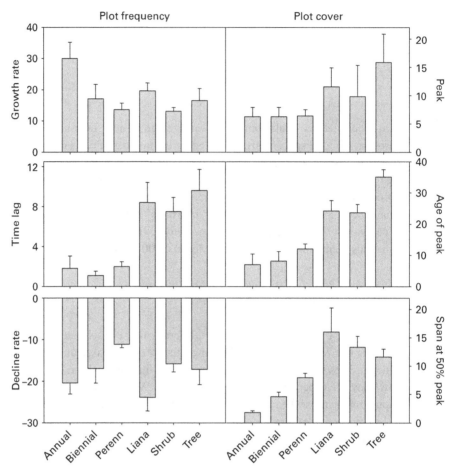

Figure 6.6 Variation among life forms in population dynamics within the BSS data. Data based on the 84 species that generated clear population dynamics during succession. Measures using plot occupancy (left) are used to illustrate the spread and decline of species during succession; plot cover is used to show patterns of abundance in succession (right).

annual and biennial species were generated by the combination of a small sample size for each form (16 annuals and 13 biennials) and the presence of late-successional, forest understory species in each. Without these woodland species, annuals and biennials would have peaked much earlier than perennials.

The span over which species were abundant ($\geq 50\%$ of maximum cover) showed very clear separation among life forms. There was a continual increase in the span of abundance in herbaceous species as life span increased. Within the woody species, lianas were abundant over longer periods than either shrubs or trees. Shrubs and lianas both began to decline in abundance as canopy closure occurred, providing a good estimate of their persistence in the community. However, most tree species had not begun to decrease within the first 50 years of succession, so their estimates of persistence were attenuated in these data.

In general, the BSS data support the presence of a general life span tradeoff generating a gradient of species along a range of live-fast–die-young to grow-slow–last-long plant strategies when aggregated at the life-form level. This broad pattern is in alignment with predicted tradeoffs of r- and K-selected species (MacArthur and Wilson, 1967; Pianka, 1970). However, population lag times and the successional age of population peaks were not directly linked to rates of spread or other population metrics across life forms. Instead they revealed a general dichotomy between woody and herbaceous species. This pattern may result from a combination of the slower dispersal of larger seeded woody species and the longer delay between establishment and reproductive maturity. Most woody species appear to have initially colonized as relatively few individuals, become reproductive, and then spread throughout the fields via local dispersal. This would have constrained woody life forms to dominate much later in succession than herbaceous forms, regardless of their ultimate growth rates.

Tradeoffs in population dynamics: plant strategies

Patterns in plant characteristics have resulted in the concept of plant strategies. These strategies are composed of suites of associated life history, and physical and physiological traits, which evolve in response to the biotic and abiotic conditions that determine plant success in communities (Westoby, 1998; Grime, 2001; Craine, 2009; Cornwell and Ackerly, 2010; Shipley, 2010). The constraints imposed by allocation patterns and other tradeoffs in these traits generate the potential suite of dynamics exhibited by plant populations during succession. Some treatments of plant strategies focus solely on physical and physiological characteristics to the exclusion of population dynamics. However, strategies should be strongly linked with the dynamics of species in communities. For example the, r- to K-selected continuum of strategies began with recognizing two diametrically opposed types of population dynamics – the traits associated with this continuum followed later. For this reason, we will begin our exploration of plant strategies by focusing specifically on population dynamics.

We can use the temporal dynamics of species in the BSS to explore the primary axes of variation in species dynamics that occur during succession. To generate a holistic view of plant population dynamics and the tradeoffs among them, a principal components analysis was conducted using all 14 of the population metrics described above. This approach allowed variation among species to be seen without the artificial categorization imposed by life form. Of the axes generated by the analysis, only the first two explained more variation than would be expected by chance alone. These two axes captured a combined 70.6% of the variation among species in their population dynamics. To visually relate the ordination to successional processes, the coordinates were subsequently rotated to correlate the time of peak cover with the first PC axis.

The ordination separated species based on their population dynamics. As expected from the rotation, the ordination exhibited a general trend from short-lived herbaceous to long-lived woody species from left to right along the X-axis (Figure 6.7). This representation highlights the variation among species within life forms, in contrast to the

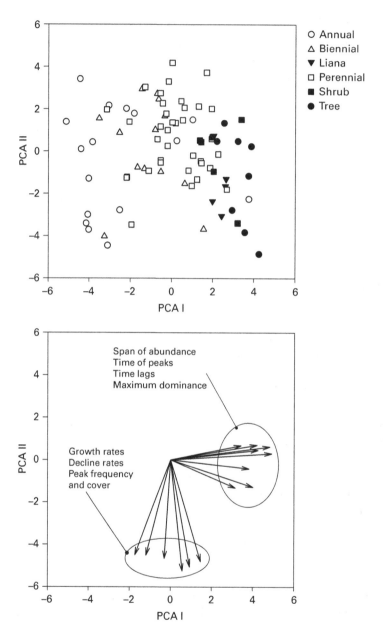

Figure 6.7 Principal components ordination of the 14 metrics of population dynamics for the species of the BSS. The ordination was rotated to maximize correlation of the first axis with the timing of peak cover. Only the first two PC axes were statistically relevant, explaining over 70% of the variation in the data. The top panel plots all 84 species in the ordination space showing relationships among species in their successional dynamics. The lower panel indicates the association of the individual population metrics with the two PC axes.

group means presented above. Though annuals and biennials were predominately on the lowest portion of the X-axis, several species occurred in other regions of the ordination. Perennial species displayed a wide diversity of population dynamics, largely overlapping with biennials, shrubs and lianas. In general, a continuum of population dynamics was exhibited by species within this system.

The relative contribution of each of the population metrics to the ordination appeared to be equivalent as all vectors were similar in length (Figure 6.7 lower panel). Unexpectedly, the ordination revealed two primary gradients along which species population dynamics were arrayed. Furthermore, the individual population metrics that generated these two gradients appeared to be separate from each other. The first set of associated population metrics were those associated with the successional age at which a species peaked. This gradient was composed of when a species peaked in succession, the time period over which the species was abundant, the length of population lag times and the maximum dominance attained by the species. All of these measures were positively associated with each other. Therefore the gradient spans from non-persistent, early successional species with short lag times to those species that dominate for long periods, peak late in succession and exhibit long population lag times. This at first glance appears to be a classical gradient from early to late successional species. However, the r- vs. K-selection dichotomy of early and late successional species would also place population growth rates as an important component of the successional transition of species. Population growth estimates were independent of the first PCA axis. It is also interesting that dominance in the community was strongly associated with the first axis, while peak cover and frequency were associated with the second PCA axis. The dominance measure clearly captured variation more directly associated with persistence in the successional community than peak cover or frequency alone.

The second set of correlated population metrics was essentially separate from those of the first axis. This axis of variation was composed of rates of population growth and decline, and the peak cover and frequency of species within the system. In general, species with fast growth rates tended to reach high peak abundances, but also collapsed relatively quickly. Species that expanded more slowly tended not to become as abundant, but also declined at relatively slow rates. It is very interesting that measures of population growth rate were not associated with how long a species persisted or when a species peaked in succession. The tradeoffs associated with population growth rate appear independent of many of the processes important to succession.

The presence of two distinct gradients of population dynamics was a bit surprising. Clearly, frequency and cover based measures of the same population behavior should be correlated at some scale. However, the two independent axes of variation in population dynamics suggest the presence of two disparate sets of population tradeoffs for species – one reflected population persistence in the community and successional status, the other reflected population growth rates. These results are also in stark contrast to the idea that succession represents a shift from r- to K-selected species (Odum, 1969). Population dynamics in the BSS separate the usual metrics along which the life-history tradeoffs are projected into two separate axes of variation. Growth rates, perhaps the fundamental component of r- and K-selection, were independent from how long a species was dominant in the community or when a species peaked during succession. Therefore, population dynamics in the BSS show two dominant gradients in life history – one associated with population growth (r) and one associated with dominance and persistence in the community (K). Examining the performance of different genotypes within a species has similarly suggested that selection may not cleanly function along an r to K gradient (Bonser and Ladd, 2011). The community-wide approach to population ecology in the BSS may be very interesting to apply to other systems to determine whether these associations are general features of co-occurring populations or an idiosyncrasy of this successional system.

We can also use the patterns of species dynamics in the BSS to address another continuum of plant strategies – the competition–colonization tradeoff (Tilman, 1983; Tilman et al., 1993; Turnbull et al., 1999). In this scenario there are two opposing plant strategies. Plant species may either possess characteristics that make them superior competitors or those that allow them to quickly colonize disturbances. Colonizing species successfully maintain themselves in the community by completing their life cycle and dispersing persistent seeds into surrounding areas before the slower to colonize competitive species displace them. Within the BSS, we can identify a few measures of species dynamics that relate to competitive or colonization ability. Competitive species would be expected to have high dominance (cover in each plot) and to persist for long periods of time. In contrast, species with high colonization ability should be able to increase quickly, particularly in frequency, when conditions are appropriate. When we look at the ordination of population dynamics, these suites of traits are not positioned along different ends of a single axis, as would be expected in a tradeoff, but rather are associated with different axes – suggesting statistical independence. The lack of association between colonization and competitive ability may reflect the inability of newly established competitive species to successfully displace established colonizing ones, regardless of their competitive superiority when adults (Yu and Wilson, 2001). The lack of pattern may also be an artifact of following only successful species in this analysis. However, the presence of a competition–colonization tradeoff in less-abundant species would suggest a fundamental shift in tradeoffs that would seem unlikely. Overall, this approach highlights the limitations of simple, two-trait tradeoffs in capturing the complexities of ecological interactions (Seifan et al., 2013).

Linking population dynamics with species characteristics

Though examining species dynamics through succession is a valuable exercise, relatively few studies will be able to directly document species dynamics over sufficient time periods to provide ecologically useful information. As the characteristics of a species should relate to their population dynamics within a community, a more practical approach may be to relate the successional role of a species to the characteristics that a species possesses. In this way it may be possible to form generalizations about successional processes that can be applied to other systems without the burden of following 50 years of species dynamics. Linking a species' traits with its successional role has been done in other studies, but has been typically limited to the contrast between early and late successional species. As successional role is clearly a continuum from early to late species, this approach may only approximate the potentially non-linear transitions that occur during succession. That being said, there is a wealth of predictions to compare the BSS data to (e.g. Odum, 1969; Walker and Chapin, 1987; Prach et al., 1997).

We will return to the idea of species traits and their role in succession later (Chapter 12). For now, we will focus on linking species traits with when their peak cover occurs during succession. For simplicity, these traits are largely based on the physical characteristics of the species. The simplest of these traits is the potential height attained by a species. While this characteristic is not the most complex to measure, it captures the ability of a species to access light (Westoby, 1998), one of the primary limiting resources in succession (Bazzaz, 1979; Bazzaz and Carlson, 1982; Tilman, 1985; Carson and Pickett, 1990; Tilman, 1990). Not surprisingly, the average potential height was greater for species that peaked later in succession (Figure 6.8). Species that peaked before 20 years of succession also tended to have less variation in maximum height. This was probably due to the intense competition for light, particularly when clonal perennials dominated the community. Species that peaked after 20 years of

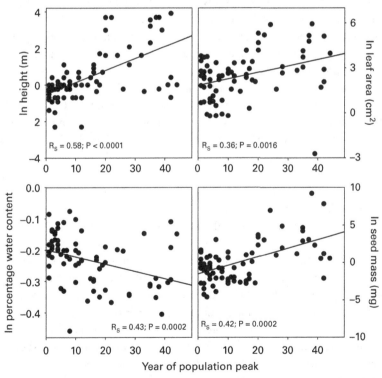

Figure 6.8 Associations between when a species peaks during succession (based on cover) and the species' height, leaf area per leaf, leaf water content and seed mass. Trends represented by best fit lines through the data. Statistical tests are Spearman rank–sum correlations.

succession exhibited a much wider range of potential heights, along with a higher mean. This variation reflects the multilayered structure of later successional communities that contain canopy trees, as well as shorter understory species.

The next suite of traits all involve the physical characteristics of leaves – leaf area, leaf water content and specific leaf area. Leaf area is related to the ability of species to capture light resources, balanced by the physiological constraints associated with heat gain and transpirational water loss. Plants that peaked later in succession tended to have larger leaf areas than earlier species (Figure 6.8). These late successional species would occur, or at least regenerate, under shaded conditions, when light capture would be more important than water loss. Leaf water content estimates allocation to the physical structure of a leaf. Though water makes up the majority of the fresh weight of all leaves, plants with lower leaf water contents have more physical structure in the leaves, making them tougher and less palatable to herbivores (Molinari and Knight, 2010). Within the BSS data, leaf water content declined in species that peaked later in succession (Figure 6.8).

Specific leaf area is the area generated by a single gram of dry leaf tissue and is associated with the relative growth rates of species (Westoby, 1998, Wright et al., 2004). As such, it has become one of the most widely used plant functional traits. While specific leaf area and leaf water content would appear to measure similar aspects of leaf structure, specific leaf area (not shown) was not correlated with when a species peaked during succession. As early successional species would be expected to have higher growth rates, the lack of association with successional status is surprising. The lack of an association is consistent with the principal components analysis of population dynamics that found growth rates were independent of successional status. Interestingly, specific leaf area appears to be quite important in capturing physiological changes in the entire plant community during succession, which we will return to in Chapter 12.

Seed mass typically varies by several orders of magnitude within most plant communities (Mazer, 1989; Westoby et al., 1992; Kidson and Westoby, 2000) and is one of the direct measures of offspring quality along the r- to K-selection gradient (Pianka, 1970). Seed mass is strongly associated with the ability of species to regenerate under shade, drought or any other limiting condition (Lord et al., 1995; Saverimuttu and Westoby, 1996; Bonfil, 1998) and is considered to be one of the primary axes of variation among species that differentiates plant ecological strategies (Westoby, 1998). Within the species pool of the BSS, there is a range in seed mass from 0.01 to nearly 1000 mg, though this does not include the dust-like seeds of terrestrial orchids, which lack endosperm. Species that peaked later in succession had significantly greater seed masses than early successional species (Figure 6.8). Despite the increase with successional position, there was still a large amount of variation among species peaking at similar times, indicating a range of species regeneration strategies occurred even within the same successional community (Leishman et al., 2000). The seed mass gradient through successional time follows predictions based on early and late successional species (Odum, 1969; Prach et al., 1997; Prach and Pyšek, 1999).

Associations with mutualists are typically thought to increase through succession. For pollination, early successional species are thought to depend on abiotic factors, while

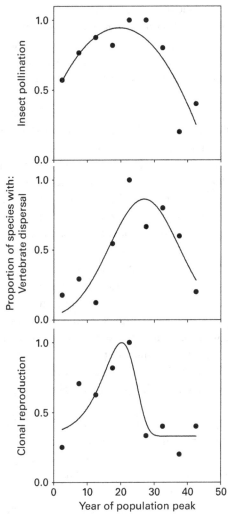

Figure 6.9 Association between when a species peaks during succession and the proportion of species that are pollinated by insects, dispersed by vertebrates or reproduce vegetatively. Data plotted are calculated proportions for each five-year age class and trends are indicated by best fit lines.

late successional species are thought to primarily rely on mutualistic associations with pollinators (Prach and Pyšek, 1999). Within the temperate regions of eastern North America, these mutualists are almost entirely insects. However, we do not see this classic early vs. late successional dichotomy in pollination mode in the BSS data. Species that peaked early in succession were predominately wind pollinated (Figure 6.9). As succession proceeded through the herbaceous perennial phases, species largely switched to the showy flowers associated with insect pollination. However, as the system made the final structural transition to a forest, the proportion of species that were insect pollinated decreased. In the young deciduous forests that developed on the site, there was a mixture of wind- and insect-pollinated species. Canopy trees were primarily, but not entirely, wind pollinated, while the lianas and shrubs were all insect pollinated. As the tree species dominant within the old-growth forest are largely wind pollinated (dominant genera – *Acer*, *Carya*, *Fraxinus*, *Quercus*), this late successional decrease in pollination via insects appears likely to persist. As we have often seen, the simple dichotomies of early and late successional species fail to capture the complexities of successional transitions.

Dispersal mode affects both the range and spatial patterning of seeds dispersing into a site. The two primary means of dispersal within the BSS data are vertebrate dispersal, chiefly by birds and small mammals, and wind dispersal. Predictions for succession are that the prevalence of vertebrate-dispersed species will increase during succession (Odum, 1969; Debussche *et al.*, 1996). There was a dramatic increase in the probability of a species being vertebrate dispersed over the first 30 years of succession, but subsequently a dramatic drop (Figure 6.9). A very low proportion of the species early in succession were vertebrate dispersed. The prevalence of biotic dispersal increased quickly as bird dispersal was the only mechanism of spread for the liana and shrub species that peaked at intermediate successional stages and displaced many of the wind-dispersed perennial herbs. Several of the early successional trees, chiefly *Cornus florida* and *Juniperus virginiana*, were also bird dispersed. Only two of the tree species studied, *Acer rubrum* and *Fraxinus americana*, were wind dispersed. However, these species are abundant in portions of the old-growth forest and are likely to persist. Even in the latest successional stages of the BSS, there was a mix of species with biotic and abiotic dispersal mechanisms, particularly when all strata were included. This mixture of canopy and understory species generated the decrease in vertebrate dispersal late in succession.

Clonality, or the ability of a species to reproduce via rhizomes, stolons or other vegetative parts, is a very important plant characteristic for both succession and invasion biology (Thompson *et al.*, 1995; Prach and Pyšek, 1999). It is a trait strongly associated with competitive plant strategies and with the acquisition of soil resources (Grime, 2001). Based on the predictions of Grime's c-s-r plant strategies (Grime, 2001), we would expect clonality to peak at intermediate stages of succession rather than exhibit a linear temporal gradient. Within the BSS data, there is a clear peak in clonality at intermediate stages of succession (Figure 6.9). While this is not a particularly rigorous test of Grime's hypothesis, we will return to this idea again in Chapter 12, when we will scale the presence/absence of a functional trait by the abundance of species during succession.

Similar to the association of traits with the successional status of species, the traits of species were also associated with their population dynamics. Using the principal component axes generated based on the population dynamics of species during succession (Figure 6.7) we correlated trait values with population dynamics (Table 6.1). In general, variables that were associated with when in succession a species reached peak cover were also associated with the first PC axis. Correlations with the second PC axis were weaker and not statistically significant once multiple comparisons were accounted for. Based on the two orthogonal groups of population metrics generated in the PC analysis, plant height, leaf area, leaf water content, vertebrate dispersal and seed mass were all positively associated with not only when a species peaked, but also the length of

Table 6.1 Correlations of species traits with the principal components axes of population dynamics (See Figure 6.7). Once multiple comparisons are accounted for using a Dunn–Sidák correction, none of the correlations with PCA II were statistically significant.

Population trait	PCA I	PCA II
Height	0.65^b	−0.20
Leaf area	0.38^a	−0.24
% water content	-0.54^b	0.06
Specific leaf area	−0.26	0.04
Insect pollination	0.02	0.08
Vertebrate seed dispersal	0.37^a	−0.22
Vegetative reproduction	0.16	0.08
Seed mass	0.42^b	−0.28

^{a} $P<0.01$
^{b} $P<0.001$

population lag times, how long species were abundant and their dominance during succession. These characteristics were independent from the population growth rates and peak abundances exhibited by the species in the community. Again, specific leaf area, a physiological measurement related to growth rates, was not associated with either PC axis, suggesting it is not simply related to species performance in this system. As pollination and vegetative reproduction have non-linear relationships with when species peak during succession, it is not surprising that these characteristics were not directly related to population dynamics as a whole.

Scaling dynamics from population to community scales

At some level, all community dynamics must result from the dynamics of the constituent populations. The BSS data provide a unique opportunity to examine community-level dynamics and decompose them into the population-level processes that generate them. While the BSS data do not follow or even identify individual plants, the population metrics discussed above allow for reasonable characterization of species temporal dynamics. Below, we will explore an example of the value of this type of data and illustrate the benefits of such an analytical approach. As non-native species invasions are a major ecological and conservation concern, we will focus on how the BSS permanent plots have helped us to understand the dynamics of invasion and how applying a population perspective to invasion impacts can translate to a mechanistic understanding of those impacts. The basis for scaling population dynamics up to the community comes from the basic premise of island biogeography (MacArthur and Wilson, 1967), that the balance of colonization and extinction rates determines the current state of the system.

First let us look at the population dynamics of *Rosa multiflora*, a non-native invasive shrub throughout eastern North America (Banasiak and Meiners, 2009). Instead of

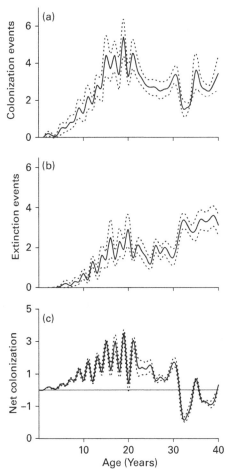

Figure 6.10 Colonization and extinction dynamics of *Rosa multiflora* throughout succession. Colonization events represent new species occurrences, whereas extinction events represent the loss of *R. multiflora* from a plot. The difference between plot colonization and extinction is net colonization, the overall change in the number of plots that contain the species. Dotted lines represent 1 SE around the mean of the 10 BSS fields. From Banasiak and Meiners (2009), With kind permission from Springer Science and Business Media.

following the overall pattern of the number of plots occupied, we can also follow colonization and extinction in individual plots (Figure 6.10). This yields more direct information on the dynamics of the invasion. Early on, colonization rates were consistently positive, showing that plots were being colonized by the species. This, in concert with overall low extinction rates, led to a net increase in the number of plots colonized. The increase in and ultimate peak in colonization rates until 20 years after abandonment would follow the availability of seeds within the site, as more individuals became reproductively mature. After the peak, colonization overall decreased and became more episodic. Extinction rates slowly increased during succession, constantly generating new opportunities for colonization. Extinction rates increased sharply once the tree canopy closed around year 30. Only then did net colonization rates drop and become negative. Following the dynamics of this species in individual plots not only explains the overall trajectory of the species, but also reveals dynamics that may help to manage this invasion. As colonization in forests is episodic, identifying the conditions that generate bursts of establishment will allow targeted intervention. Furthermore, the continued colonization of new plots even after the species had begun to decline in abundance indicates the sustained potential of this species to increase rapidly if conditions change. As the increase in plot extinction rates followed canopy closure, canopy disturbances would be expected to lead to an increase in this species that may then persist for decades.

The invasion of non-native species represents a management concern because of the perceived impacts that these species may have on communities (Hager and McCoy, 1998; Parker et al., 1999; Pimentel et al., 2000; Huston, 2004; Vilà et al., 2011). Any impacts of an invader on a community must also be reflected in the population dynamics that generated the change in community structure. The population processes altered by the invader then are the mechanisms of those impacts (Yurkonis and Meiners, 2004; Yurkonis et al., 2005). When four non-native species invasions were examined using the BSS data, a single mode of impact was found. To alter species richness at the plot scale, the scale at which plant species would interact, there must be a shift in the balance of local colonization and extinction rates with invasion. The study found no changes in local extinction rates for the species studied, but colonization rates were depressed as the magnitude of invasion increased (Figure 6.11). Mechanistically, this suggests that invasion impacts are not driven

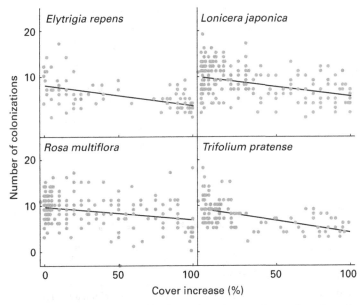

Figure 6.11 The influence of four non-native species invasions on the number of species that colonize plots in the BSS. Data plotted are the number of new species that colonize a plot as a function of the increase in cover by a non-native species over the same time period. From Yurkonis et al. (2005) with permission from John Wiley and Sons, Inc.

by competitive displacement of established plants in the community, but rather the continued depression of colonization rates and interactions at the seed and seedling life-history stage. As species turnover is constant in ecological communities, the depression of colonization rates would inhibit new species from entering the local community as openings in the plant canopy became available. This also suggests that the rate of invasion impacts should be strongly tied to the rate of community turnover.

We can take the population approach to invasion impacts one step further. If local colonization rates are depressed by invasion when looking across all species, we may also expect invasion to alter the colonization dynamics of associated populations. To test for impacts on individual species, the population dynamics of species that were common during each invasion were related to the magnitude of invasion (Yurkonis and Meiners, 2004; Yurkonis *et al.*, 2005). As in the larger-scale analysis, the impacts of invasion on individual species were largely through reductions in local colonization rates (Figure 6.12). At this scale there were increases in local extinction rates for some

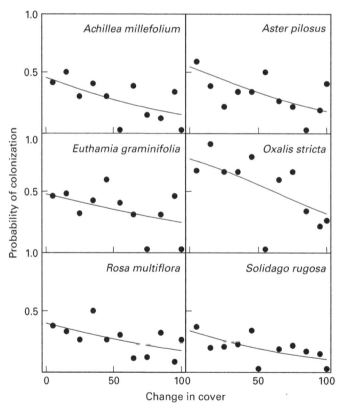

Figure 6.12 Influence of the non-native species *Lonicera japonica* on the colonization rates of common species. Points represent the proportion of plots colonized in 10% cover increments of *L. japonica*. Lines are the results of logistic regressions that predict the probability of colonization as a function of *L. japonica* cover. From Yurkonis and Meiners (2004) with permission from John Wiley and Sons, Inc.

species, but this was much less common than influences on colonization rates. These increased extinction rates were masked when all species were examined together, but reveal the variation in species–species interactions that should be expected in most communities. The presence of a predominant mechanism of invasion impact may indicate the equivalent of an assembly rule for invasion impact.

By examining the population dynamics that underlie the expansion and impact of non-native species invasions, we see the mechanisms that generate the patterns seen in the population and the community. However, this approach is not mechanistic at the most reductionistic level. These data do not determine whether competition for nutrients or light, propagule pressure or some other process ultimately drives dynamics. However, this approach does suggest the population stages where these mechanisms are operating. This is a critical step in the mechanistic hierarchy as it forms the linkage between the processes, such as competition for nitrogen and the structure of communities, that may be limited by that resource.

Succession as a population process

Though the theories on the processes that generate successional dynamics were originally focused mechanistically at the community level, this approach was largely rejected as the Clementsian paradigm fell out of favor. Later reductionist approaches dealt with mechanisms that were based solely on the ecology of the constituent populations (e.g. Connell and Slatyer, 1977; Peet and Christensen, 1980). Now that we have explored both the community and population dynamics of this system, where do the BSS data fall with regards to this dichotomy?

Clearly the community-level patterns in life-form transitions are directly composed of the individual population dynamics shown in this chapter. Transitions in life form were generated by integration across individualistic species dynamics. Similarly, we see strong linkages between the characteristics of the constituent species and how they perform (i.e. their population dynamics) during succession. We must remember, however, that we can only really examine the population dynamics of the most abundant species over time. While the remaining species do not contribute much in the way of cover, they represent the majority of the diversity within the system. The population perspective is therefore useful, but does not completely address the diversity of processes and patterns in succession. Similarly, associations between species characteristics, such as seed mass and dispersal mode, and their population dynamics are informative, but as we will see later (Chapter 12) do not capture the full functional dynamics of succession.

So, is succession a population process? At a certain scale the answer must be yes. The species that compose the community generate the compositional changes seen through successional time. The characteristics of the building blocks necessarily constrain the characteristics of the whole. Populations colonize, establish, reproduce, and if conditions change to those which prevent regeneration, collapse during succession. The BSS data provide excellent, direct observations of species in all of these population phases.

The difficulty in this generalization is that all of these processes are occurring in the context of a larger community. Competition is typically thought to be one of the major forces that structure successional communities (Tilman, 1985; DeSteven, 1991b; Tilman and Wedin, 1991; Myster and Pickett, 1992a; Inouye *et al.*, 1994; Clay and Holah, 1999; Royo and Carson, 2006; Wright and Fridley, 2010). In its simplest form, interspecific competition is by definition a community-level process, because it involves interactions among species. Add this to all of the other potential interactions among species, and you generate a very complex interaction landscape where the characteristics of each species function to generate population dynamics. Species dynamics during succession are the net result of the ecological sorting of traits and life histories within this heterogeneous interaction landscape – the biotic complex within Tansley's ecosystem concept (1935). Population dynamics of species isolated from a community bear little mechanistic relationship to dynamics within a community. Succession is quite simply a population process in the context of a changing community. Separation of these two ecological scales is both artificial and uninformative. This is why a hierarchical, upwards and downwards view of causality is useful.

7 Impacts of drought and other disturbances on succession

> "Vegetation is dynamic – an ever-changing complex now appearing quiescent and in complete equilibrium with the habitat, now displaying obvious evidence of change. I believe that failure to recognize the dynamic aspect of vegetation is a primary cause of differing concepts."
>
> E. L. Braun 1956. The development of association and climax concepts: their use in interpretation of the deciduous forest

While explicit discussion of disturbance as a driver of vegetation did not arise until later, we clearly see that Lucy Braun espouses the dynamic view of ecology championed by Clements, though with much less metaphorical baggage. Her work primarily focused on examining stands of old-growth forests in the eastern United States at a time when they yet persisted in abundance. Her choice of study system explains her focus on defining climax communities and their relationships. The major vegetation change at that time was the loss of chestnut (*Castanea dentata*) from the southern Appalachians. Most of the forests that she examined would certainly fall towards the more stable end of vegetation as her work pre-dates the increased nutrient deposition, overabundant deer populations and non-native species invasions that threaten those forests today. While we do not currently accept all of the connotations (deserved or undeserved) of the climax concept, we see in Dr. Braun's writings a clear projection of the fluid nature of vegetation – both successional and climax.

Succession and disturbance are clearly linked processes. However, most discussion about the relationship between these two processes has revolved around the linkage between the characteristics of the succession-initiating disturbance and the successional processes which follow. Stand initiating factors may be the dominant relationship between disturbance and succession, and are a primary feature of our conceptual organization, but disturbances occur at many spatial and temporal scales. The successional processes of the BSS have been continually influenced by disturbances of varying intensities, spatial extents and temporal scales. At fine spatial scales, these disturbances may include trampling by deer, burrowing activities of small mammals or the localized collapse of woody species. Influences at broader spatial scales would include periodic fluctuations in rainfall, hurricanes and, at least historically, fires (Buell *et al.*, 1954; Small 1961).

We have explicitly dealt with the influences of the stand-initiating disturbances of the BSS in our discussion of assembly rules (Chapter 9). Here we will cover the continuing disturbances that occur within succession and their influences on the dynamics and structure of the system. Specifically we will discuss the nature of disturbance events, their impacts on the structure and composition of the plant community, recovery and the implications of these disturbance events in influencing successional transitions. We will largely focus on fluctuations in rainfall as they are the most common, large-scale disturbance within the system and can be directly addressed by the BSS data. As there has also been extensive experimental work at the BSS site on small-scale disturbances, we will draw information from experimental work to supplement the BSS data. Our goal is to develop an integrated view of disturbances within the context of succession with particular emphasis on how disturbance may influence succession within multiple classes of successional drivers. Along the way, we will confront some successional predictions with the BSS data and find some surprising deviations from traditional successional wisdom.

Characteristics of disturbances

Disturbances are incredibly varied, as are their impacts on plant communities. To account for the variation, ecologists characterize disturbances in a variety of ways. These metrics typically focus on the frequency, intensity, size and timing of the events as these measures capture largely independent aspects of disturbance (Connell and Slatyer, 1977). The frequency of disturbances determines the period of time between events during which the community can recover. Frequent disturbance events may reduce the abundance of species whose regeneration time exceeds that of the disturbance return time in favor of more short-lived species. The importance of disturbance frequency to plant communities is highlighted within Grime's scheme of plant strategies, where the frequency of disturbance is one of the two primary axes that define potential plant habitats (Grime, 2001). More intense disturbances tend to generate greater influences on the plant community and on resource availability than less intense ones. For example, soil disturbance from small-mammal digging will have greater influences than the disturbance caused by trampling, though each may have a similar spatial extent. Similarly, we should expect a much greater impact of a season-long drought than one that spans a period of weeks.

The size of a disturbance influences the speed, and perhaps even the process by which a community recovers. Small forest gaps, such as those caused by the fall of a single tree or a few branches may fill in from the extension growth of surrounding trees, while larger events may require the establishment of new individuals (Bazzaz, 1996). At fine spatial scales, small soil disturbances in herbaceous communities may be colonized by the clonal expansion of adjacent plants, while larger disturbances may provide opportunities for the seedling establishment other species (Goldberg and Werner, 1983; Tilman, 1983). At very coarse scales, the size of the stand-initiating disturbance is critical in regulating succession (Golley et al., 1994). The largest fields of the BSS have taken longer to accumulate tree cover as there is dispersal limitation from the adjacent forest (Chapter 9). These larger fields have developed simultaneously patches of young forest and patches of perennial herbs characteristic of mid-successional communities. Such limits to dispersal and regeneration have been used to justify harvesting of forests in narrow strips to speed recovery (Gorchov et al., 1993; Gorchov et al., 2013).

Finally, the timing of a disturbance in relation to the phenology of the community has large influences on not only regeneration patterns but also on the severity of the impact. For example, prairie fires occurring during the growing season have much different impacts on the composition of grasslands than fires that occur when the dominant grasses are dormant. Dormant season fires lead to the enhanced growth and performance of warm-season grasses to the detriment of forbs. In contrast, fires that occur while the grasses are still active tend to break the competitive dominance of grasses and allow the recruitment of subordinate species (Biondini et al., 1989; Copeland et al., 2002). Within the BSS, we can see similar phenological effects on composition occurring in fields that were last plowed in the fall, generating a pulse of winter annual establishment (Chapter 9).

Variation in the frequency, intensity, size and timing of disturbances within successional communities will interact to generate both the immediate impacts expressed by the community and the dynamics that follow. The term used for community dynamics resulting from a disturbance is often recovery. As we have discussed previously in the context of succession, the dynamics generated by disturbance may or may not generate a community similar in composition or structure to the pre-disturbance condition and may alter successional trajectories in unpredictable ways (Kreyling et al., 2011). The BSS data provide examples of recovery in the traditional sense, where composition returns to pre-disturbance conditions, and instances where disturbances generate a compositional transition in the community.

Characteristics of droughts within the BSS

All of the interacting characteristics of disturbances are important in evaluating the role of drought in a community. While the spatial extent of drought is certainly important in determining movement patterns and food availability for consumers at broad spatial scales, the BSS fields occur at a fine enough spatial scale that rainfall, or lack thereof, is essentially consistent across the site. For this reason we will not deal with the size of disturbance in this section. We will, however, return to this idea in reviewing the role of other disturbances in succession.

Throughout the first 50 years of the BSS study, rainfall has fluctuated frequently, with extremes in both rainfall and drought (Figure 7.1). Droughts have occurred throughout succession with typically less than eight years between events that reduced rainfall for the entire growing season by 25% or more. The first 50 years of the BSS contained 10 such events. This averages to a return cycle of 5 (50/10) years for droughts of at least moderate intensity. Similarly, periods of growing-season rainfall that were 25% or more above long-term averages occurred nine times, with a maximum period between events of 12 years. This periodicity reflects a return time of heavy rainfall of every 5.6 (50/9) years. An interesting feature of the temporal pattern of dry and wet years is that they often occur sequentially, with large changes in moisture from year to year. We will see

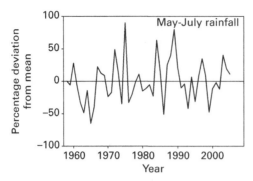

Figure 7.1 Periodicity of drought throughout the BSS study period. Data plotted are percentage deviation from 50-year averages for the months of May through July. These months occur before the annual vegetation sampling and are most strongly associated with vegetation changes at the site. All weather data are from the New Brunswick weather station, approximately 9 km from the study site.

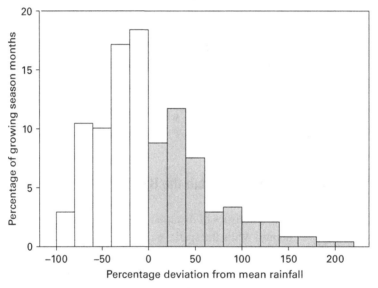

Figure 7.2 Distribution of variation from mean monthly rainfall for the growing season (May–September). Months with greater than average rainfall indicated by shaded bars, months with lower than average by open bars.

later that these fluctuations, as well as absolute rainfall amounts are important in shaping the plant community.

As may be expected, individual months within the growing season varied much more widely than did the growing season as a whole (Figure 7.2). When examined individually, 17% of months during the growing season had less than 50% of normal rainfall, and 5% of months received less than 75%. When these extreme events align with the reproductive or germination phenology of a species, then there may be large reductions in the abundance or spread of that species. Similarly, extremely wet years were also

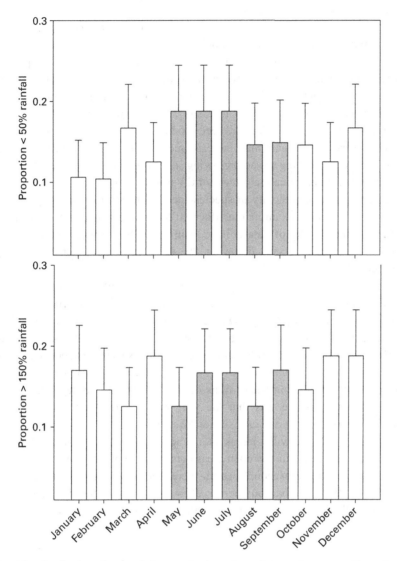

Figure 7.3 The distribution of wet and dry months throughout the growing season. Data plotted are the proportion of wet and dry months during the BSS based on deviation from 50-year monthly averages. Shaded bars represent rainfall during the growing season of May through September.

relatively common. During the growing season, 15% of months exceeded average rainfall by 50% or more, and 11% exceeded the average by at least 75%. Again, the timing of these wet periods may differentially influence the performance of individual species, particularly those whose germination, growth or reproductive phenology coincided with high moisture levels.

The last aspect of drought disturbance to be addressed is that of timing. While the analyses will largely focus on dry and wet years that occur during the growing season, these events can occur thoughout the year (Figure 7.3). Across all months, 15% had rainfall that was less than 50% of the long-term average, and 16% had rainfall amounts that exceeded the long-term average by 50%. Statistically, there was no difference among months in the probability of being either wet or dry. While the distribution of wet and dry months throughout the year was even, the biological impact of those events would clearly differ greatly. Droughts occurring late in the growing season may reduce seed production in those species that bloom during the summer. Wet springs may increase the establishment success of seeds produced in the previous year and increase the growth, flowering and seed production of species that bloom early in the growing season. Wet years may increase seed production, leading to an increase in colonization rates during the next growing season (Banasiak and Meiners, 2009). Wet periods may also enhance the growth of C3 species relative to C4 grasses (Fay et al., 2002). Due to the complexity of species-specific interactions with weather patterns, we will not explicitly examine the effects of the timing of drought events in the data. Because of the timing of data collection in the BSS, we will specifically focus on drought events that occurred during the growing season, but before vegetation sampling.

The characteristics of drought are difficult to quantify as are their long-term impacts. As an example, the most recent severe drought at HMFC occurred in the summer of 1999. The drought was quite strong in June and July (Yurkonis and Meiners, 2006). At the time of vegetation sampling, many of the remaining goldenrods (*Solidago* spp.) were dead and there were very few surviving subordinate plants and seedlings. Following the late-July sampling, hurricane Floyd made landfall on September 16, producing heavy rains of over 17 cm in two days. As a whole, the result of this hurricane was that the year was slightly wetter than average. The combination of drought and excess moisture generated some interesting dynamics in the system, which we will address later. The BSS fields were largely forested at the time of the drought, with few open areas persisting. In other areas at HMFC that were still dominated by herbaceous species, there were much greater effects on the plant community. While herbaceous perennials were largely reduced by the drought, the moist fall allowed a large number of plant species to germinate and establish. These species, as well as those which established in the more open community the following spring generated large shifts in composition. Large populations of the winter-annual *Barbarea vulgaris*, the short-lived perennial *Chrysanthemum leucanthemum* and the cool-season grass *Danthonia spicata* developed in one of the HMFC fields (Meiners, unpublished data). There were even a number of blooming *Spiranthes* orchids mixed in the community, which had not been previously observed in that field. This anomalous weather pattern generated opportunities and episodic reproduction that may continue to influence the plant community for decades. Seeds produced during this period may remain dormant for long periods only to germinate following soil disturbance or a subsequent drought (Oosting and Humphreys, 1940; Roberts and Vankat, 1991; Leck and Leck, 1998; Plue et al., 2010). While the community of established plants may recover over a relatively short period, the persistent and perhaps unseen effects of disturbance events may linger.

Impacts of rainfall fluctuation on community structure

The impacts of disturbances on communities are typically measured through alterations in either the diversity or biomass of the system (Runkle, 1982; Armesto and Pickett, 1985; Petraitis et al., 1989; Myster, 2003). Here we will begin our discussion of rainfall impacts with total cover, our surrogate for biomass, and species richness. One interesting feature of the BSS data was that the total change in rainfall from one year to the next was much more important in determining the change in community structure than was the deviation from normal rainfall in the current year. In other words, a year with only 50% of the long-term average rainfall had a greater effect when it followed a wet year than when it followed an average year. For this reason, the analyses presented within this section will focus on the total change in rainfall deviation from one year to the next (Figure 7.1). When analyses were conducted on deviation from the long-term average alone, the directions of the effects were the same, but the associations weaker. We also present the response variables as the change from one sampling year to the next. As both cover and richness change over successional time (Chapter 5), these long-term changes in structure need to be accounted for. Examining the net changes in community responses removed the effects of successional age and allowed focus on interannual changes in the community.

Plot cover responded as would be expected for any type of disturbance (Figure 7.4). As the change in rainfall became more negative, reflecting a shift from wetter to drier conditions, there was a concomitant decrease in cover. When weather patterns shifted towards wetter years, total plot cover also increased. The plant community appears to track annual fluctuations in rainfall, even while accruing cover over succession. The greater importance of the change in rainfall from one year to the next over absolute rainfall amount suggests that at least some species are able to capitalize on temporary increases in moisture availability. However, the continual turnover in species composition through succession means that the identity of these species also changes over time. The depression in cover seen during drying periods may result in more opportunities for the species colonization (Yurkonis and Meiners, 2006). However, rainfall fluctuation was not related to the amount of bare soil in the community ($R = 0.074$; $P = 0.19$). The lack of relationship between rainfall variation and bare soil suggests that in the BSS, openings in the plant community were not environmentally generated or were ephemeral.

The association between cover and change in rainfall was also reflected in species richness at the plot scale. As weather patterns shifted from drier to wetter, plot richness also increased. Likewise, the more rainfall decreased between consecutive years, the greater the loss of species richness. Gains in local species richness would reflect both the establishment of new individuals from seed and the clonal expansion of established plants into sample plots. Similarly, local loss of species would represent both the mortality of individuals and the contraction of plants with the majority of their biomass outside the plots. The local loss of species and cover in plots in dry periods should result in the increased availability of light and soil resources, perhaps generating temporary opportunities for the colonization of new species (Burke and Grime, 1996; Davis et al., 2000). Though species richness at the plot scale did not systematically change

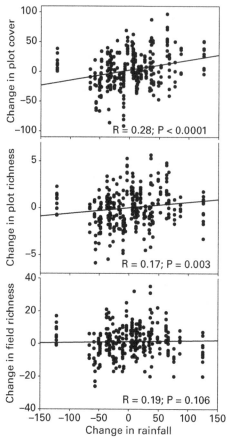

Figure 7.4 The relationship between annual change in rainfall ($Rain_t - Rain_{t-1}$) and changes in community structure. Data plotted are change in average plot cover (top), plot richness (middle) and total species richness (bottom) for each field. Change in rainfall is calculated as the annual change in deviation from normal rainfall. Statistical tests are Spearman rank–sum correlations.

throughout much of succession (Chapter 5), richness was still dynamic over time. The successional stability of local richness, despite continual fluctuations generated by rainfall, argues even more strongly for the existence of local limits to the number of species in herbaceous communities (Gross et al., 2000).

Species richness at the field scale responded much differently to fluctuations in rainfall than did local community structure. There was no correlation between the overall change in rainfall and changes in species richness (Figure 7.4). Therefore, fine-scale processes within plots do not scale up to the entire field. Many of the local colonizations and extinctions were likely of species present elsewhere across the field and would not necessarily alter field richness. When the direction of the change in rainfall was ignored (deviation from normal only), a very different response to rainfall fluctuation was found. Field richness was negatively associated with the absolute value of the change in rainfall (Figure 7.5). In other words, consecutive years with similar rainfall tended to increase slightly in species richness, while those years that differed

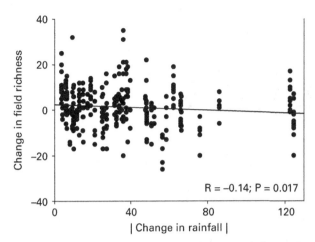

Figure 7.5 The relationship between the absolute change in rainfall ($|Rain_t - Rain_{t-1}|$) and the change in field-scale species richness ($Richness_t - Richness_{t-1}$). Results from Spearman rank–sum correlation are shown.

dramatically in rainfall, with either an increase or decrease, tended to lose species. Again, this pattern is superimposed on the general increase in species richness that has occurred through succession. The ecological consequences of consistency is an interesting and somewhat surprising result as most research focuses on the importance of heterogeneity, not homogeneity, in maintaining diversity (Tilman et al., 1993; Vivian-Smith, 1997; Wagner et al., 2000; Lundholm 2009; Eilts et al., 2011). An association with absolute change in rainfall was not seen in either plot cover or richness, suggesting that heterogeneity, or the lack of it, may function differently across scales.

Community recovery from drought

The ability of successional communities to recover from drought is complicated by the frequency of moderate and strong drought events contained within the BSS data. The frequency of these events often does not allow sufficient time to assess recovery before the next event occurs. The dynamic nature of the system also provides challenges as successional changes must be separated from drought effects. In early successional communities of the BSS, where turnover occurred rapidly and droughts were frequent, full separation of these two concurrent processes would be impractical.

While the dynamic nature of the system precludes discussion of specific drought-induced changes for much of succession, we can explore the underlying population dynamics that generate shifts in composition. Drought events generated pulses of extinction events in herbaceous communities at the plot scale (Bartha et al., 2003; Yurkonis and Meiners, 2006), linked with a loss in total cover (Figures 7.6 and 7.7). These extinction events occur as distinct increases set against a background level of species turnover. Following these events are pulses of local colonization that exceed the

Impacts of drought and other disturbances on succession

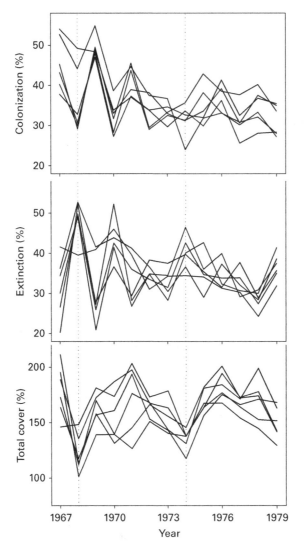

Figure 7.6 Synchronicity among fields in plot cover, colonization and extinction rates for six of the BSS fields (C3, C4, C5, D1, D2 and D3). This period contained two droughts, 1968 and 1974, indicated by vertical dashed lines. Percentage colonization and extinction is the number of species lost or gained scaled by the number of species in the plot during the previous year. Data from Bartha et al. (2003). Used with permission from John Wiley and Sons.

background colonization rates characteristic of succession. Both plot extinction and colonization rates return to background levels quickly after the drought has passed. These effects appear to be relatively transient influences of drought. Again, the dynamics associated with droughts are consistent with the hypothesis that fluctuating resources regulate invasibilty (Burke and Grime, 1996; Davis et al., 2000).

These transient pulses of local extinction and colonization generate very different effects on community composition depending on when the drought occurred. In the early

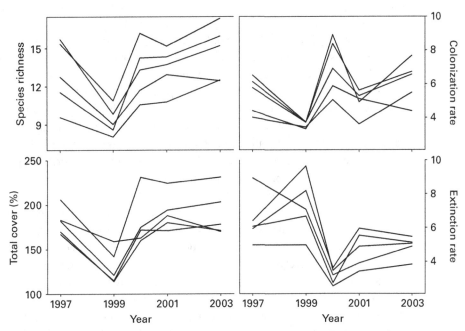

Figure 7.7 Changes in species richness, cover and species turnover during the 1999 drought. Data presented are field averages ± 1 SE. Data from Yurkonis and Meiners (2006). Used with permission from Springer.

to mid-successional communities studied by Bartha et al. (2003), these pulses of turnover generated lasting effects on composition, as they initiated changes in dominance within the BSS. The droughts appeared to hasten the replacement of species during succession. Species in the process of being displaced via differential performance would be more susceptible to the stresses imposed by drought as they would be competing with their successional replacements. Early work by Billings (1938) found that as succession proceeded, plant communities had much deeper root systems. Based on this, differential impact of drought on early successional species would seem likely. Losses of earlier species would allow more opportunities for the expansion and colonization of later-successional species and speed successional transitions. As these droughts largely occurred within perennial communities, any establishment of ruderal species would be temporary as they would be quickly displaced as the community closed. However, pulses of recruitment may also allow non-native invaders to colonize.

The role of disturbance as a successional accelerator seems counter-intuitive, as disturbance is typically thought of as a re-setting function for plant communities. However, the response of the community is probably related to the intensity of the disturbance. Even though the droughts resulted in a significant loss of cover and richness, the majority of the established perennial vegetation would have survived and recovered once conditions allowed. If, however, the drought resulted in the mortality of a large proportion of the plant community, it is likely that an overall resetting of successional processes would occur. As the range of drought severity contained within the BSS is relatively large, events of sufficient magnitude to move succession backwards would need to be very severe and would appear to be quite rare.

Drought impacts that occurred while major structural transitions were still developing within the system are in stark contrast to those that happen once the forest canopy closed. The BSS fields were largely forested when a major drought occurred in 1999. Despite the severity of the drought, the forest canopy remained intact, with little mortality of canopy individuals. As the canopy layer was relatively unresponsive to the drought, analyses focused solely on the herbaceous understory, where the majority of the diversity occurred. As shown in the earlier droughts, there was a decrease in understory cover and a pulse of local extinction events (Figure 7.7). Following the drought, plot colonization rates increased, rapidly replacing the cover lost during the drought. However, the 1999 plant community was not in the process of transitioning to another successional phase. The species that recolonized and expanded following the drought were largely the same species that dominated before the drought. This consistency generated understory plant communities with very similar compositions within three years of the drought (Figure 7.8). Relatively few species were lost from the system during the drought. The dynamics in response to the severe drought generated an overall recovery of plant community structure and composition late in succession.

Though the 1999 drought largely generated transient dynamics within the understory plant community, this does not mean that all species responded similarly to the drought. The invasive species *Alliaria petiolata* (Garlic mustard) had been an increasing component of the BSS data since it invaded in 1982. Prior to 1999, *A. petiolata* was increasing in frequency and cover across the site (Figure 7.9). During the drought, the frequency of *A. petiolata* continued to increase, probably a result of the previous year's seed production and dispersal (Figure 7.9). Cover of *A. petiolata* declined precipitously during the drought, but rebounded dramatically in the following year. As the fall following the drought was wetter than average, the first year rosettes of the biennial would have been able to establish at high rates in the temporary openings within the understory community. The increased establishment led to a temporary population explosion of this invasive species. While the abundance of this species quickly returned to pre-drought levels within the community, the pulse of seeds generated from this event will likely persist for a long period of time (Pardini *et al.*, 2009). Therefore, though the drought generated a pulse of recruitment in this species, the resulting seed bank may spread the influence of the drought over a much longer period of time and may help to maintain populations of this invader in the forests of the BSS. The inhibitory effects of *A. petiolata* on mycorrhizal associations and understory diversity would magnify the importance of such episodic population dynamics (Roberts and Anderson, 2005; Hale *et al.*, 2011; Lankau, 2011; Castellano and Gorchov, 2012).

Succession and community stability

The successional development of communities has long been conceptually linked with the stability of the system. As one of his predictions for succession, Odum (1969)

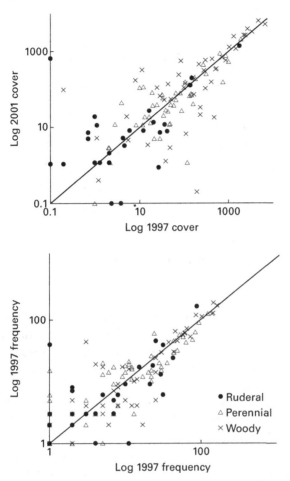

Figure 7.8 Recovery of community composition following the 1999 drought. Data plotted are total cover across all fields (top) and frequency across all plots (bottom) for each species. Species are identified by life form: ruderal – annuals and biennials, perennial – herbaceous perennials and woody – lianas, shrubs and trees. One-to-one lines represent equivalence in abundance between pre- and post-drought communities. Data from Yurkonis and Meiners (2006). Used with permission from Springer.

suggested that stability should increase with succession. The basic argument for this prediction is in the fundamental differences between early and late successional species. Though the dichotomy between early and late successional species strategies is clearly an artificial division along a gradient of species strategies, we will use it for the simplicity of argument. Early successional species are adapted to completing their life cycle in the periods between disturbance events. The strategy of early successional species is typically to allocate large amounts of their resources to seed production. This strategy maximizes reproduction in the short term, and maximizes the long-term persistence of the species in communities with unpredictable availability of suitable regeneration sites. The high allocation to reproduction necessarily results in lesser allocation to

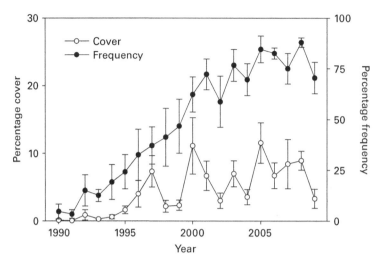

Figure 7.9 Population dynamics of *Alliaria petiolata* in response to the 1999 drought. Data are pooled across all fields.

the development and maintenance of structures that would allow these organisms to persist through drought events. Most important to this strategy would be the lack of deep root systems or tissues for the storage of carbohydrate reserves that would allow the species to either continue growing or to go dormant and re-grow after the drought subsided (Billings, 1938; White et al., 2001; Navas et al., 2010). While allocation patterns allow early successional species to expand rapidly and to take advantage of openings within the plant canopy, these same characteristics would inhibit the ability of these species to withstand stressors such as drought.

Late successional species, in contrast, allocate a much greater proportion of their resources to below-ground structures and to the development of energetic and nutrient reserves. This is accomplished at the expense of rapid rates of growth and early reproduction. Once sufficient size has been attained, allocation may then switch to reproduction at the expense of further growth (Quinn and Meiners, 2004). Allocation to growth and maintenance allows late successional species to establish, grow and persist in plant communities without large openings and to integrate across poor- and high-quality resource years. Based on this generalized strategy, later successional species would be expected to survive droughts at greater rates and to be less impacted by the drought.

In addition to the characteristics of the species themselves, forest communities tend to ameliorate environmental conditions, which may also reduce the influence of disturbances such as drought. Forest interiors tend to be more humid (Matlack, 1993; Chen et al., 1995), which may mitigate the effects of brief droughts, particularly for forest understory species. Canopy trees may also be involved in the hydraulic lifting of water during dry periods (Horton and Hart, 1998; Liste and White, 2008), which can increase water availability to other species. Finally, connections among trees, either through root grafts or mycorrhizal networks, may allow the movement of resources (Simard et al.,

1997a, 1997b; Fraser et al., 2006) to lessen stress on individual plants. Together, the evolutionary, physiological and environmental characteristics of late successional species and communities generate the expectation of increased stability in response to perturbation.

As has been shown earlier in this book, the BSS data often do not meet the theoretical expectations of successional dynamics. The results with regard to Odum's linkage between stability and succession are also mixed. To address whether succession influenced the susceptibility of the plant community to environmental variation, we separated the successional sequence into the following age categories: 1–10, 11–20, 21–30 and 31+ years after abandonment. This separation resulted in groups of field ages at various stages of successional development that contained a range of rainfall amounts. In the first 10 years following abandonment, there was a strong positive correlation between the annual change in rainfall ($Rain_t - Rain_{t-1}$) and both the change in plot cover and richness (Figure 7.10). This fits with Odum's predictions, as early successional communities should be less stable and therefore strongly influenced by changes in water availability. The direction of the association also makes sense as the greater the decrease in rainfall, the greater the net loss of cover and local richness. Early successional species would be expected to do well in periods of relatively high resource availability and low stress. Similarly, these early successional communities contained greater areas of bare soil, which would provide opportunities for establishment and expansion of populations when sufficient moisture was present.

During the next two periods, years 11–20 and 21–30, annual changes in rainfall were unrelated to changes in both species richness and cover. The lack of association in these mid-successional communities fits with predictions, as later successional species should be better able to deal with the stresses generated by drought. The lack of association in the 11–20 year old fields is a bit surprising as these communities were still relatively young and contained a large proportion of early successional perennials. The diversity of these mid-successional communities may allow for complementarity among species in their responses to fluctuations in rainfall. Variation in species drought responses may generate stability within communities, particularly in additive measures of structure such as cover or richness (Tilman and Downing, 1994; Doak et al., 1998; Yachi and Loreau, 1999). However, average species richness at the plot scale was stable throughout this time period, so changes in local diversity did not generate the shift in drought response across years.

During the last period examined, when the fields were all largely forested, a significant departure from Odum's predictions was seen. A relationship re-appeared between change in rainfall and change in plot species richness, and nearly so in total cover. This is the time period that would be expected to show the most stability and the least responsiveness to interannual changes in rainfall based on the strategies of the species and the buffering capability of the forested environment. The re-appearance of weather-generated variation in the plant community may be related to the development of a multilayer community, where most plot diversity was located in the understory. This layer was composed primarily of herbaceous species and seedlings or small stems of woody lianas, shrubs and trees. This layer may have been more responsive to

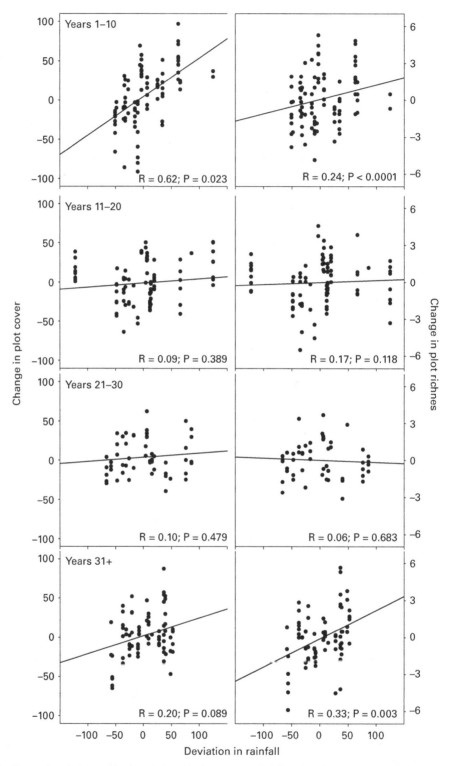

Figure 7.10 Successional changes in the relationship between Δrainfall and both Δcover and Δrichness at the plot scale. The data were separated into four time periods and the Spearman rank–sum correlation calculated for each.

environmental fluctuations than deep-rooted, established woody plants. The young forests also contained remnant individuals of species that had dominated the perennial phases of succession before the canopy had closed. Droughts may function to remove some of these species from the understory. However, community composition recovered from a major drought during this period, with most species quickly regaining their pre-drought abundance (Yurkonis and Meiners, 2006). Whatever the mechanism, succession and community stability are not as simply linked as has been hypothesized. It may be that as the young forests of the BSS age and shift more consistently towards forest species, stability will re-emerge.

Other disturbances

Other disturbances within successional communities are much more limited in spatial extent than are the influences of rainfall. They typically involve damage to or death of one or a few individuals within the community. The resulting community dynamics are therefore much more spatially constrained than the processes that have been previously discussed in this chapter. The sources of disturbance also shift from those external to the system, such as variation in rainfall, to those that arise from within the system, such as trampling or digging by mammals or tree-fall gaps. Despite the limited size of these disturbance events, they can be critical in maintaining the populations of some species and may allow the colonization of new species within an otherwise closed community (Goldberg, 1987; Carson and Pickett, 1990; Lavorel *et al.*, 1994; Collins *et al.*, 2001; Wilby and Brown, 2001; Gibson *et al.*, 2005). Excluded from this discussion are system-wide influences such as grazing, mowing or fires. The first two represent the resumption of agricultural activities, while the last would be catastrophic to the recovery of forest. Though fire was certainly a part of the forests of the region historically (Buell *et al.*, 1954), fires within an herbaceous-dominated successional community would prevent the establishment of woody species and would result in a large-scale re-setting of succession to a herbaceous community (Collins and Adams, 1983). For this reason fire is not a common occurrence or management practice for successional systems with a deciduous forest endpoint until a mature forest canopy has developed.

Disturbances within successional old fields are largely a result of the foraging or digging activities of mammals, resulting in patches of bare soil (Goldberg and Gross, 1988) or trampling (McCarthy and Facelli, 1990). The smallest of these disturbed patches will likely be colonized by the local clonal expansion of plants that border the disturbance. As clonal expansion is spatially constrained by the rate of vegetative spread into a disturbed patch, disturbances that exceed the spread of clonal parts, at least temporarily, can be colonized by seed recruitment. Plants which capitalize on such ephemeral opportunities for establishment may ultimately be outcompeted by the surrounding plants. Those species able to reproduce at least once before being displaced will produce and disperse seeds that will be able to colonize later soil disturbances. Effectively, these species escape competition as they complete their life cycle before they are displaced by competitively superior species. The difference between competitive and opportunistic plant strategies has led to the dispersal–competition tradeoff (Petraitis *et al.*, 1989; Tilman *et al.*, 1993; Fynn *et al.*, 2005). Interestingly, this tradeoff is also a tradeoff between differential species availability and differential performance. Within the BSS, species population dynamics do not appear to support a tradeoff between local dominance and rate of spread throughout the community (Chapter 6).

Disturbance and its influence on the plant community have been experimentally studied in a number of successional systems, including the fields of HMCF that surround the BSS. Disturbance has been experimentally manipulated by clipping or mowing vegetation, pulling back the plant canopy to increase light penetration and by turning over the soil with a trowel or mechanical tiller. All of these manipulations serve to increase the amount of resources available to colonizing species. The abundance of early successional species, particularly annuals and biennials, are typically greatly increased by disturbance, regardless of how the manipulation was carried out (Hils and Vankat, 1982; Armesto and Pickett, 1985; Goldberg, 1987; Vankat and Carson, 1991; Collins *et al.*, 2001; Wilson and Tilman, 2002; Gibson *et al.*, 2005; Honu *et al.*, 2006). As a consequence of increased colonization and disruption of competitive interactions, disturbance typically increases the diversity of the plant community. The results from experimental manipulations follow the effects of naturally occurring soil disturbances (Tilman, 1983; Goldberg and Gross, 1988). Germination opportunities provided by disturbance may not directly translate into increased diversity, as post-germination mortality may be quite high (Goldberg, 1987; Wilby and Brown, 2001). Disturbance may interact with resource availability (Wilson and Tilman, 1991; Huston, 2004; Gibson *et al.*, 2005), leading to contingencies in the impacts of disturbance. For example, disturbances paired with adequate water availability may lead to much greater rates of seedling establishment and survival (Goldberg, 1987; Carson and Pickett, 1990; Lavorel *et al.*, 1994). Relative to the influence of resource availability and leaf litter, fine-scale disturbances in successional communities may play a minor role in community structure (Carson and Pickett, 1990; Bartha, 2001).

Within the BSS data, we see patterns that are consistent with the role of soil disturbances as increasing local diversity. We can quantify soil disturbance based on the amount of bare soil within each plot. Though bare soil would clearly be available for colonization regardless of its origin, the data yield no information on the source or spatial extent of each soil disturbance. For example, a plot with 6% bare soil may have a single bare patch, or several, and may be generated by digging, plant mortality or trampling. As the spatial extent of each patch, as well as the bordering vegetation, is clearly important in the re-vegetation of patches (Goldberg and Werner, 1983), the BSS data are clearly limited in assessing recovery below the scale of resolution in the data. Using the BSS data, the association between local species richness and the amount of bare soil was determined for three different field ages (4, 13/14 and 29/30). These periods reflect large differences in the amount of bare soil available (Chapter 5). The amount of bare soil was positively associated with the number of species in the same year (Kendall's Tau-b, age 4 – R = 0.129; age 13/14 – R = 0.137; age 29/30 – R = 0.155; all P < 0.001). However, the amount of bare soil in a plot was not related to the number of species in that plot the following year. This suggests that openings within the community are brief (Goldberg and Gross, 1988; Rogers and Hartnett, 2001) and that colonizing species often do not persist for more than one year. Bartha (2001) separated bare soil from soil disturbed by digging mammals and found patches of soil without leaf litter to be much more important in regulating community structure. While we cannot determine the source of bare ground in the BSS, our results suggest that the condition of the soil surface is important in producing regeneration niches for some species.

Seed banks are often implicated in the response of plant communities to disturbance, as this is the most immediate source of new propagules to colonize the patch. Seed banks frequently contain high numbers of early successional species (Oosting and Humphreys, 1940; Roberts and Vankat, 1991; Lavorel et al., 1994; Leck and Leck, 1998; Gibson, et al., 2005). The bias towards early successional species within the seed bank makes them particularly important in disturbance recovery early in succession (Roberts and Vankat, 1991; Vankat and Carson, 1991). As succession proceeds, the density of the seed bank decreases and becomes less directly linked with the post-disturbance vegetation and with the surrounding vegetation (Oosting and Humphreys, 1940; Vankat, 1991; Vankat and Carson, 1991; Leck and Leck, 1998). As succession proceeds and generates a young forest, soil disturbances in the absence of a canopy gap are probably unimportant for the colonization of structurally dominant species, but may allow the colonization of understory species. When a large canopy gap occurs within a forest, recruitment of trees may come from the seed bank (Marks, 1974), from established seedlings (Marks and Gardescu, 1998) or from newly dispersed seeds (LePage et al., 2000). The diversity of sizes of canopy gaps within forests generates a range of conditions within the gap and maintains canopy diversity (Runkle, 1982; Sipe and Bazzaz, 1994, 1995), similar to the effects of soil disturbances in early succession. The BSS fields, now young forests, have experienced relatively few canopy gaps, though these will certainly become very important to the continued dynamics of the system.

Conclusions

In general, we can say that disturbance is important for the continued development and structure of successional vegetation. It is more difficult to pin down exactly what those effects are, as the system is driven by contingencies. Fluctuations in rainfall may influence the cover and richness of fields, though this depends heavily on when during succession the variation occurs. Severe drought events may usher in key transitions in dominance, or may result in rapid recovery of the community to pre-drought conditions – again dependent on when during succession the drought occurs. Not only may disturbance affect community recovery, but the recovery of the community may mitigate its ultimate response to disturbance. Differential responses of species to the drought will mitigate the impacts on the system and set the stage for subsequent recovery. One key feature of all of the disturbances discussed is their effects on resource availability (within differential site availability and conditions). Effects on this primary level of community drivers clearly cascade through the other drivers of community change. Differential species availability within the seed bank or local dispersal generates the pool of species that may respond, which are then filtered through differentials in performance. As the resource base changes during succession, as well as the pool of species within the vegetation and seed bank, we may also expect responses to disturbance to change.

8 Dynamics of diversity

> "Early in the history of a community we may suppose many niches will be empty and invasion will proceed easily; as the community becomes more diversified, the process will be progressively more difficult...In this way a complex community containing highly specialized species is constructed asymptotically."
> George Evelyn Hutchinson, Homage to Santa Rosalia or why are there so many kinds of animals? (1959)

Ecologists have been wrestling with understanding patterns of diversity for decades, with no evidence that interest is waning. Incorporated into this preoccupation is a desire to understand controllers of diversity, patterns of coexistence, evolutionary processes and the importance of diversity to communities, landscapes and ecosystems. Ecologists have developed many ways of addressing, describing and predicting diversity, each with a unique conceptual perspective (Scheiner and Willig, 2005). Like many topics in ecology, scale is inherently intertwined with our understanding of diversity. Local drivers of diversity that occur at the scale of species interactions or opportunities for local establishment (e.g. Pickett, 1980; Tilman *et al*., 1993; Gross *et al*., 2000; Myers and Harms, 2009) may have little bearing on diversity at coarser ecological scales (Levine, 2000; Ricklefs, 2008). The varied diversity perspectives that have been generated each have an appropriate scale of operation, though this is not always clearly defined. Changing scales likely leads to changes in the underlying processes involved. Not recognizing scale dependency may lead to confusion when integrating studies.

While there are a plethora of diversity patterns that may be investigated, there is less known about the dynamics of diversity. The G. Evelyn Hutchinson quote that begins this chapter deals more specifically with the development of a community (think succession), and is therefore temporal in nature. His words reflect much of niche-based theory that was developing when the address from which the quote was taken was presented to the American Society of Naturalists in 1959, but it is also reminiscent of much of what is still discussed in invasion biology today. Interestingly, this text also holds a strong prediction – that community development will be asymptotic. We shall return to this idea later in the chapter.

The general lack of a temporal perspective in diversity studies is likely a reflection of the temporally limited nature of the data that are often available. Temporal duration, of course, is the strength of the BSS, as we can explicitly follow the dynamics of the system over time. Understanding the dynamics that generate diversity patterns is a critical linkage to the mechanisms that may be generating these patterns. Fortunately, succession does have a clear predictive framework for understanding the dynamics of diversity – the intermediate disturbance hypothesis (Loucks, 1970; Grime, 1973; Connell, 1978; Petraitis *et al.*, 1989; Wilkinson, 1999) with a relatively large body of literature that we can draw upon. We will discuss this hypothesis in detail later in the chapter, along with some interesting variations.

Further complicating the field is the (pardon the repetition) diversity of ways to represent diversity. Diversity may be thought of as the number and evenness of species in a community alone or in combination, may represent within- and between-site variation, and may be represented graphically or distilled into a number. Assessing the value of each approach to diversity is well beyond the scope of this chapter. Instead, we have chosen to reflect the varied approaches to diversity in our examination of the BSS data. By approaching diversity in multiple ways, we hope to pull together the disparate approaches so that linkages may be highlighted.

While we have introduced some of the basic diversity patterns as a part of describing the baseline community changes earlier in this text (Chapter 5), we will explore these themes in much greater detail here. Inherent in all of these approaches are issues of scale and temporal dynamics. First we will begin by dissecting some of the basic patterns of diversity to determine their responses to spatial and taxonomic scale. We will then use Whittaker's dominance–diversity curves to describe community changes in succession, following this up with a discussion of dominance and rarity in succession. This will lead us to a quick detour into the temporal scale of turnover in diversity. Building on this material, we will conclude with a detailed analysis of the intermediate disturbance hypothesis. Overall, our goal is to generate a dynamic and integrative view of diversity in succession to place it into the broader context of community dynamics. Following one of the primary themes of this text, we believe this approach will also help to develop a broader understanding of diversity in all communities.

Scale and patterns of diversity

Scale is always a critical issue in dealing with diversity. We have briefly looked at the differences between the plot and field scale for species richness in Chapter 5. Briefly, species richness at the plot scale did not vary over successional time, while fields continued to accumulate species for nearly 35 years (Figure 5.5). However, spatial scale is not the only scale of operation for diversity – we may also examine taxonomic scales. This approach may yield information about functional diversity, as species in the same taxonomic group are typically more functionally similar than species from disparate taxa (Silvertown *et al.*, 2001; Webb *et al.*, 2002; Cadotte *et al.*, 2009), within the context of their phylogenetic constraints (Silvertown *et al.*, 2006). When we look at the entire pool of species in the BSS, the ratio of species: genera: families was approximately 4.8: 2.7: 1. This ratio gives us a base expectation of the distribution of taxonomic diversity within the system, assuming random draws from the larger species pool.

In examining diversity at the 1 m^2 plot scale over successional time, there was again no real temporal change in richness for any taxonomic level (Figure 8.1). As is

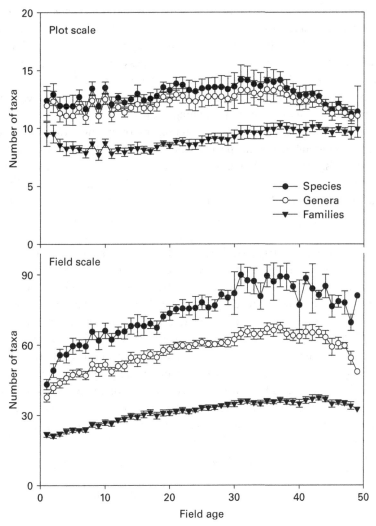

Figure 8.1 Successional trajectory of species, genus and family richness at the 1 m^2 (top panel) and field (bottom panel) scales. For each interval the data plotted are means ± SE.

necessarily the case, there was a drop in richness from species to genera to families (average richness 11.4, 11.1 and 9.1, respectively). However, the decrease in richness between species and genera was very small, with genera richness averaging 95% of the richness of species. This is quite different from the 44% decrease that would be expected based on the number of taxa in the system. The near equivalence of species and generic richness suggests that at small spatial scales there are limitations to coexistence, so that on average, each genus is represented by one species. Such an effect may be evidence for limiting similarity in controlling local species richness and species performance (MacArthur and Levins, 1967; Pacala and Tilman, 1994; Emery, 2007; Pillar et al., 2009). In comparing the richness of genera to families, there was again a clear decrease in family richness. Though a decrease in richness would be expected at a higher level of aggregation, the shift was again less than the expected decrease suggested by the taxonomic structure of the total species pool. Family richness was 82% of generic richness in the temporal data, whereas family richness was only 37% of generic richness in the species pool. Again, there were on average fewer family co-occurrences in the expressed community than the species pool would predict. This further suggests that there may be limiting similarity functioning at local scales.

At the field scale there was a much clearer separation among taxonomic scales, as well as a strong temporal pattern in species richness (Figure 8.1). The temporal pattern was strongest for species and genera, and somewhat muted in families. In contrast to the variation seen at the plot scale, there was a much larger shift in richness among taxonomic levels at the field scale. In both genera and families, there were still fewer co-occurrences than the species pool would predict, but these were complicated by the temporal pattern of richness in these groups. Very early in succession, species and generic richness were nearly equivalent, but rapidly separated over a few years. This separation over time suggests that early successional communities may be colonized by only one member of a genus, but that other members quickly colonize and spread. The overall temporal pattern of reaching a sustained maximum richness 30–40 years after abandonment was the same for all taxonomic levels. Interannual variation and variation among fields also decreased at coarser taxonomic levels. The reduction in variation suggests that there may be complementarity within genera and families, where different taxa may be present in each field, but still represent the same higher taxonomic level.

Differences between plot and field scales largely follow expectations based on the spatial scale of species interactions. Competitive interactions predominately operate between adjacent individuals, which are best represented by the plot scale in the BSS data (Meiners *et al.*, 2001; Yurkonis *et al.*, 2005). At that local scale, we see the strongest sorting of taxa in the generic and familial relationships of species. In fact, throughout succession the plot data approached the one species, one genus threshold, the strongest pattern of sorting possible. In contrast to the plot pattern, there was much less evidence for sorting at the field scale. Clearly, at scales coarser than the scale of direct species interactions there will be more coexistence mediated by spatial heterogeneity, priority effects and dispersal limitation (Fitter, 1982; Armesto and Pickett, 1985; Huston, 1994; Tilman, 1994; Vivian-Smith, 1997; Fukami *et al.*, 2005). Competitive interactions among even closely related taxa will not occur often enough within a field to generate significant patterning during the constant vegetation changes that occur in succession.

Partitioning of diversity

The relationship between scale and diversity may also be placed into a landscape context. While the vast majority of diversity in this system is contained at the field scale, measurement of the number of species in each field does not take into account differences in composition among the fields. Diversity across a landscape is addressed in the concept of α and β diversity, where α represents the diversity within individual communities and β represents turnover among communities. β diversity is often represented as a multiplicative factor that represents the accumulation of species from individual communities (α diversity). Another approach is to partition diversity among α and β components in an additive manner (Veech et al., 2002; Veech and Crist, 2009). From a practical standpoint, the additive approach places both components of diversity into a proportional scale, making the results more intuitive. Such partitioning may also be useful in placing diversity into broader spatial contexts and in informing conservation efforts (Wagner et al., 2000; Gering et al., 2003).

How might we expect succession to relate to α and β diversity? To answer this question, we must re-visit our conceptual framework, particularly the roles of differential site availability and conditions, and differential species availability. Differential performance would not be expected to have strong effects on diversity at larger scales as it would be limited to fine-scale species interactions. Performance would also not be expected to change dramatically across the BSS site because of the overall similarity in soils and climate. Experimental variation in how the BSS fields were abandoned was a direct manipulation of differential site availability and conditions. Pre-abandonment treatments generated persistent compositional differences among fields, particularly early in succession (Chapter 9). The impacts of abandonment conditions on species composition may lead to differences in α diversity among fields and may also generate much greater β diversity across fields with different successional starting points. Initial differences in composition should lead to β being the dominant diversity component early in succession. As succession proceeds and the importance of those initial conditions decreases, β diversity would be expected to decrease as fields converge compositionally.

We may also generate similar expectations for α and β diversity based on differential species availability (Christensen and Peet, 1984). The ability of species to colonize a newly abandoned site is a function of reproductive output, position of reproductive individuals in the landscape and the dispersal ability of the species propagules. For species that can form seed banks or have other persistence strategies, there is the potential for site history to also mitigate the ability of species to persist within the site and then establish following abandonment. Based on these factors, we expect community composition early in succession to be a stochastic combination of species based on local availability (del Moral, 1998; Chazdon, 2008). Early dispersal limitation should be mitigated over time as fields accumulate species from the landscape. Again, this mechanism would generate early heterogeneity among fields and dominance of the β component of diversity. As species disperse and colonize across the system, differences in composition should lessen, increasing the relative importance of the α component of diversity. However, species may begin to assort themselves along local environmental gradients late in succession, increasing β diversity (Christensen and Peet, 1984; Harrelson and Matlack, 2006). As the BSS fields are all still relatively young, in close proximity and have very similar soils, we do not expect dramatic differences based on local environmental gradients.

In partitioning diversity, we again find that the BSS data do not fully follow our initial expectations. As was predicted, early in succession β was the dominant diversity

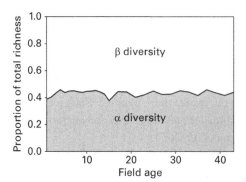

Figure 8.2 Successional changes in the partitioning of diversity between α (field) and β (site) components. Additive partitioning was done using Partition V3 (Veech and Crist, 2009) and is expressed as a proportion to remove the influence of overall successional changes in richness.

component, comprising roughly 60% of the diversity in the system (Figure 8.2). However, the relative importance of α and β diversity components did not change with succession. This is despite the dramatic accumulation of species in fields as succession and turnover occurred. The nearly doubling of richness in the BSS fields did not change the relative importance of the β component, which remained at an average of 57% with very little fluctuation (range 54–62%). The disparity from expectations is likely related to where the majority of diversity resides in this system. Most species are present as only a few scattered individuals, so the accumulation of diversity in succession is mostly an increase in rare species, which we will discuss more fully in the following section. With most species being rare and necessarily in only one or a few fields, the β component remained high. What was completely unexpected and is difficult to explain is that the increase in diversity in rare species was proportionately matched by increases in diversity within fields to maintain a constant ratio of α and β diversity. The stability of diversity components suggests that there was some pattern of assembly that regulated diversity across the system and did not vary with succession. Continued dispersal limitation seems a likely candidate, though the process would be expected to be more stochastic than the pattern seen in the BSS data.

Dissecting patterns in dominance-diversity

One of the graphical ways in which diversity is approached is in the generation of rank–abundance or dominance–diversity curves (Whittaker, 1965). Early work with rank–abundance curves focused on evaluating models of niche partitioning within a community, well beyond our goals here. However, there is utility in dissecting these curves to understand the underlying dynamics that generate patterns in diversity. We presented examples of these curves in our treatment of community patterns in succession (Chapter 5). Succession typically leads to an increase in richness and a lessening of the steepness of the rank–abundance curve (Figure 5.7; Bazzaz, 1975).

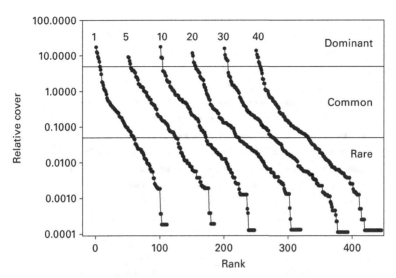

Figure 8.3 Example of rank–abundance curves generated to divide species into dominant (>5% relative cover), common (5% to 0.05% relative cover) and rare (<0.05% relative cover) within each age. Species designations were then used to follow changes in the abundance of each group over successional time (Figures 8.4 and 8.5).

Following the work of Murray et al. (1999), we used the position of species in a rank–abundance curve to categorize their role in the community. To determine patterns of species abundance and the change in these patterns over time, we generated a rank–abundance curve for each BSS field at each successional age. Species abundances within each age/field were then categorized as dominant (>5% relative cover), common (5% to 0.05% relative cover) or rare (<0.05% relative cover) based on natural breaks in the relative cover of each species at the site (examples plotted in Figure 8.3). Because of the temporal nature of the BSS data, we then looked at the dynamics of these abundance groups over time and the dynamics of species within those groups.

The temporal dynamics of dominant, common and rare species varied dramatically (Figure 8.4). The number of dominant species in each age was quite low, as would be expected, with an overall average of four dominant species. The number of dominant species did not change dramatically with succession. In contrast, common and rare species were both more numerous than the dominants and responded temporally to succession. The number of common species per field was lowest in year one (30 species), peaked around year 30 (42 species) and then slightly decreased to 38 common species per field in year 50. This group represented the majority of diversity in the BSS fields for most of succession. The number of rare species increased significantly over time, from 15 species in year 1 to 40 species in year 50. Based on these analyses, the majority of the successional increase in the species pool is generated by the addition of subordinate species rather than the loss of dominant species in the system. Though the categorization and analyses presented here are structurally different than the core satellite hypothesis (based on cover rather than on frequency), these subordinate species would be largely analogous to satellite species (Collins et al., 1993; Gibson et al., 2005).

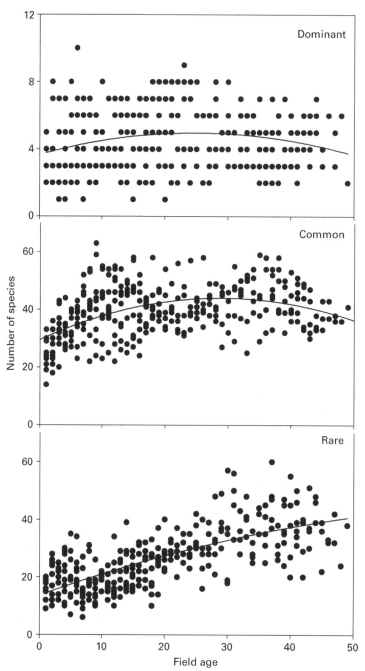

Figure 8.4 Temporal changes in the number of dominant, common and rare species within each of the BSS fields. Lines represent best fit quadratic functions. Regression information: Dominant – $F_{2,316} = 6.10$, $P < 0.0025$, $R^2 = 0.04$, model $S = 3.67 + 0.104(age) - 0.002(age)^2$; Common – $F_{2,316} = 38.09$, $P < 0.0001$, $R^2 = 0.19$, model $S = 29.54 + 1.002(age) - 0.017(age)^2$; Rare – $F_{2,316} = 166.4$, $P < 0.0001$, $R^2 = 0.51$, model $S = 13.77 + 0.769(age) - 0.005(age)^2$. Except for the quadratic term for rare species, all parameter estimates were significant.

To examine species transitions among dominant, common and rare abundance groups, a second set of analyses was conducted. In this analysis, a single rank–abundance curve was generated for each age by pooling data for all 10 fields and the component species were categorized as before. The complexity of variation among fields may be interesting, but would inhibit our ability to clearly categorize species based on abundance. We then followed all species throughout succession and tracked the amount of time that they were considered rare. Nearly all species (98.1%) were rare at some point during their successional trajectories. This intuitively makes sense since our definition of rarity was based on relative cover. Species will generally have low cover at two times during their successional trajectories, when first colonizing and when being displaced from the community. Surprisingly, 44.5% of species were always rare, persisting at low abundance in the community for up to 26 years. The remaining species all increased in abundance and became common species, with a few becoming dominants. The high frequency of rare species would be a large component of β diversity, though the temporal patterns of rare species would suggest that diversity partitioning should also change with succession.

In following the dynamics of rare species, there was a strong linkage between the number of years that a species was present in the site and the proportion of time that the species was considered rare (Figure 8.5). As a species' residence time increased, the proportion of time spent as rare decreased dramatically. This suggests that plant species tend not to persist as rare. Instead, they either increase and become a more abundant component of the community, or go locally extinct. In fact, many of the species that

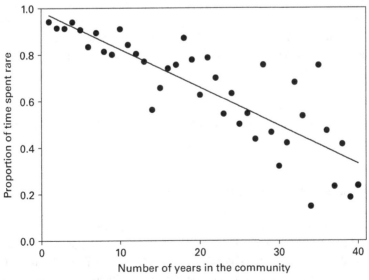

Figure 8.5 Relationship between the number of years a species was present in the BSS data and the proportion of that time spent as a rare species (Figure 8.3). Each data point represents the mean of all species with that residence time. Regression information: $F_{1,38} = 96.3$, $R^2 = 0.72$, $P < 0.0001$, $P_{rare} = 0.99 - 0.016(age)$.

Figure 8.6 Colonization of one of the BSS fields by the aggressive non-native species, *Microstegium vimineum*. Photo taken during the annual sampling in July, 2005.

persisted as rare may represent repeated failed colonization events rather than the persistence of a few individuals. These species may continue to colonize the site from the surrounding landscape until local conditions change or the species arrives in sufficient numbers to establish a population (Rouget and Richardson, 2003; Lockwood *et al.*, 2005). As an example, the invasive non-native grass *Microstegium vimineum* initially invaded several fields between 1982 and 1988, but then disappeared from the data until 1995. This second invasion started in the old-growth forest on the far side from the BSS and swept across the forest understory, moving into and through the BSS fields like an advancing wave (Figure 8.6).

Spatial turnover in diversity

Though studies of diversity often focus on static patterns or coarser spatial scales, the BSS allows us to address local variation in diversity and the stability of that variation. Small-scale studies have the potential to assess the roles of both local interactions and environmental conditions in generating patterns of invasion (Stohlgren *et al.*, 1998; Levine, 2000; Shea and Chesson, 2002; Brown and Peet, 2003). If local diversity is regulated by processes such as nutrient availability, seed availability or local heterogeneity, then we may expect that patterns of diversity will be constant over time. In other

Table 8.1 Association of plot species richness over time. Values plotted are Spearman correlations calculated on data ranked within fields to place all fields on an equivalent scale. Positive correlations indicate that the spatial pattern of richness is consistent between years – i.e. species-rich plots in year 1 are species-rich plots in year 2. Significant correlations are indicated in bold. As we are interested in pattern only, corrections for multiple comparisons were not carried out.

	Year 10	Year 15	Year 20	Year 25	Year 30	Year 35	Year 40	Year 45
Year 5	**0.11**	0.03	−0.01	0.04	0.07	−0.01	0.05	−0.10
Year 10	–	**0.16**	0.02	0.05	−0.02	−0.01	−0.03	−0.08
Year 15		–	**0.36**	**0.20**	**0.14**	0.06	−0.02	0.01
Year 20			–	**0.50**	**0.31**	**0.12**	0.06	0.02
Year 25				–	**0.52**	**0.26**	**0.15**	**0.10**
Year 30					–	**0.44**	**0.27**	**0.12**
Year 35						–	**0.39**	**0.23**
Year 40							–	**0.50**
Year 45								–

words, if a particular plot has conditions favorable to diversity, then we would expect that plot to remain higher in diversity as long as those conditions persist. However, if local diversity is more a function of species interactions and stochastic dispersal events, we would expect that successional turnover should lead to the breakdown in the spatial pattern of diversity.

Earlier, we discussed the idea that very early successional communities result from the combination of stochastic processes and historical contingencies (more on this in Chapter 9). Based on this, we would not expect plant communities in the first few years of succession to sort along environmental gradients or to be strongly structured. Therefore, we started our analyses with data from year 5, and examine each five-year period after that. We then simply correlated species richness in each plot with that plot's richness at other time periods. This allowed us to look at patterns in diversity and their persistence. Overall we see that diversity was consistent in adjacent time periods and that the patterns break down as the time interval increases (Table 8.1). Species richness in year 5 was only associated with year 10, whereas year 10 was associated with richness in years 5–15, etc. The time span over which richness was positively correlated increased during succession, suggesting a linkage between turnover in diversity and the life span of the organisms. We also see that patterns of diversity slowly broke down. If we examine the correlations for year 45, we see that plot richness in year 25 was weakly correlated with richness in year 45, 20 years later. The strength of correlation increased as the time interval shortened.

The spatial inconsistency of species richness and linkage with the life spans of the constituent organisms suggests that species richness at the plot scale is primarily driven by species interactions rather than resource availability or other qualities of the plot itself. This is consistent with the idea of local limitations on the number of species that can coexist in a 1 m^2 plot (Harrison, 1999; Gross et al., 2000). Species richness at the plot scale did not change throughout succession (Figure 5.5), despite dramatic increases in richness across fields. Such biotic limitations to coexistence, probably driven by competitive interactions, would mitigate the ability of local diversity to respond to other controllers of diversity.

Intermediate disturbance hypothesis

The intermediate disturbance hypothesis specifically relates disturbance to the maintenance of diversity in communities (Loucks, 1970; Connell, 1978; Huston, 1979; Petraitis et al., 1989). Simply put, the hypothesis states that at intermediate levels of disturbance there should be a peak in diversity. This peak is driven by the opposing community drivers of disturbance and competition. Without disturbance, or another opportunity for colonization, competition will lead to the loss of early successional species that colonize following disturbances (Huston, 1979; Shea et al., 2004). On the other hand, disturbances that occur too frequently will lead to the elimination of late successional species with much slower life histories. The optimum between these two drivers is achieved when disturbance is frequent enough to maintain early successional species, but not so frequent as to lose late successional species (Glitzenstein et al., 1986; Collins et al., 1995; Roxburgh et al., 2004). Of course, the frequency that actually qualifies as intermediate will depend on the life history of the species involved and the type of disturbance (Collins et al., 1995). It should also be noted that while the hypothesis is known as the intermediate disturbance hypothesis, there has been a range of diversity–disturbance patterns described. Given the prominence of the intermediate disturbance hypothesis as an ecological concept, it is surprising that unimodal peaks comprise a minority of the patterns found (Mackey and Currie, 2001; Shea et al., 2004). While the intermediate disturbance hypothesis has been criticized for its apparent lack of applicability on empirical and theoretical grounds (Fox, 2013), it represents a fundamental way in which people have thought about the relationship between succession and diversity.

Over the years there has been dramatic variation in the ways in which the intermediate disturbance hypothesis has been applied to systems (Mackey and Currie, 2001; Shea et al., 2004). Some researchers examine only patterns of species richness, the original focus of the hypothesis, while others incorporate a broader suite of diversity measures. Perhaps more importantly, some researchers focus on the maintenance of diversity in landscapes, while other researchers focus on within-site processes. At the landscape scale, a certain frequency of disturbance will lead to local patches at various stages of succession and therefore maintain species diversity within the landscape (Denslow, 1980; Roxburgh et al., 2004). In this scenario, any individual patch will likely not contain all species. For researchers that focus on within-site dynamics, succession following disturbance leads to the gradual increase in late successional species and loss of early successional species. Between the initiation of succession and eventual dominance of late successional species, there will be a peak in diversity when both types of species are present in the community (Loucks, 1970; Collins et al., 1995; Roxburgh et al., 2004). It is this second application of the intermediate disturbance hypothesis that we can address with the BSS data.

To determine whether the BSS data support the intermediate disturbance hypothesis, we will re-visit some of the data presented earlier in our treatment of community dynamics (Chapter 5). Instead of looking at changes in mean diversity with successional age, as was done previously, we will this time examine all fields individually and fit a quadratic function through the data. Regressions of species richness, evenness and Shannon–Wiener diversity all show significantly humped diversity responses to successional age (Figure 8.7). However, they differ markedly in the intermediate age that they identify as the diversity peak. Species richness, the most common measure used in testing the intermediate disturbance hypothesis, showed the latest peak in diversity at 39 years after abandonment. The other two diversity measures had smaller successional responses and peaked much earlier in succession than did species richness. Evenness peaked at 22 years after abandonment, whereas Shannon–Wiener diversity peaked at 25 years. The difference among measures reflects the continued accumulation of rare species late in succession. These species would contribute little cover to the system and would have minimal contribution to indices that include equitability.

The dependence of the diversity–disturbance relationship on the particular diversity measurement used is common to many systems and suggests that care should be taken in selecting individual metrics for study (Svensson *et al.*, 2012) and in applying the concept to conservation. The species-richness response follows the original usage of the intermediate disturbance hypothesis and addresses species coexistence. Measures of diversity incorporating the abundance distribution of species reflect within-site heterogeneity in composition. These two approaches to diversity represent different conservation goals and successional optima. Management intervention will not be able to maximize both evenness and richness measures and so therefore must necessarily select one view of diversity over another as the target.

The mechanism used to explain intermediate disturbance is the coexistence of early and late successional species. Rather than forcing a dichotomy on the species pool of the BSS and force species into being either early or late successional categories, we identified the decade of succession that each species peaked in. We then plotted the richness of each of these groups in response to successional age and examined their overlap as a measure of coexistence (Figure 8.8). There are two features of interest in these data. First, the total richness of each of these groups varied greatly. Species that peaked in the first 10 years of succession were the most species rich, followed by species that peaked years 11–20, 21–30 and finally 31–40. However, the latest successional category, species that peaked years 41–50, was more species rich than the previous two intermediate ages. Of course we know that many of the species in late successional communities are quite rare (Figure 8.4).

The second feature of the data is that each of the successional groups exhibited strong temporal patterns. The earliest successional species decreased in richness throughout most of succession. Similarly, each of the later groups peaked in richness at intermediate ages, coinciding with when their cover peaked. Even those species that peaked in cover 31–40 years after abandonment decreased late in succession, though this was a very small suite of species. What does all of this mean with regards to the coexistence mechanism at the heart of the intermediate disturbance hypothesis? Even 10 years into succession there is the loss of many early successional species in the vegetation. They likely remain dormant in the seed bank and so may not be completely lost to the system. The period 20–30 years after abandonment appears to have the most successional groups

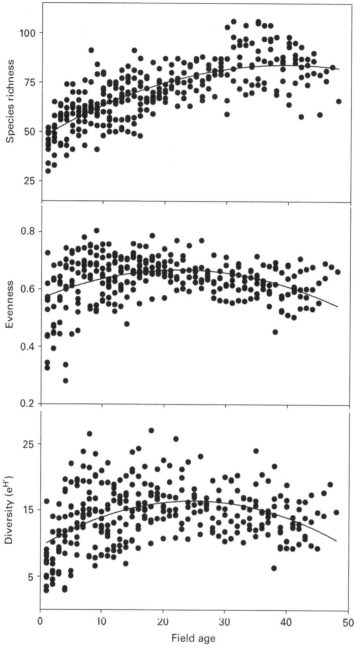

Figure 8.7 Successional dynamics of species richness, evenness and Shannon–Wiener diversity. Each data point represents an individual field and the line is a best-fit quadratic function. Regression information: Species richness – $F_{2,311} = 205$, $P < 0.0001$, $R^2 = 0.57$, model $S = 47.1 + 1.85(\text{age}) - 0.023(\text{age})^2$; Evenness – $F_{2,311} = 24.1$, $P < 0.0001$, $R^2 = 0.13$, model Evenness $= 0.568 + 0.009(\text{age}) - 0.0002(\text{age})^2$; Diversity (expressed as species equivalents) – $F_{2,311} = 32.5$, $P < 0.0001$, $R^2 = 0.17$, model $e^{H'} = 9.5 + 0.56(\text{age}) - 0.011(\text{age})^2$.

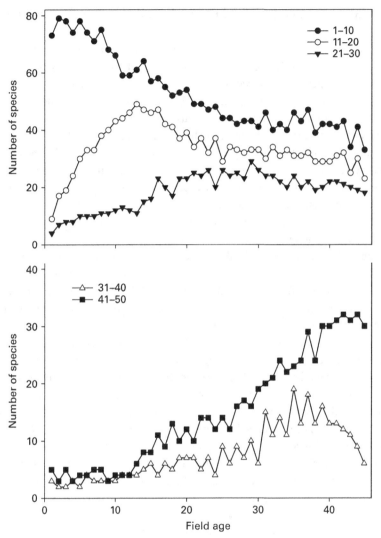

Figure 8.8 Coexistence of species from different successional stages as hypothesized by the intermediate disturbance hypothesis. Temporal changes in the richness of species categorized by the decade during which they peaked in succession. Data presented are from all fields pooled together.

represented, though the latest successional species are probably not reproductive by that age. This is the phase identified by evenness and Shannon–Wiener diversity as the most diverse. As the richness of species in the first 30 years of succession remains relatively stable past year 30, the increasing richness of the latest successional stage drives the peak in diversity identified by species richness. Therefore, coexistence of species from different successional stages is verified as a primary driver of the intermediate disturbance hypothesis. Whether this coexistence represents a tradeoff in colonization and competitive ability (Petraitis et al., 1989; Tilman, 1990) is not clear, but is certainly consistent with the BSS data.

Non-native species and the intermediate disturbance hypothesis

One component of the plant community that is rarely dealt with explicitly in analyses of the intermediate disturbance hypothesis is the role of non-native species (Catford et al., 2012). These species may directly reduce the richness of local communities in areas of heavy invasion (Meiners et al., 2001; Yurkonis and Meiners, 2004; Yurkonis et al., 2005). Non-native species invasions are also often linked with disturbance, so interplay with the intermediate disturbance hypothesis may be expected. The overall abundance of non-native species remained constant in succession (Figure 5.4), but how should their diversity respond? As many of the non-native species in the region are disturbance adapted and shade intolerant, we may predict that non-native species will have a clinal response, decreasing in species richness with time since disturbance.

When native and non-native species were separated, there were two disparate temporal patterns, though both could be considered intermediate disturbance responses (Figure 8.9). Native species exhibited the same general pattern as was exhibited in the whole community, though the peak in richness was then projected to occur several years earlier, at 34 years after abandonment. The model fit was also much improved over the model that incorporated both native and non-native species. When non-native species were examined separately, their temporal response was unimodal, but concave. Richness of non-native species was greatest early in succession, when the community was dominated by disturbance-adapted agricultural weeds, and late in succession, when a suite of woodland invaders was colonizing the young forests of the BSS. The minimum richness of non-native species occurred at year 22, though this was not a dramatic decrease in the number of taxa. The mechanism behind this depression in richness was not clear. It could be generated by an overall lack of mid-successional non-natives in the species pool when compared with agricultural weeds and forest invaders. The temporal response could also be generated by the non-native species themselves. The minima in richness coincides with the dominance of the non-native community by a few species, predominately *Lonicera japonica* and *Rosa multiflora*. Both of these species generated local losses in species in heavily invaded plots (Meiners et al., 2001; Yurkonis et al., 2005) and so may have effectively prevented the colonization of other non-native species. Whether this temporal pattern in non-native species is a feature unique to the BSS or a more general pattern is unknown. However, the overall muted response of non-native species, in contrast to the clear peak in native species, suggests that different ecological processes were regulating their richness.

The temporal dynamics of diversity

As is often the case, the more deeply we delve into a system, the more complexity we uncover. The BSS data clearly show evidence of patterning that suggests strong sorting at the local ($1m^2$ plot) scale. This occurred in examining both taxonomic hierarchies (limiting similarity) and in generating a consistent level of species richness throughout

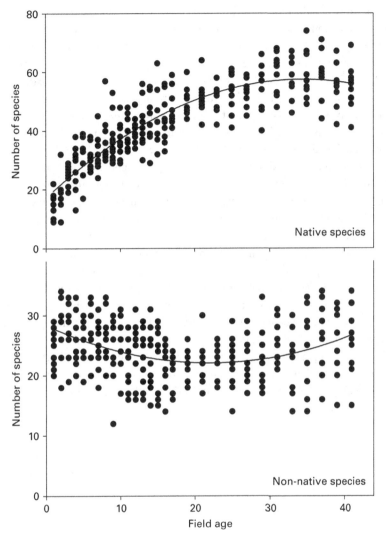

Figure 8.9 Species richness at the field scale separated out into native and non-native species. Regression information: Natives – $F_{2,283} = 489$, $P < 0.0001$, $R^2 = 0.78$, model $17.2 + 2.34(\text{age}) - 0.034(\text{age})^2$; Non-natives – $F_{2,283} = 20.9$, $P < 0.0001$, $R^2 = 0.13$, model $28.3 - 0.57(\text{age}) + 0.013(\text{age})^2$.

succession. Diversity at coarser spatial scales was much more responsive over succession, revealing patterns consistent with the intermediate disturbance hypothesis. However, we also uncovered large differences in the peak of diversity based on the metric used, whether dominant, common or rare species were examined, and between native and non-native species. So, the question of when diversity peaks is contingent on the composition of the community, metric used and scale examined.

The original hypothesis posed by Hutchinson, of asymptotic community assembly is not borne out by the data. The evidence for peaks in diversity is clear. This is perhaps not a clear conviction of Hutchinson's ideas, as his work was based largely on niche theory and therefore would rely on strong and direct species interactions. At the plot scale, where interactions should be the most intense, we saw that the ultimate size of the community was constrained. Similarly, patterns of richness over time suggest that species interactions generate patterns of diversity rather than abiotic factors. Hutchinson's ideas only break down at coarser spatial scales, where spatial heterogeneity and stochasticity would be important in limiting the strength of interactions.

The utility of addressing all of these diversity issues within the same ecological system allows us a broader perspective on the linkages among diversity patterns and places them into a temporal context. At the broadest scale, the rules for determining fine-scale assemblage composition from the larger species pool do not change with the age of the community. The remaining analyses all deal with how those assemblages assort themselves and generate gradients in species abundance – an outcome of differential performance. The sorting of species in space and time leads not only to rare, common and dominant species, but also to the compositional transitions characteristic of successional dynamics. All of the diversity measures that incorporated aspects of differential performance responded to successional time. Differential performance manifests itself in the BSS data both in generating local abundances and in regulating assemblage membership following colonization (differential species availability). By incorporating the dynamics of diversity into this larger effort to understand plant community dynamics, we also build a context for these changes in diversity. Too often, diversity studies focus on patterns of diversity removed from the larger context of the ecological system. This is a dangerous proposition, as diversity – both static and dynamic – is an emergent property of the underlying interactions and processes that generate community structure and dynamics.

Part 3

Integrative themes

This section of chapters continues the focus on the BSS data initiated in the previous section. However, the chapters presented here are centered on the conceptual framework that guides our successional thoughts. For each of the themes presented, we have used the conceptual framework to guide our exploration and discussion of the data to provide an integrative and holistic view of the system. The depth provided in each of these chapters will be much greater, though the range of chapters explored is not comprehensive of all that could be addressed with the BSS data. We selected research themes here based on their importance to our understanding of dynamic systems and based on our personal research interests.

Beyond the specific ideas explored in each chapter, there are two goals to this section. First, we present these chapters as case studies, examples, of the utility of our conceptual framework in addressing ecological issues in a complex system. We argued for the necessity of conceptual frameworks early in this book – this section is where we hope to build support for that position. Using the conceptual framework has forced us to address each issue in additional ways that we may not have otherwise have done, resulting in a much more complete view of the system.

Our second goal for this section re-visits one of our primary motivating forces – that of integration. This is where we have practiced it. The conceptual framework has not only organized our approach to each research theme, but has also allowed integration across individual drivers and processes to form a broader view of dynamic communities that encompasses many of the contingencies inherent to ecological systems. The application of a single conceptual framework to such disparate ideas as community assembly, species invasions and heterogeneity allows a further level of integration – that of understanding the relative importance of each class of driver across research themes. There is also value in training ourselves to examine research questions following a conceptual framework. In doing so, we will be less likely to immediately focus on a single driver of the system, but will rather adopt a broader, contingency-laden starting point to our research. This is the true value of integration to ecology.

9 Convergence and community assembly

> "The phenomena of nature are fluent[sic], not rigid, and no set of pigeonholes will entirely contain them. It is our task to give verbal expression to constantly changing phenomena in a way that will parallel their mutations as closely as is humanly possible. If we are to succeed in this, it is essential that we reduce the skeleton structure to a bare minimum..."
>
> William Skinner Cooper, The fundamentals of vegetational change (1926)

It may seem a bit odd to start a chapter addressing assembly rules with a quote that suggests that ecological dynamics are difficult to cleanly define and describe. The term assembly *rule* would certainly imply that processes are predictable, whether or not they result from deterministic or probabilistic mechanisms. Cooper's quote is part of a larger treatment of dynamic ecology that re-visited the foundations of vegetation science in order to evaluate the individual components, the "stones" of the foundation. Within this text, Cooper rejects the tendency of ecologists to treat each ecological system as different and distinct, with its own unique processes. Instead, he pushes for a more synthetic view of dynamic ecology that examines the commonalities among systems – what we may call rules today. He argues for this need based on the universality of change in vegetation; all communities are changing, have changed in the past or may begin to change in the future. Therefore all types of vegetation, regardless of community type, rate of change or driver of that change, may be studied under a single conceptual umbrella. To illustrate his thoughts, Cooper used an analogy of a braided stream, with the current state of vegetation being a cross section through the multiple channels. Time was represented as the flow within a channel and compositional change as the movement of the channel's position over time. The structure and dynamics expressed in a particular location would be a combination of the current environmental and biotic drivers as well as the historical context of the site. Over time vegetation may converge or diverge in response to changing drivers and their interactions, leading to the merging or splitting of stream channels in Cooper's metaphor.

Cooper's sermon against treating each ecological system as disparate and unique is appropriate as we turn to assembly rules. The definitions of assembly rules for ecologists vary as dramatically as the types of communities they study (e.g. Wilson *et al.*, 1996; Weiher and Keddy, 1999; Wilson, 1999; Cingolani *et al.*, 2007; Hille Ris Lambers *et al.*, 2012). Assembly rules in their simplest form are descriptions of patterns – increasing numbers of species on larger islands, changes in the number of species with soil fertility, or changes in life form as fire frequency changes. Other researchers maintain that assembly rules must be more specific – a decrease in water level of 5 cm will increase the abundance of species A by X%, or for every year since fire, the population of species B will decrease by n individuals. Some assembly rules are broad treatments addressing numbers of species only; other rules are much more mechanistic, linking species traits with environmental drivers. All rules attempt to understand how local communities are assembled from the broader species pool. This may involve assembly based on vagility, coexistence based on species similarity, or changes in abundance based on the interaction of physiological processes with the environment. All of these mechanisms were espoused by Diamond (1975) in his initial discussion of assembly rules, though his treatment of allowed and forbidden species combinations that generate checkerboard patterns of occurrence has received the most attention. As succession inherently represents the (re-)assembly of a community following disturbance, it has been an active area of research in assembly rules in many systems (Wilson *et al.*, 1996; Harrelson and Matlack, 2006; Ruprecht *et al.*, 2007; del Moral *et al.*, 2012; Raevel *et al.*, 2012).

In Cooper's language, the barest skeleton would be assembly rules that function at the broadest, least community-specific level and would represent the commonalities of dynamic ecology. At this level, we suggest that the conceptual framework set forth as the guiding principle for this book is an assembly rule. Simply put – the vegetation of an area will be determined by variation in site availability and conditions, variation in species availability and variation in species performance. As our framework is hierarchical, this structure allows one to focus in on details appropriate to individual systems with their unique suite of available species, ecological drivers and environmental constraints (Chapter 3). This structure allows the contribution of individual drivers, or rules, to be studied and perhaps to add more detail and predictability to the generalizations that may be formed.

In this chapter, we will focus on patterns within the BSS as they relate to assembly rules at the broadest scale of operation – differentials in site availability, species availability and species performance. The assembly rules that we will address are those that often come up in successional studies. To provide an appropriate context for these rules, we will place them into our guiding conceptual model. These rules will assess the impact of agricultural history on populations and successional processes and combine both temporal and spatial aspects of assembly. Though we will deal with the specifics of our data, we will constantly keep in mind Cooper's admonition to avoid focus solely on the details unique to our system. At the end, we will re-visit the call for predictive assembly rules (Weiher and Keddy, 1999) and propose some rules that may be tested in other systems. By explicitly linking pattern and process, the formation of assembly rules can be easily transferred to the restoration of a diversity of systems (Temperton and Hobbs, 2004).

Convergence: an expectation of assembly rules

Despite the complaint that succession can be unpredictable at local scales, successional processes are often treated as being deterministic, generating a more or less expected progression of successional stages. One of the resulting expectations of that determinism is the convergence of successional systems over time (Christensen and Peet, 1984). Convergence in composition or structure relies on the consistency of successional dynamics across a landscape. Consistency in dynamics would suggest that there are in fact overarching rules to how successional communities in an area assemble. If the rules governing assembly changed substantially across sites, then we would expect that successional dynamics should also vary and convergence not occur.

Successional convergence to varying degrees has been found in both primary (Lichter, 1998; del Moral, 2009; Kuiters et al., 2009) and secondary successional systems (Christensen and Peet, 1984; Frelich and Reich, 1995; Sheil, 1999; Brown and Peet, 2003). However, a lack of successional convergence has also been seen in a number of systems (Collins and Adams, 1983; Facelli and D'Angela, 1990; Myster and Walker, 1997; Blatt et al., 2005). As the composition and dynamics of early successional communities may be driven largely by stochastic processes that regulate the initial arrival of species into a community (del Moral et al., 1995), these communities would be expected to show marked variation in composition. As locations accumulate species, variation among local species pools should also decrease. With the development of a somewhat uniform species pool, more deterministic processes, such as competition should influence composition, and convergence should be expected (del Moral et al., 1995; del Moral, 2009). Successional drivers may also function to inhibit convergence. For example, convergence may be prevented when stochastic processes such as herbivory or local disturbance occur at high enough rates to disrupt species interactions and prevent displacement (Blatt et al., 2005). Specific to secondary successional systems, variation in land-use history may generate persistent legacies that maintain vegetation differences among sites (Olsson, 1987; Foster, 1993; Standish et al., 2006; Flinn and Marks, 2007). These land-use legacies are often driven by alteration and slow recovery of soil properties (Standish et al., 2006; Walker et al., 2010), so we would expect convergence to happen more quickly in less impacted soils or when soil recovery can happen quickly (Cramer et al., 2008). Agricultural landscapes also generate heterogeneity in species availability based on the location, disturbance history and vegetation composition of adjacent habitats. Lands abandoned in a matrix of pasture will exhibit quite different composition and dynamics than fields adjacent to remnant habitats or corridors.

Convergence may also differ with the scale of observation. Successional processes may lead to convergence at coarse scales, but microhabitat-associated divergence at fine scales (Frelich and Reich, 1995). Other systems have found convergence to occur at both fine and coarse scales (Sheil, 1999) or to vary among sites based on dispersal limitation (del Moral, 1998; Dzwonko and Loster, 1990). If dispersal limitation is important, perhaps even an assembly rule in its own right, compositional convergence may be an unrealistic expectation. For example, Collins and Adams (1983) found no compositional convergence in grassland succession, but found that all sites became dominated by woody species – structural convergence. Such structural convergence was a primary expectation in Clements' more deterministic view of succession (Clements, 1916). It should be noted that a lack of convergence does not necessarily suggest the absence of useful assembly rules, but may instead reflect the same rule expressed under different environmental conditions or with a dissimilar local species pool.

Functional convergence, an increase in similarity of the functional composition of a community, is a much less studied aspect of succession. Two communities may differ dramatically in composition, but still be composed of similar functional groups. In one clear experimental test of compositional and functional trait convergence, Fukami *et al.* (2005) found significant functional convergence in experimental plots, despite persistent compositional differences. These results suggest that trait-based community assembly rules are important in generating successional dynamics. Compositional convergence during succession may result from functional convergence, but is not a necessary result of successional dynamics. The importance and prevalence of functional convergence in all successional communities is not clear and remains an important gap in our knowledge of succession.

The BSS data provide us with a unique opportunity to directly examine successional dynamics for evidence of convergence throughout all of the major structural transitions of succession from abandoned agricultural fields to deciduous forest. This is in contrast to most studies, which must rely on space-for-time substitution. In exploring community dynamics (Chapter 5) we have already seen the overall structural changes that occur with succession. Patterns in life-form transition are largely consistent among fields, though there is some variation, which we will discuss later. Again, consistency in assembly rules that govern species–environment and species–species interactions should generate convergence in composition. Spatially stochastic rules such as herbivory, disease outbreaks and dispersal limitation should inhibit convergence (Chazdon, 2008).

As is often the case, the BSS data reveal much more complex dynamics than the answer to the simple question, is there convergence? Using data from the ordination of succession shown in Chapter 5 (Figure 5.8), we have examined temporal changes in compositional similarity among fields (Figure 9.1). Composition data show two distinct

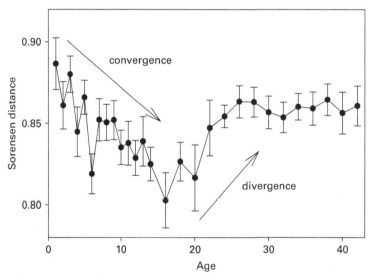

Figure 9.1 Temporal patterns in compositional variability across the 10 BSS fields. Data plotted are the average (± 1 SE) Sorensen distance across fields. Compositional data were the same as used in the NMS ordination of successional dynamics for Chapter 5 (Figure 5.8). Composition data were relativized prior to analysis to equalize the contribution of all species to compositional similarity.

phases of dynamics in similarity among fields – convergence and divergence. During the first year of succession, there was substantial compositional variation among fields. This is expected as fields were abandoned in different years and the fields were subjected to different agricultural treatments pre-abandonment. During the herbaceous phases of succession, composition across the 10 BSS fields became more similar, reaching the greatest convergence (lowest Sorensen distance) 16 years after abandonment. Thereafter, compositional similarity decreased for a 10-year period then stabilized, beginning in year 26.

What does this say about assembly in this system? We see evidence of both deterministic processes that generate convergence, and stochastic processes that generate variation among fields. Despite the initial variation among fields generated by differences in pre-abandonment agricultural treatment (dealt with in detail in the next section), these fields became more similar to each other as succession proceeded when herbaceous species dominated. This suggests that compositional differences initiated by abandonment treatment, crop history, local seed availability and even variation in weather at the time of abandonment (1958–1966) were insufficient to generate permanent shifts in the successional dynamics of this system.

The period of divergence coincided with the increase, spread and ultimate dominance of woody species in the BSS fields (Figure 5.1). It is during this period that we see the potential importance of dispersal limitation. If we assume that seed mass is negatively correlated with a species' dispersal ability, we can generate tests for dispersal limitation within the data. Since the shift in convergence coincided with an increase in woody species, we separated species into herbaceous and woody taxa for analysis. Variability among fields in species abundance was measure as the minimum coefficient of variation (CV) in cover during succession for each species. Variability in species' abundance among fields would generate variation in composition among fields. Individual species were typically most evenly distributed across the BSS fields near their peak abundance (Chapter 11). In general, herbaceous species had a much broader range of CV across fields than woody species; while shrubs, lianas and trees had lower variation across fields. However, these species were also more likely to be structurally dominant in the data. Variation in abundance among fields was not related to seed mass in herbaceous taxa (Figure 9.2). However, variation in abundance among fields increased with seed mass in woody species. Small-seeded woody species were much more evenly distributed among BSS fields than those with large seeds. For example, species such as *Toxicodendron radicans* and *Rosa multiflora* became equally abundant in nearly all fields, while large-seeded species such as *Juglans nigra* were restricted to only a few fields. Dispersal limitation in these taxa led to variation in the tree canopy composition of each field. As canopy trees have the potential to live much longer than any other life form, this divergence has persisted. This is not to suggest that all of the observed compositional divergence was a function of differences in the forest canopy. Following canopy closure, the BSS fields began to be colonized by a suite of herbaceous species characteristic of forest understories. These forest species were also very patchily distributed, primarily occurring in only one or two of the fields, and they had not spread rapidly. Given sufficient time, these fields may again undergo compositional convergence as a smaller set of shade-tolerant canopy trees

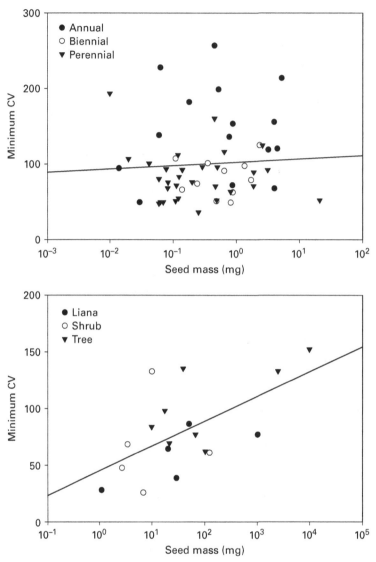

Figure 9.2 Dispersal limitation in herbaceous (top panel) and woody (bottom panel) species in the BSS. Data plotted are the minimum coefficient of variation in cover across the 10 fields for each species. Species that were evenly distributed across fields would have a low minimum CV, indicating a lack of dispersal limitation. Species with uneven distribution among fields would have a high minimum CV and would appear to show evidence of dispersal limitation. Spearman rank–sum correlations: Herbaceous – $R_S = 0.10$, $P = 0.45$; Woody – $R_S = 0.48$, $P = 0.045$. Seed mass data came from literature sources supplemented with on-site collections.

replace the existing canopy and as understory species disperse and establish throughout the site (Harrelson and Matlack, 2006).

What does the evidence of sequential periods of convergence and divergence mean with regards to assembly rules? In the original treatment of assembly rules, Diamond

(1975) dealt with both the vagility of species and their requirements in understanding the patterning of island bird communities. Dispersal ability is clearly a key species characteristic in determining the assembly of species within each field. Variation in species availability from local species pools necessarily constrains the species that colonize abandoned land and, subsequently, constrains the potential range of composition that may be generated through species–species or species–environment interactions (Foster *et al.*, 2011; Houseman and Gross, 2011). Many treatments of assembly rules, specifically those that attempt to identify specific physiological mechanisms, focus solely on interactions. As not all species are equally available to local communities, interaction-based assembly rules must be contingent on dispersal-based rules. In successional systems, and perhaps most community types, assembly rules must account for both classes of drivers to realistically capture community dynamics. The contingency of interaction-based assembly rules on dispersal-based rules will likely inhibit the ability to generate predictive rules that would pass the most stringent definitions of an assembly rule.

Historical impacts on assembly: abandonment conditions

Originally, the BSS was designed to test the effects of pre-abandonment condition on successional dynamics. Consequently, four fields were abandoned following hay production and the remainder all abandoned following row crops. The fields also varied based on the season of abandonment and whether the soil was left intact or plowed one last time (Chapter 2). These original treatments provide an opportunity to assess assembly rules operating at the level of site availability and conditions. These rules necessarily tend to be toward the descriptive end of the spectrum of assembly rules, but the patterns suggest mechanisms that may be tested experimentally.

Grass abundance – Four of the fields were planted in *Dactylis glomerata* hay before abandonment, and these fields are, therefore, expected to have higher abundance of this and other grass species, at least initially. The higher abundance of grasses in planted hay fields persisted for seven years before grass cover collapsed (Figure 9.3). In contrast, row crop fields had initially low grass cover that increased dramatically, while grass cover declined in the former hay fields. The mechanism of this successional replacement appears similar, though the timing of the transition was offset by nearly 10 years. Grasses in the hay fields were largely replaced by taller perennial forbs, well before these were in turn replaced by woody species. Grasses in the row crop fields were fairly slow to establish after abandonment and would have been competing with other early successional species during their establishment. This competition, in combination with the slower life history of perennial grasses, would have generated the delay in grass abundance relative to hay fields. Grasses in row crop fields peaked in abundance in year 10 and then consistently declined until year 20, as taller perennial forbs and woody species replaced them. It appears, therefore, that the agricultural establishment of grasses in the hay fields allowed a successional transition between grasses and taller forbs to be detected. In the row crop fields of the BSS, the successional replacement of grasses by forbs occurred simultaneously with the transition to woody species.

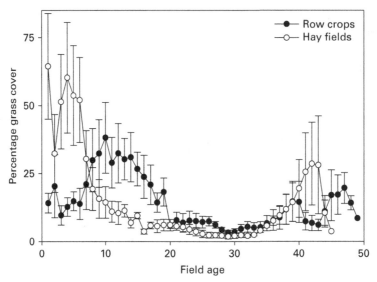

Figure 9.3 Abundance of grasses during succession in fields that were abandoned as hay fields and as row crops. Plotted data are means (± 1 SE) for the hay fields (N = 4) and row crop fields (N = 6).

Planting the fields in hay also altered the development of associations within successional communities (Meiners et al., 2002a). The richness of native and non-native species was positively associated in nearly all years within hay fields. In contrast, the pattern took 10 years to consistently appear in the row crop fields. This difference suggests that some level of community structuring occurred in the hay fields prior to abandonment. The same pattern eventually developed in the row crop fields, but only after a period of colonization and community organization following abandonment. Thus, the same assembly rule applied to fields that differed in cropping history, but the timing of this association was contingent on site history.

Short-lived species early in succession – The germination, growth and reproduction of short-lived species such as annuals or biennials are strongly linked with the availability of appropriate microsites within the community, typically the availability of areas free from competitors (McConnaughay and Bazzaz, 1987). These species may also respond strongly to environmental conditions and herbivory, leading to shifts in community dominance (Pickett and Bazzaz, 1978; Wilson and Tilman, 1991; Collins et al., 2001; Wilby and Brown, 2001; Honu et al., 2006). For these reasons, we may expect short-lived species to be particularly sensitive to variation in abandonment conditions among fields. We may also expect short-lived species to differ dramatically in response to variation in weather conditions over the nine years (1958–1966) that fields were abandoned. The narrow successional window over which annual species are abundant also means that their responses to agricultural history should be brief.

Within the BSS, there are two sets of short-lived species that we may expect to respond quickly to initial site conditions. Members of the Brassicaceae, the mustards, are strongly adapted to disturbance. Many of these species are weedy in agricultural systems and likely had well-established populations prior to abandonment. Most mustards germinate when soil temperatures are cool – either early in the spring (true annuals), or in the fall and over winter as vegetative rosettes (winter annuals). In both cases, they flower and produce seeds early in the growing season. The BSS contains a suite of early successional mustards composed predominately of *Barbarea vulgaris* (non-native, 65% of mustard cover), *Raphanus raphanistrum* (non-native, 23%) and *Lepidium campestre* (native, 12%). The BSS also contains a diverse pool of early successional non-mustard annuals that tend to germinate later in the growing season. These species include (in order of decreasing relative abundance): *Ambrosia artemisiifolia* (native, 30% of annual cover excluding mustards), *Erigeron annuus* (native, 24%), *Bromus racemosus* (non-native, 16%) and *Digitaria sanguinalis* (non-native, 9%). Similar to the mustards, these species are also weeds in agricultural systems, but tend to be more problematic for crop production, as they persist later into the growing season, competing with crops for longer time periods. These species also likely had established populations when the fields were actively being farmed.

It is reasonable to expect a priori that row crop fields would develop greater populations of short-lived species compared to fields with established grass cover. Early successional species depend on openings in the plant community for establishment, which should be more available following row-crop agriculture. The continuous perennial grass cover developed in the hay fields, and which had been maintained by mowing, lasted for many years after abandonment and would not appear to provide adequate opportunities for short-lived species to establish. However, the only consistent pre-abandonment treatment that was associated with the cover of mustards and annuals was whether the fields were plowed immediately prior to abandonment. Fields that were plowed after the final harvest and abandoned as bare soil consistently had greater cover of annuals or mustards (Figure 9.4) than any other treatment. Fields C4 and C5 were both row-crop fields that were abandoned as bare soil. These two fields developed greater cover of mustards than fields abandoned with intact crop stubble. Interestingly, field C7, a hay field that was plowed at abandonment, also had a high cover of mustards. These three fields also maintained the highest abundance of mustards into the second year after abandonment.

There appears to be some complementarity in the short-lived species pool. The two other fields abandoned following plowing, D3 (row crop) and E2 (hay) did not produce high mustard cover, but instead had the greatest abundance of non-mustard annuals (Figure 9.4). Again, the presence of a well-developed hay crop did not appear to affect whether the community became dominated by annuals. Even though plowing only temporarily disrupted the cover of grasses in the hay fields (Figure 9.3), this disruption allowed the establishment of annuals with some persistence into year 2. Key to this pattern appears to be the presence of disturbed soils, most likely with little plant-litter cover. Fields abandoned following row crops would certainly have had more open area, but would have retained crop litter that may have physically inhibited seedling emergence or reduced soil temperature fluctuations that serve as germination cues (Marks, 1974; Facelli and Pickett, 1991b; Facelli and Facelli, 1993; Bosy and Reader, 1995; Wilby and Brown, 2001).

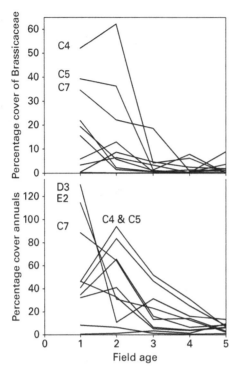

Figure 9.4 Variation among fields in the successional abundance of mustards (Brassicaceae) and non-mustard annuals in the 10 fields of the BSS. Data plotted are field averages. Only fields with high abundances of mustards or annuals are labelled.

The responsiveness of short-lived species to pre-abandonment treatments only generated vegetation effects during the first few years following abandonment. However, seeds produced early during succession may persist in the seed bank for long periods of time. These seeds may then serve as the source of colonists following soil disturbances, drought or other establishment opportunities that may take place years later (Marks, 1974; Roberts and Vankat, 1991; Leck and Leck, 1998).

Woody species early in succession – The establishment of plants often occurs in relatively discreet temporal windows (Egler, 1954; Johnstone, 1986). In succession, such periods of establishment may generate major compositional transitions (Bartha *et al.*, 2003). These windows have been noted in trees in the BSS and other successional studies (Rankin and Pickett, 1989; Debussche and Lepart, 1992; Peroni, 1994). Establishment windows for woody plants may be closed by the development of continuous herbaceous cover (Debussche and Lepart, 1992) or may be episodic opportunities generated by fluctuations in seed and seedling predators (Manson *et al.*, 1998). In the BSS, woody plants become established early in succession, but do not dominate until much later in succession. Therefore, factors that alter establishment windows for woody species early in succession may continue to alter community structure much later in succession. In the BSS data, we see the potential for pre-abandonment conditions to regulate woody plant establishment. However, each life form (shrubs, lianas and trees) was responsive to a different pre-abandonment treatment.

Trees compete heavily with grasses and other herbaceous species for water and light, leading to reduced regeneration (Petranka and McPherson, 1979; De Steven, 1991b; Gill and Marks, 1991; Davis *et al.*, 1998; Li and Wilson, 1998). Therefore, the presence of an established dense cover of perennial grasses in the hay fields would be expected to reduce tree regeneration, and likely the regeneration of other woody species. Though depression of tree growth by grasses has been documented in many successional systems, the temporal change in tree cover was not different between row and hay fields (Figure 9.5). Likewise, shrub cover was not significantly different over time. Lianas were the only woody group to respond to crop history with cover significantly higher in former row-crop fields. While the increase in liana cover in the absence of an established competitive plant community is easily understood, the lack of a response for shrubs and trees is difficult to explain. Competition is clearly not the only way in which herbaceous vegetation may affect woody plant regeneration. Established plant cover may increase seed predation (Reader, 1991; Reader and Beisner, 1991; Myster and Pickett, 1993; Hulme, 1994), but may also increase emergence (De Steven, 1991a). The net effect on regeneration would be the combination of all effects, beneficial and detrimental, and may generate an overall neutral association between hay fields, and tree and shrub regeneration. Why only lianas should respond is not clear, though the response is consistent among liana species.

Season of abandonment only affected the temporal increase in abundance of shrubs, with greater cover developing in those fields abandoned in the spring. While trees varied dramatically in seed mass and dispersal mode, all of the shrubs and lianas that became dominant in the BSS were bird-dispersed. Shrubs in general had smaller seeds than those of the lianas, though there is substantial variation in seed mass (Figure 9.2). Shrub cover, more than any other life form, was dominated by a single species – *Rosa multiflora*. The association with season of abandonment may be generated wholly by specific requirements of *Rosa multiflora* rather than a more general shrub characteristic. The next most abundant group of shrubs, *Rubus* spp., did not vary with season of abandonment, but rather had increased abundance in fields formerly used for hay production. *Rubus* spp. are commonly found in pastures and are strongly clonal, so this association is consistent with patterns seen in other successional study sites. An association with hay fields may suggest that *Rubus* spp. was established in hay fields prior to abandonment and persisted as clonal fragments. However, *Rubus* were first identified within the permanent plots in year 4, so any individuals that persisted in the hay fields must have been in very low abundance. Alternatively, grass cover may have facilitated germination and establishment of *Rubus* from the surrounding landscape.

Only tree cover was affected by whether the crop residue (crop stubble or established hay grasses) was left intact, or plowed immediately prior to abandonment. Tree cover increased more rapidly in those fields that were left intact, rather than disturbed. Seedlings that were plowed would have likely been killed or at least uprooted. The lack of plowing would have allowed any trees that had become established in the year before abandonment to survive undamaged and have a one-growing-season advantage

Convergence and community assembly

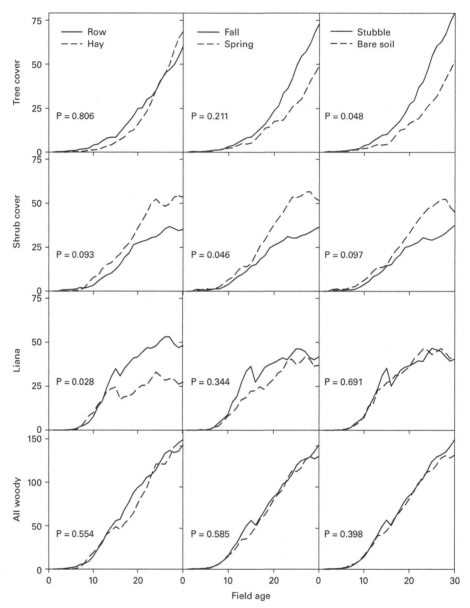

Figure 9.5 Temporal patterns in woody life forms and total woody cover in response to pre-abandonment treatments. Average percentage woody cover across all fields is plotted for each treatment. Only the first 30 years of data were examined as following this period, trees dominated all fields and other life forms decreased. P values come from RMANOVA analyses of the effect of each abandonment condition.

over their counterparts in plowed fields and associated herbaceous plants. In hay fields, it is possible that some tree seedlings persisted in the crop despite being mown annually, and these plants may have been quick to expand despite interactions with the established grasses. The benefits of an extra season's growth would extend to all woody species, so it is not clear why only tree species would have responded to this treatment. It is also possible that the final plowing increased germination and success of new tree seedlings. Whatever the mechanism responsible, the diversity of regenerating trees in the site suggests that this was a general tree response to site history.

When all woody species were examined collectively, the temporal change in cover was statistically indistinguishable for all of the pre-abandonment treatments. Woody species maintained the same successional trajectory across all fields. Therefore, none of the treatments altered the establishment and growth of woody plants or influenced the overall advantage of woody species in successional communities. To generate consistent rates of woody expansion, shrubs, lianas and trees appear to have responded complementarily to initial field conditions. Each life form as a whole only responded to a single agricultural treatment, different from the other woody life forms. Abandonment conditions only shifted the relative abundance of woody life forms in mid-successional communities. After year 30, all woody life forms except trees experienced rapid declines and differences generated by abandonment conditions disappeared.

Spatial assembly rules: the impacts of forest edges on succession

Succession is typically discussed as a temporal phenomenon, though it is an inherently spatial one as well. Site location within a landscape will determine the flux of plant species into the site, as well as the use of the site by vertebrate dispersal vectors and herbivores (Johnson and Adkisson, 1985; McClanahan and Wolfe, 1993; Matlack, 1994; Lawson *et al.*, 1999; Cadenasso and Pickett, 2000; Cadenasso and Pickett, 2001; Myers *et al.*, 2004). During succession, patches of vegetation originate, spread and come into contact with each other (Harrison and Werner, 1982; Luken and Thieret, 1987; Ben Shahar, 1991; Burton and Bazzaz, 1995). As vertical woody structure develops within an herbaceous community, avian seed dispersers are attracted, altering the spatial pattern and magnitude of seed dispersal (McDonnell and Stiles, 1983; McDonnell, 1986; Robinson and Handel, 1993, 2000).

Within many successional systems there is another spatial feature – an adjoining remnant habitat. Habitat edges may be critical to succession because they may serve as a source population for colonizing species, particularly those of late succession (Hughes and Fahey, 1988; Johnson, 1988; Hughes and Bechtel, 1997). However, edges generate much more heterogeneity than just seed dispersal (Cadenasso *et al.*, 2003a; Matlack, 1994). Forest edges adjoining successional habitats generate microclimate gradients (Cadenasso *et al.*, 1997), alter spatial patterns of seed and seedling predation (Myster and Pickett, 1993; Manson *et al.*, 2001; Meiners and LoGiudice, 2003), and alter the spatial pattern of herbivory (Myster and Pickett, 1992b; Inouye *et al.*, 1994; Cadenasso and Pickett, 2000; Meiners *et al.*, 2000). This complex gradient of edge effects interacts with patterns of seed availability to generate gradients in vegetation (Myster and Pickett, 1992b; Meiners and Pickett, 1999; Meiners *et al.*, 2002b).

Mechanistically, the potential influence of edges on successional dynamics resides within two of the major drivers of succession – differential seed availability and differential performance. Spatial patterning of seeds generated by the presence of a forest edge can come about through deposition of locally produced seeds or through attracting dispersal agents from the surrounding landscape. Once seeds are present within the site, then differential species performance can generate further patterning via plant–environment interactions, spatial responses of seed and seedling predators, and spatial variation in herbivore activity. These two classes of drivers may generate different patterning in the community. To fully understand assembly across edges it is necessary to separate these two processes and their role in generating vegetation gradients.

Separating differential availability from performance is somewhat challenging within the BSS, as individual plants are not followed. In general, we can look at rates of plot occupancy as a measure of species availability and cover of species as a measure of differential performance. This approach is not perfect, as plot occupancy integrates species availability and establishment, while greater rates of dispersal may generate greater cover in small, non-clonal species. As the initiation of the BSS preceded interest in spatial patterning in plant communities, plots in most fields do not capture vegetation directly adjacent to the old-growth forest edge. Despite this limitation, the data have been useful in understanding spatial dynamics (Myster and Pickett, 1992b). We will use the BSS data to understand how the forest edge influences successional dynamics and try to separate dispersal from performance mechanisms.

Tree regeneration – Previous work with the BSS (Myster and Pickett, 1992b) documented strong spatial association of trees with the forest edge early in succession, but this pattern disappeared as succession proceeded. Dispersal mode and the palatability of the trees to herbivores also played a role in generating spatial patterning. If we take a broad view of succession, the establishment of tree cover is the primary structural transition that leads to the loss of shade-intolerant species and an increase in forest understory species. For this reason, we will explore tree species as a whole to understand whether the edge constrains tree regeneration and therefore succession from old field to forest.

We will explore the spatial pattern of tree regeneration in one of the BSS fields, C4. We chose this field because it has a single border with the old-growth forest and has a large range of plot distances from the forest edge to increase our ability to detect edge effects. The closest plots in this field are located 26 m from the forest edge, though there would be forest canopy overhang well beyond the edge. As the number of tree stems is recorded in the BSS data, we can follow the number of individuals over time. From year 15, when trees first started to expand in the field, to year 41, we see a rapid increase in the number of tree stems (Figure 9.6). In year 15, plots averaged less than 0.25 stems per plot, increasing to over five stems 15 years later. There is no indication of an edge response in tree density at the distances covered by the plots. However, there is dramatic spatial variation in years 31 and 41, and a collapse in tree density in year 35. The densities attained were more than sufficient to generate a continuous forest canopy.

Not surprisingly, tree cover also increased over the same time period, but did so in a much more consistent manner. Trees expanded rapidly after year 25, with tree cover averaging around 90% 41 years after abandonment. An edge effect was evident in years

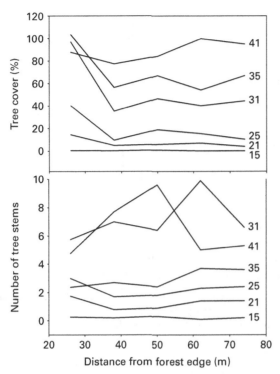

Figure 9.6 Spatio-temporal patterns in tree cover (top panel) and stem density (bottom panel) for field C4. Data represent the influence of the old-growth forest edge on the spatial pattern of trees in six successional time periods.

25–35, with a marked depression in tree cover between 26 and 38 m from the edge. At greater distances into the field, cover remained constant, accumulating a consistent amount of cover over time. By year 41, tree cover did not vary with distance from the edge. The edge constrained tree regeneration early in succession, but this does not appear to have been driven by the establishment of trees, as seedling density was spatially constant and sufficient to generate a closed canopy forest. Plots closer to the old-growth forest generated a closed canopy 10 years earlier than those farther from the edge. As seedling density was constant, tree growth appears to have been greater closer to the forest. This pattern may have resulted from a variety of mechanisms, including selective foraging by browsers (Myster and Pickett, 1992b), availability of appropriate mycorrhizal symbionts (Mason et al., 1983; Allen and Allen, 1984; Johnson et al., 1991) and changes in microclimate that directly or indirectly affected tree growth (Meiners et al., 2002b).

Mechanistic analyses of species dynamics – Plant species are commonly noted to increase or decrease in association with forest edges. However, the processes that generate these spatial responses are rarely examined. These processes are the critical step to understanding how successional communities assemble themselves across edges. To explore the mechanisms that generate edge responses, a suite of 75 species that became abundant in at least one of the six BSS fields that shares a single border with the

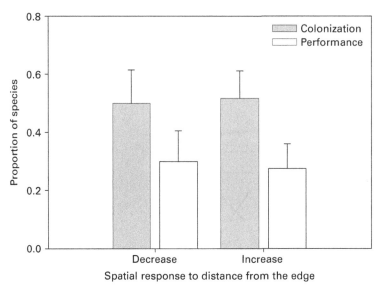

Figure 9.7 Proportion of species with spatial responses to the forest edge generated by differential colonization and performance. Species are separated into those that increased with distance into the field and those that decreased. Lang and Meiners, unpublished data.

old-growth forest were analyzed (Lang and Meiners, unpublished data). To begin, the spatial pattern in response to the edge was determined at the peak cover of each species. Of the species examined 20 (27%) decreased in abundance with distance into the fields, while 29 (39%) increased with greater distance into the fields. This variation is not surprising as species commonly have variable edge responses and this species pool contained species from all successional stages and life forms.

For each species with an edge response, spatial patterns of plot colonization (a combination of dispersal and early establishment) and cover were examined to determine the underlying dynamics that generated edge patterns. To partially separate spatial patterns in cover from those of colonization, only plots which had been colonized by a species were included in analyses of cover data. These two measures serve as surrogates for differential dispersal and performance in generating individual species' edge responses. Overall, 51% of species, regardless of the direction of their edge response, had spatial patterns influenced by differential colonization (Figure 9.7). In contrast, analysis of cover data showed that 29% of species with edge responses had patterns generated by variation in how abundant species became in colonized plots. Many species showed evidence of both processes. Together, these analyses argue that dispersal and early seedling establishment are much more important than subsequent growth in generating edge-related changes in vegetation. This result is similar to analyses of the BSS data that showed that the majority of non-native species impacts came from alteration of species colonization rates (Yurkonis et al., 2005; Yurkonis and Meiners, 2004) Having said that, differential performance was still an important mechanism in generating edge responses.

BSS assembly rules

The BSS data suggest some process-based assembly rules. While these rules will not satisfy those requiring specific predictive ability to qualify as an assembly rule, the rules presented here are more likely to translate to other systems than much more specific ones. While any of the patterns shown in this chapter may be considered an assembly rule at some scale, we have chosen to focus on a few more integrative rules. These rules form hypotheses that may be tested in other systems. When the underlying rules for assembly differ among systems, this represents an opportunity for developing a broader understanding of community dynamics (Cramer et al., 2008) and for restoration (Temperton and Hobbs, 2004).

- *Successional communities go through a period of species accumulation driven by stochastic processes, followed by the ecological sorting of species through more deterministic processes.*

Succession effectively represents a mechanistic shift from factors that determine species arrival to a site towards those that determine how the species assort themselves ecologically (Pickett et al., 2011). Immediately following abandonment is a period when the community is essentially open and susceptible to local variation in seed, rain and abandonment conditions. While the majority of individuals will be adapted for such disturbances, the community will also contain a random assemblage of most life forms. As available space becomes filled, interactions such as competition will increase, colonization windows will close and species will become assorted by their environmental requirements. The increase in species–species and species–environment interactions will lead to a concomitant development of structure within the community and the ultimate displacement of less successful species.

- *Species availability constrains species interactions.*

As a logical starting point, if a species is not present in a community then it cannot possibly interact with other species. While many herbaceous species became evenly represented among the fields, woody species were much more spatially limited. This difference among life forms resulted in a period of convergence in herbaceous communities as a subset of species rose to dominance in all fields, but then a period of divergence as dispersal-limited woody species came to dominance. In this context, convergence can only be expected when sites are similar environmentally and share a common species pool with dominants available to all sites. The timing of convergence will depend on the rate at which dominant species arrive to, and spread within, all sites.

- *Site history and arrangement can generate short- or long-term shifts in composition, but little change in vegetation structure.*

Despite being a primary driver of successional dynamics, variation in abandonment conditions did not generate dramatic variation in the community that persisted throughout succession. Similarly, spatial constraints associated with edges did not alter the successional transition to forest. Beyond initial high abundances in short-lived species in some fields and the dominance of grasses in hay fields, overall successional trajectories of life forms did not vary among fields. Despite temporary advantages to lianas or shrubs in some fields, all were eventually replaced as dominants by trees. There is evidence for neither alternative successional endpoints nor alternative successional pathways to forest in this system. The overall consistency in environment and species pool across the site constrained succession to pass through the same structural transitions.

- *Colonization is more important in generating spatial structure than subsequent growth.*

Factors that affect species dispersal and early establishment appear critical to determining the majority of species' responses to edges or in generating impacts of non-native species. The seed–seedling transition is very susceptible to environmental conditions, herbivory, pathogens and competition (Grubb, 1977; Fenner, 1987). Once this critical threshold is crossed and the seedling established independent of seed reserves, species are much less responsive. This pattern suggests that focusing purely on interactions between adult plants misses a critical stage for many species.

Coda

In the search for assembly rules, we need to recall the earliest origins of the idea. Diamond (1975) formulated his ideas in the context of island biogeography, species dispersal ability and resource utilization. He also had the benefit of temporal data on islands where all bird species had been removed by volcanic or tidal wave activity. These earliest rules dealt with island area (site conditions), species vagility (availability) and habitat requirements, as a baseline – and then focused on a subset of species to understand structuring based on competitive interactions (performance). What should this tell us about the search for assembly rules? Assembly rules need to encompass all of the dominant mechanisms that structure a community in order to be useful. Failure to do so ignores important contingencies and may generate unnecessary debate in ecology.

10 Successional equivalence of native and non-native species

> "Perhaps the most notable thing about the weeds that have come to us from the old world, when compared to our native species is their persistence, not to say pugnacity... Our native weeds are for the most part shy and harmless..."
>
> John Burroughs, *A Year in the Fields* (1896)

> "We may enquire, whether weeds have any common characteristic which may give them advantage, and why the greater part of the weeds of the United States, and probably of similar temperate countries, should be foreigners."
>
> Asa Gray, The pertinacity and predominance of weeds (1879)

Even before the field of invasion biology was codified in Charles Elton's seminal book *The Ecology of Invasions by Animals and Plants* (1958), ecologists and other observers of the natural world had noticed the spread and potential impacts of the new plant species that they saw springing up in their surroundings. As those first observations were made, so too were the first generalizations. The two quotes that begin this chapter are examples of the early formation of the conceptual dichotomy between our native plant species and those that have been introduced. In fact, Gray's comments are in response to another, even earlier discussion of the differences between native and non-native plant species (Claypole, 1877). Based on the success of European weeds in the new world and the general lack of reciprocal invasions, these papers argue that these floras are inherently different, with the European flora more adaptable and therefore more successful on both continents. Clearly, this cannot be the case, as the following century was characterized by a large number of North American plant species that became established and even invasive in Europe.

More recently, some have argued that the separation of invasion biology from the rest of ecology is not beneficial to the field because that division is an artificial one (Davis *et al.*, 2001). We may also question whether the conceptual separation of native from non-native species is also an artificial one. Should we really expect that plant species and the factors which constrain their dominance in communities are inherently different just because they evolved in two geographically disjunct areas?

Biological invasions are wonderful opportunities to study ecological processes in action (MacMahon *et al.*, 2006). If for no other reason than people tend to notice the spread of new species, we know much more about the movement patterns of non-native plant species throughout landscapes than we could ever hope for the historical and contemporary movements of natives. In many cases, this increased awareness of non-native plants is a function of the legitimate management concerns that many of these species pose. From a purely theoretical standpoint, biological invasions are excellent model systems in which to examine the controllers on species abundances, the impacts of species on each other and the selective processes that continually shape populations. As all of these processes are general ones common to all species, we must ask whether non-native species are convenient case studies representative of all plant species or do they represent something inherently different from our native flora?

The dichotomous approach to non-native plant species is at least partially a dichotomy generated by the scale and mechanistic focus of the two approaches. Researchers who treat non-native species as different from the natives have tended to focus specifically on a single mechanism of success, such as escape from natural enemies or competitive superiority. There is also a tendency to focus on species determined to be invasive. While concentrating on problematic species is clearly appropriate from a conservation viewpoint, application to the understanding of invasion biology as a whole is problematic. When summaries of invasion are generated from such studies, the results are highly skewed by focusing on only the most egregious invaders and likely do not represent the non-native plant assemblage as a whole. The desire to explain and potentially mitigate the invasions of non-native species has made this approach by far the more common one. The difficulty in this approach is in determining how it improves our understanding of biological invasions as a whole. The conclusion that two species are regulated by different ecological processes, perform differently under similar conditions or have other dissimilarities, speaks more to the individualistic nature of species and their evolutionary history than to invasion biology as a whole. The examination of ecologically equivalent groups is therefore critical to addressing the equivalence of native and non-native species (van Kleunen *et al.*, 2010)

Those researchers that view native and non-native species as a distinction in evolutionary origin only, often use a quite different approach. These researchers tend to focus on larger data sets that involve many species without using subjective criteria imposed by invasiveness or any other qualitative categorization of species. The argument behind at least initially assuming that native and non-native species are equivalent is based in the universality of the fundamental tradeoffs that constrain all plant species. These common constraints should lead to native and non-native species pools that are functionally equivalent (e.g. Huston, 1994; Thompson *et al.*, 1995; Grime, 2001; Tecco *et al.*, 2010). Clearly, an individual species, regardless of origin, may possess characteristics that allow it to succeed in a particular recipient community. However, native and non-native species, as a whole, should not differ in these characteristics or in their overall success. From this null position, we may expect native and non-native species to respond similarly to the environment and to have similar suites of characteristics. Deviation from equivalence would represent differences in the ecological characteristics of the native and non-native species of an area and would be directly informative to invasion biology.

In this chapter we will examine the ecology of a broad suite of native and non-native species. The direct utility of such a community level approach in determining the equivalence of native and non-native plants in succession is clear. The approach removes the necessity of all qualitative non-native species categorization based on invasiveness, naturalization, dominance or other criteria. All communities are composed of a gradient of species abundances from those that are the structural dominants of the community to those that are subordinate and contribute relatively little biomass to the system

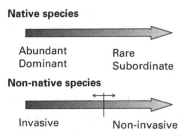

Figure 10.1 Conceptualization of the ways in which native and non-native species are categorized based on abundance and perceived interactions. While the gradient of species abundances is widely recognized in studies of native communities, non-native species are typically separated into invasive and non-invasive categories. As the separation of invasive species from the larger non-native pool is subjectively placed along this gradient, we suggest that comparisons should be based on the entire range of species present within native and non-native assemblages.

(Whittaker, 1975). Any dividing line placed along such a gradient is inherently a qualitative separation (Figure 10.1). Furthermore, if criteria were developed that separated out the most successful non-native species, the invasives, these criteria must also be applied to the native species assemblage. This process would ensure the comparison of ecologically equivalent groups of species (van Kleunen et al., 2010). Our approach here is not to select species by their dominance in the community, but to functionally group them based on basic life-history characteristics. As life history is important in determining the general constraints on a species, incorporating life history into analyses should ensure that we are comparing ecologically similar groups of species. This approach also allows us to identify and account for any bias in the species pool generated by selective species introduction. As plant introductions in eastern North America have largely occurred from mesic temperate areas of Eurasia, we would expect that non-native species in the region would have evolved with the same basic set of challenges and constraints as natives. Non-native species should therefore show some level of pre-adaptation to their introduced range (Schlaepfer et al., 2010; van Kleunen et al., 2011).

If a community-level perspective is important in understanding plant invasions, why then should we look to successional systems as being useful in addressing these questions? There are several attributes which make successional systems useful in addressing biological invasions (Meiners et al., 2009). First, successional systems worldwide are heavily invaded, typically by a diversity of non-native species (Inouye et al., 1987; Rejmánek and Drake, 1989; Omancini et al., 1995; Bastl et al., 1997; Meiners et al., 2002a; Matthews and Spyreas, 2010; Tognetti et al., 2010; McLane et al., 2012). The diversity of non-native species in successional communities is in direct contrast to other community types, which may contain many fewer non-natives. The non-native assemblages of successional habitats also contain a range of species life forms, generating a broad basis for statistical comparison and generalization. Second, successional systems exhibit community dynamics over relatively short time periods compared to other community types. This is not to say that other communities are not dynamic, but that successional systems turnover at rates that allow direct observation of community and population dynamics. The speed of successional dynamics has made these plant communities important model systems for testing theories of community ecology (e.g. Tilman, 1984; Stevens and Carson, 1999). For invasion biology, rapid rates of turnover mean that the colonization, spread and ultimate decline of a plant species can be directly observed. This results in a complete and dynamic view of plant invasions and their impacts.

The last, and perhaps most important, beneficial feature is wholly unique to successional systems. The succession-initiating disturbance constrains all species, native and non-native alike, to the role of invader. In the BSS, the annual plowing or mowing maintained the communities perennially in a disturbed state until the cessation of agriculture. With the exception of the planted hay grasses, all species had to initiate populations after abandonment. Therefore, all species have had to go through population phases of colonization (i.e. invasion), spread and, in most cases, decline. The constraints imposed by the agricultural disturbance means that the dynamics of species can be compared directly and on an equivalent basis. In contrast, studies of invasion in other communities must compare the invading suite of species with those species already established within the system. The resident populations may be essentially stable, while the newly invading populations would be continuing to increase. Factors that limit the resident populations, such as pathogens, microbial feedbacks or specialist seed predators would also be well-established. The appropriate comparison for the invader in this case would be the dynamics of the resident species when it first spread throughout the community and encountered similar escape from biotic limitation. This type of information is relatively rare in ecological studies. The BSS data provide a unique opportunity to compare equivalent and simultaneous dynamics within a community.

This chapter will explicitly address whether native and non-native assemblages are equivalent using a variety of different approaches. In all cases we will employ a community level perspective, assessing as many species as possible. If we relate the conceptual structure that frames this book to plant invasions, two classes of community drivers can be most directly addressed with the BSS data: differential species availability and differential species performance. In this case, species availability is largely determined by the history of species introduction to the area. The species pool is set by regional introductions then functions to constrain the types of interactions and dynamics that generate differential performance. Differential site availability and conditions provides the backdrop for all species interactions and dynamics. We will deal with the differential availability of species first, then address differential performance directly by assessing population dynamics, and indirectly by assessing variation in species traits. Finally we will address community-level processes that may be altered by plant invasions. After presenting the lessons that the BSS data offer on invasions, we will return to our overarching conceptual framework and develop it specifically in the context of biological invasions.

The non-native species assemblage of the BSS

Of the species found within the BSS, 120 are non-native. The vast majority of these species are Eurasian in origin, as are most plants that have been introduced to North America.

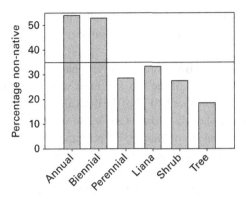

Figure 10.2 Representation of non-native species among life forms in the BSS data. The horizontal line represents the overall proportion of non-native species across all life forms, 35%. Non-native representation varied significantly among life forms ($\chi^2 = 25.8$, df = 5, $P < 0.0001$).

While non-native species composed 35% of the total BSS flora, there was significant variation in non-native representation among life forms (Figure 10.2). Annual and biennial species were heavily biased towards non-native species, with over 50% being non-native. Annuals and biennials also made up 48% of the entire pool of non-native species. In contrast, only 23% of woody species at the site were non-native species. The bias towards short-lived herbaceous species in the non-native assemblage reflects the overall agricultural nature of the landscape at the time of abandonment and the close association of many non-native species with agriculture (Mack, 2000; Mack et al., 2000; Pyšek et al., 2003). Based on the introduction bias alone, we would expect that non-native species would be shifted towards weedier plant strategies relative to the native species. The dominance of the non-native species pool by short-lived annuals and biennials makes accounting for life history critical in all subsequent comparisons of native and non-native species.

While we will deal with all non-native species collectively, it is worthwhile to note that the BSS data contain a good representation of non-native species considered to be invasive or otherwise problematic (Meiners et al., 2009). The hay fields were originally planted in *Dactylis glomerata*, a species considered invasive in many habitats. Two short-lived herbaceous plants, *Alliaria petiolata* and *Microstegium vimineum*, are regionally invasive in forest understories. Several herbaceous perennials are considered invasive, including: *Chrysanthemum leucanthemum*, *Elytrigia repens*, *Hieracium caespitosum*, *Plantago lanceolata*, *Poa compressa* and *Rumex acetosella*. There was also a diversity of invasive shrubs present – *Berberis thunbergii*, *Elaeagnus umbellata*, *Lonicera maackii*, *Rubus phoenicolasius* and *Rosa multiflora*. *Rosa multiflora* dominated mid-successional communities and populations of *L. maackii* and *R. phoenicolasius* have expanded dramatically during the past 10 years. Similarly, the lianas contained *Celastrus orbiculatus* and *Lonicera japonica*, both regionally problematic invaders of forests. Within canopy trees, two non-native species qualified as regionally invasive, *Acer platanoides* and *Ailanthus altissima*. After 50 years of succession, *A. platanoides* had established only a few individuals within the BSS fields, though these appear to have become reproductively mature around 2007 based on the presence of seedlings.

Ailanthus altissima was a locally important canopy tree within the BSS. Both invasive tree species have been established for a long time in the adjacent old-growth forest of HMFC (Ambler, 1965).

It is also interesting to note that the BSS contains several native species that have become invasive in Europe. In a survey of the top 125 most widespread non-native species in Europe (Lambdon *et al.*, 2008) we find 11 of the most abundant native species within the BSS. These include the short-lived herbs *Ambrosia artemisiifolia*, *Conyza canadensis*, *Erigeron annuus* and *Oenothera biennis*; the perennial herbs *Aster lanceolatus*, *Oxalis stricta*, *Solidago canadensis* and *S. gigantea*; the liana *Parthenocissus quinquefolia* and the trees *Acer negundo* and *Quercus rubra*. Another tree, *Prunus serotina* has also become locally invasive in Europe (Chabrerie *et al.*, 2010) The presence of these North American invasives suggests that comparisons of the native and non-native flora of the BSS do involve ecologically equivalent species with the same potential to be invasive. Of course, invasiveness may not necessarily be constrained to non-native species (Falk-Petersen *et al.*, 2006; Simberloff, 2011). Within North America, native plants such as *Juniperus virginiana*, *Solidago* spp. and *Vitis* spp. have been considered invasive (Werner *et al.*, 1980; Fike and Niering, 1999; Briggs *et al.*, 2002), though invasiveness in natives if often associated with some sort of disruption in the community (Simberloff *et al.*, 2012).

Population dynamics

One of the most direct ways to determine the functional equivalence of species is to examine their population dynamics. The dynamics exhibited by a population incorporate all of the characteristics of a species and the interaction of these characteristics with other species and the environment. This is by far a superior approach to looking for differences in the physical or physiological characteristics of species. Interpreting differences in the characteristics of species assumes that these characteristics are important in determining the performance of the species in the community. For example, two groups of species may differ in herbivore defense/tolerance. If the two groups of species exhibit the same dynamics and patterns of abundance within communities, the functional meaning of the difference in herbivore response is unclear. In our treatment, we will explore the population dynamics of native and non-native species before addressing their characteristics. In this way, we will generate a context in which to place the mechanistic analyses. It should be noted that while the data on population dynamics will be composed of the 84 core BSS species used previously, all physical characteristic data will include as broad an array of species as possible.

As the analyses addressed in Chapter 6 revealed large effects of life history on the population dynamics of species, we must include life form in all analyses to account for basic variation among species. This approach ensures that equivalent species are being compared and addresses the bias in the species pool towards short-lived non-native species. All of the analyses presented have the general form of testing for effects of species origin, life form and their interaction. However, life forms were aggregated to

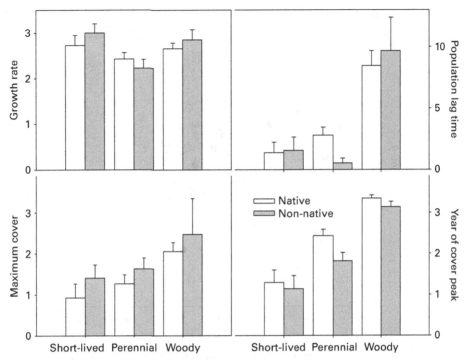

Figure 10.3 The influence of life form and species origin on the population dynamics of species during succession. ANOVAs for each measure: ln of growth rate (plots/year; Life form $F_{2,78} = 5.03$, $P = 0.009$; Origin $F_{1,78} = 0.24$, $P = 0.63$; LF × Origin $F_{2,78} = 0.99$, $P = 0.37$); Population lag time (in years; Life form $F_{2,78} = 18.16$, $P = 0.001$; Origin $F_{1,78} = 0.08$, $P = 0.78$; LF × Origin $F_{2,78} = 1.29$ $P = 0.28$); ln maximum plot cover (Life form $F_{2,78} = 3.71$, $P = 0.029$; Origin $F_{1,78} = 2.04$, $P = 0.16$; LF × Origin $F_{2,78} = 0.03$, $P = 0.98$); ln year of cover peak (Life form $F_{2,78} = 21.68$, $P = 0.001$; Origin $F_{1,78} = 1.91$, $P = 0.17$; LF × Origin $F_{2,78} = 0.59$, $P = 0.56$). For details of the measures of population dynamics, see Chapter 6.

ensure adequate sample size when splitting them into native and non-native components. Annuals and biennials were combined into a single short-lived herbaceous group and all woody species were combined.

For simplicity, we have chosen to focus on four of the population metrics likely to be key in understanding invasion and that represent both of the fundamental axes of variation found in population dynamics (Chapter 6). In analyses of all four metrics, life form was significant in determining population dynamics, but neither species origin nor the origin × life form interaction were significant (Figure 10.3). We use population growth rate based on plot occupancy as our basic measure of the ability of species to move throughout a site. Population growth rates were higher in both short-lived herbs and woody species, than in perennial herbs. While woody species were slow to dominate in cover, they were overall quite capable of colonizing plots across the site very rapidly. Similarly, population lag times were strongly related to life form, with woody species exhibiting much longer periods between when they first colonized the site and when they began to expand than either herbaceous group. Native and non-native species showed

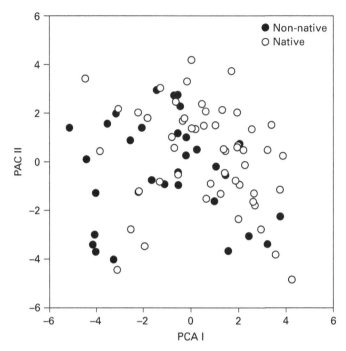

Figure 10.4 Principal components ordination of 14 measures of population dynamics indicating the origin of the 84 species. MANOVA results using Wilk's λ: (Life form $F_{4,154} = 16.04$, $P < 0.001$; Origin $F_{2,77} = 2.48$, $P = 0.09$; LF × origin $F_{4,154} = 0.45$, $P = 0.77$). For details of the ordination and the component population measures of population dynamics, see Chapter 6.

very little separation in these measures, though we might have expected the non-natives to spread more quickly based on their propensity to spread in landscapes. The two cover-based measures also revealed little separation between native and non-native species. Longer-lived species tended to reach greater peak cover and peak later in succession, as would be expected based on their successional positions and length of dominance in the community.

While only 4 of the possible 14 population metrics were presented here, all show similar patterns of variation (Meiners et al., 2009). However, individual measures of population dynamics may not sufficiently capture the patterns of variation for life forms and species origin. If we revisit the ordination of population dynamics from Chapter 6, we can examine the distribution of native and non-native species using a multivariate approach (Figure 10.4). In the ordination, the range of non-native species population dynamics is nearly identical to that of the native species. There is no indication that any individual non-native species exhibited fundamentally different population dynamics than those of the native species. When accounting for life form and species origin in a MANOVA of the two principal components axes, we again find that life form was key in determining population dynamics, but that species origin had no influence.

Overall, we see that native and non-native species are functionally equivalent in succession with regards to their dynamics. Of critical importance in evaluating the hypothesis of equivalence was accounting for the basic ecological differences among life forms. As the non-native species pool was dominated by short-lived species and relatively depauperate in trees, analyses ignoring life-form biases in the species pool would certainly find differences based on species origin. However, the difference in species pool was generated not by the biology of the species, but instead by the vagaries of species introduction. Of biological significance is that short-lived native species appeared equivalent to short-lived non-native species, woody natives were equivalent to woody non-natives, etc. We would not expect short-lived herbaceous species to be equivalent to woody species, so it makes no sense to pool across life forms and incorporate that comparison.

Functional characteristics of native and non-native species

Functional characteristics, or traits, are used to represent the ability of a species to disperse, establish, grow and reproduce within a community (Keddy, 1992; Weiher et al., 1999; Westoby and Wright, 2006; Lavorel and Grigulis, 2011). The interest in functional ecology is based in the hope that by understanding the characteristics of species, we may gain insights into how communities are structured that are not possible when species identity is the focus of study. While we will deal specifically with the functional ecology of succession later (Chapter 12), functional ecology also provides a template under which the equivalence of native and non-native species can be evaluated.

The functional traits that we have selected to explore are all characteristics that are fundamental to both succession and invasion biology. The physical traits – seed mass, vegetative reproduction, dispersal mode, specific leaf area, were selected to specifically capture aspects of dispersal, regeneration, growth and physiology. Seed mass reflects dispersal ability, as small-seeded species are more likely to disperse greater distances; seed longevity, as smaller seeds persist in seed banks for longer periods of time; and regeneration ability, as larger seeds confer advantages that allow regeneration under competition or other stresses. Vegetative reproduction is a key competitive trait characteristic of many mid-successional dominant herbs (Prach et al., 1997; Grime, 2001) and invasive species (Thompson et al., 1995; Sutherland, 2004). Dispersal mode, specifically here the dependence on vertebrates for dispersal, is linked with late-successional plant strategies (Odum, 1969; Prach et al., 1997; Grime, 2001) and with spread of woody invasive species (Rejmánek, 1996, Rejmánek and Sandlund, 1999; Aronson et al., 2007). Specific leaf area is tightly correlated with relative growth rates in a variety of systems (Westoby, 1998; Weiher et al., 1999; Wright et al., 2004) providing us with a measure of the physiological activity of each species. Finally, we will address herbivore palatability as a potentially important regulator of both invasion and succession (Shure, 1971; Brown, 1985; Brown and Gange, 1992; Brown, 1994; Keane and Crawley, 2002; Wolfe, 2002).

Physical characteristics – As in previous analyses, the structural characteristics of the species revealed strong variation among life forms, but no relationship to species origin within life forms (Figure 10.5). Seeds in all herbaceous species were much

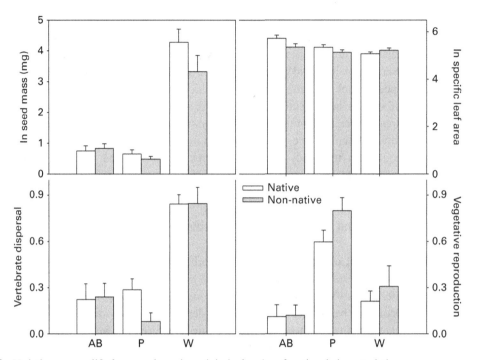

Figure 10.5 Variation among life forms and species origin in four key functional characteristics. ANOVA results: ln (seed mass +1) in mg (Life form $F_{2,137} = 65.42$, $P < 0.0001$; Origin $F_{1,137} = 1.92$, $P = 0.17$; LF × Origin $F_{2,137} = 1.42$ $P = 0.24$); ln specific leaf area in $cm^2 \, g^{-1}$ (Life form $F_{2,131} = 5.07$, $P = 0.008$; Origin $F_{1,131} = 2.13$, $P = 0.15$; LF × Origin $F_{2,131} = 2.30$, $P = 0.10$). CATMOD results: proportion of species with vertebrate dispersal (Life form df = 2, $\chi 2 = 36.3$, $P < 0.0001$; Origin df = 1, $\chi 2 = 0.98$, $P = 0.32$; LF × Origin df = 2, $\chi 2 = 2.61$, $P = 0.27$), proportion of species with vegetative reproduction (Life form df = 2, $\chi 2 = 33.45$, $P < 0.0001$; origin df = 1, $\chi 2 = 0.14$, $P = 0.23$; LF × Origin df = 2, $\chi 2 = 0.72$, $P = 0.70$).

smaller than those of woody species. The greater seed mass in woody species is also reflected in the much greater proportion of species that were dispersed by vertebrates. As both seed mass and vertebrate dispersal are expected to increase in succession (Odum, 1969; Debussche *et al.*, 1996; Prach *et al.*, 1997), variation among life forms makes sense. Woody species often become established under existing plant canopies and may require larger energetic reserves to become established. The mass of increased seed reserves also make vertebrates an effective dispersal strategy. Though the linkage of vertebrate dispersal with species invasion has been made (Rejmánek, 1996; Aronson *et al.*, 2007), if anything, the BSS data show a depression in vertebrate dispersal of non-native perennial herbs and near perfect equivalence in the other groups.

Specific leaf area was remarkably invariant among life forms and with origin, suggesting the overall equivalence of relative growth rates across groups. Though non-native species are typically thought to grow more quickly than their native counterparts, this was not seen in the BSS data as measured by leaf characteristics. We do, however, see the expected trend of decrease in specific leaf area, and presumably growth rate, as the life span of the species increases. The leaves of woody species had the lowest specific leaf area and would also have the lowest relative growth rates because of their allocation to woody tissues (Grime, 2001).

Vegetative reproduction is critical in the establishment of new plant populations and the acquisition of resources for both succession and species invasions. Clonality and other competitive mechanisms are predicted to peak in mid-successional communities (Prach *et al.*, 1997; Grime, 2001), so it is not surprising that herbaceous perennials have the greatest incidence of vegetative reproduction. Both short-lived herbaceous and woody species had a much lower prevalence of vegetative reproduction. Non-native species tended to have more vegetative reproduction in both woody species and herbaceous perennials, but this was not sufficient to be statistically significant. As with all other characteristics, vegetative reproduction was unable to identify differences between native and non-native species once basic variation among life forms was accounted for.

Overall, we see little separation between native and non-native species and large differences among life forms in physical traits. Trait separation among life forms reinforces the utility of this basic functional classification. Similarity in native and non-native species suggests that they draw from the same pool of traits and that species are species, wherever they come from.

Herbivore palatability – Herbivores are important in regulating species abundances and the dynamics of communities. Their activity, or rather lack of it, may also be critical to explaining many plant invasions (Elton, 1958; Blossey and Notzold, 1995; Wolfe, 2002; Colautti *et al.*, 2004). Most of the mechanistic studies of escape from herbivores revolve around the activity of specialist consumers, those that feed on only one or a few taxa of plants. However, most plant consumers are not specialists, but generalists. How generalists relate to biological invasions is not clear (Keane and Crawley, 2002; Agrawal and Kotanen, 2003; Parker *et al.*, 2006; Schaffner *et al.*, 2011), though their feeding activity may be much more important to the majority of plant species. Exploring variation in palatability to specialist herbivores would be non-informative, as specialists should always prefer their host taxa. Instead we will focus on differential palatability to a generalist herbivore as an assessment of defensive chemistry across species. If successful invasion is linked with absence of herbivore damage, or at least reduced damage relative to native taxa, we would expect non-native species to be less palatable than natives on average.

Molluscs have proven useful in comparing plant palatability in a variety of systems (Hanley *et al.*, 1995; Fenner *et al.*, 1999; Peters *et al.*, 2000; Elger and Barrat-Segretain, 2004). Therefore, we used a widespread snail, *Anguispira alternata*, as our generalist herbivore. A suite of 38 plant species, chosen to equally represent native and non-native species as well as a range of life forms, were used in feeding trials (Grime *et al.*, 1993). Snails were fed filter paper discs saturated with extracts from each plant species and consumption relative to discs saturated with extracts from a reference food item was measured (lettuce – *Lactuca sativa* cv "Iceberg"; Nott and Meiners, unpublished data).

Palatability to herbivores differed from the physical traits in that there were clear differences based on species origin, but none based on life form. As would be expected,

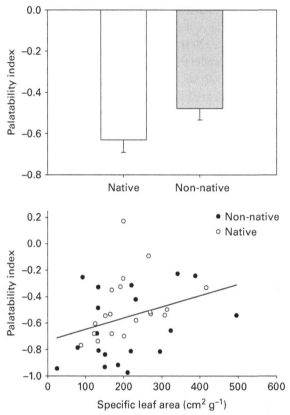

Figure 10.6 Palatability analyses of native and non-native species. Palatability index expressed as relative to a control food item (iceberg lettuce) using the relative impact index of Armas et al. (2004). Negative preference values indicate that the test item was less preferred than the control item (equal to 0). Native species were significantly less preferred than non-native (Top: Kolmogorov–Smirnov test, $Z = 1.46$; $P = 0.028$). There was also a positive association between specific leaf area and palatability (Bottom: $R = 0.36$, $P = 0.026$).

nearly all species tested were less preferred than the lettuce reference, leading to overall negative palatability measures (Figure 10.6). Native species were overall less preferred, that is had a lower relative palatability, than non-native species on average. This is in sharp contrast to the idea that non-native species' success is derived from the avoidance of herbivore damage (Agrawal and Kotanen, 2003). Of course, this test examined leaf chemistry in isolation from physical toughness, leaf pubescence or other non-chemical feeding deterrents that may alter feeding preferences for intact tissues. However, palatability was related to the physical characteristics of the leaves with higher specific leaf area species were being preferred. This is in alignment with the idea that species with high specific leaf area do not allocate much energy into either supporting or protecting their photosynthetic tissues as leaves are quickly replaced. Therefore, damage from herbivores is expected to be higher in high specific leaf area species (Westoby, 1998). The association between palatability and a leaf character suggests that the herbivore trials measured an ecologically useful aspect of species biology. However, the difference detected was not in favor of non-native species, as predicted, but suggested a net benefit to native species from a native herbivore. Interestingly, palatability may be positively linked with decomposition rates, linking this plant trait to ecosystem-level processes (Grime et al., 1996, Diaz et al., 2004, Mason et al., 2010).

Comparative community structure and dynamics

One of the major predictions of the equivalence of native and non-native species is that both groups should generate similarly structured communities. This is in sharp contrast to the often observed ability of non-native species to dominate local communities that is cited as a mechanism of their impacts. Within the BSS, the population dynamics and functional ecology of native and non-native species appear largely identical. For this reason, we should expect that the vast majority of non-native species should become relatively innocuous members of the plant community (Ortega and Pearson, 2005; Tecco et al., 2010) and exhibit no more dominance than seen in native species. As all species appear to be constrained by the same suite of tradeoffs and draw from a single pool of traits, we should also expect that native and non-native species should show similar responses to environmental gradients. Species abundance distributions are also considered an assembly rule (Wilson, 1999), so equivalence here would suggest similar forces are structuring the communities.

We have previously (Chapter 5) explored the dynamics of rank–abundance curves during succession. In general, all communities were dominated by a relatively small number of species, with a much larger number of subordinate and rare species. When we examine the BSS data to compare the abundance distributions of native and non-native assemblages, we see nearly identical distributions (Figure 10.7). The shape of the distributions changed over time, but native and non-native assemblages were statistically indistinguishable during each period examined. While the BSS data clearly contained invasive species likely to dominate the community, the abundance of these species was similar to the abundance of dominant native species. More importantly, non-natives contained a range of species with lesser abundance, many of which were fairly rare in the community. The species pool of non-native species was clearly much smaller than that of the natives, but this would be a function of species introduction history and not any inherent differences. The equivalence of abundance distributions argues strongly that the factors that control community structure in this system act similarly on both native and non-native species.

Another prediction of community structure based on native and non-native equivalence is that both groups should respond to determinants of diversity in similar ways. While the mechanisms that regulate local- and field-scale richness are not clear, we can address the spatial patterning of richness. A spatial association in the richness of native and non-native species would suggest similar responses to the underlying drivers of local richness. In many systems the richness of native and non-native species is often positively associated at spatial scales that exceed the scale of species–species interactions (Planty-Tabacchi et al.,

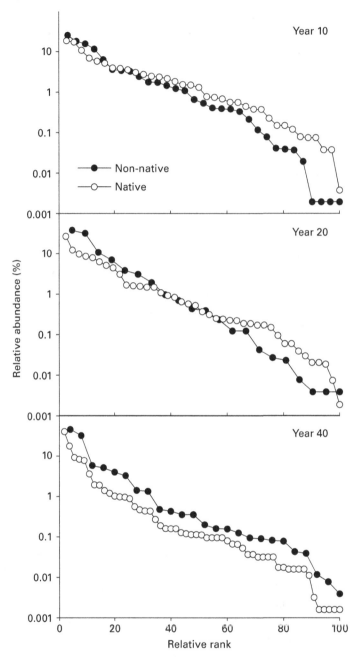

Figure 10.7 Equivalence of abundance distributions for native and non-native species during succession. Rank–abundance curves for field C3 with the X-axis standardized to visually account for the greater richness in native species. Differences between distributions were non-significant in all years (Kolmogorov–Smirnov test, all $P > 0.34$).

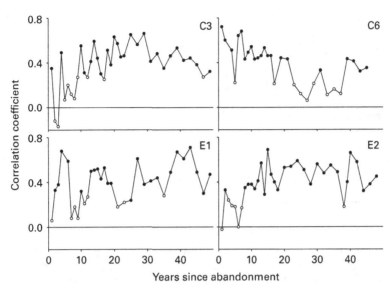

Figure 10.8 Temporal pattern of association between native and non-native species in four of the BSS fields. Values presented are Pearson correlation coefficients of plot data. Non-significant correlations (P > 0.05) are indicated by open symbols, significant by closed.

1996; Stohlgren et al., 1998; Lonsdale, 1999; Stohlgren et al., 1999, 2001). At more local scales, the relationship between richness of native and non-native is more variable, possibly reflecting the combination of environmental responses, propagule availability and direct species interactions (Wiser et al., 1998; Levine, 2000; Brown and Peet, 2003; Gilbert and Lechowicz, 2005; Bennett et al., 2012).

Within the BSS data, we can explore the spatial association of richness at the scale of both local plots and entire fields. We can also examine temporal dynamics of these associations to help to understand the processes that generate them. At the 1 m^2 plot scale, the richness of native and non-native species was positively associated in fields during most of succession (Figure 10.8). The degree of association varied temporally, but was relatively consistent in direction. There was dramatic variation in the strength of association, particularly during the first 10 years of succession. Correlations shifted from strongly positive to non-significant and back again. The strength of association was characterized by periods of increasing positive association between native and non-native species. These periods were followed by events that disrupted the association. The cause of these disruptions isn't completely clear, but appears to be a combination of successional transitions in dominance and disturbances caused by drought. The rapid fluctuations early in succession in both fields suggest that the life span of the constituent species is important in regulating the association. Whatever the source of temporal variation, native and non-native species were strongly positively associated over time, suggesting the existence of a single set of diversity regulators. This fine-scale positive association is somewhat surprising, given the ability of some non-native species to

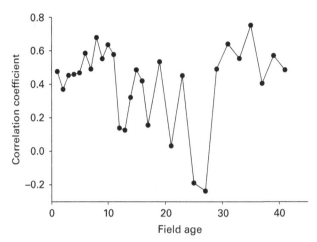

Figure 10.9 Temporal pattern of association between native and non-native species at the field scale. Values presented are Pearson correlation coefficients. Following year 17, data were combined in two-year intervals to account for the shift to alternate year sampling.

depress local richness in the BSS (Meiners et al., 2001; Yurkonis and Meiners, 2004; Yurkonis et al., 2005).

The ability of the BSS data to detect associations at the coarser field scale is limited by the small sample size of 10 fields. However, the general patterns revealed are consistent with those seen at the plot scale. In the majority of years, the richness of native and non-native species was positively associated (Figure 10.9). The early successional fluctuations in associations exhibited at the plot scale were largely dampened at the field scale. However, the period of 10–30 years post-abandonment exhibited much greater variation in association. This temporal variation was partially driven by the influence of individual climatic events on multiple ages of fields – a drought in a single year can influence richness in fields varying in age by up to 8 years (Bartha et al., 2003). These patterns are consistent with other site-level surveys, which find the richest native communities also tend to have the richest non-native communities (Planty-Tabacchi et al., 1996; Lonsdale, 1999; Stohlgren et al., 1998,1999; Gilbert and Lechowicz, 2005).

With the association between the diversity of non-native and native species, we might also expect that the dynamics of both groups were regulated by similar processes. Local species richness over short time periods is controlled by the balance between local colonization and extinction rates (MacArthur and Wilson, 1967). However, species invasions have a complex relationship with local diversity (Meiners and Cadenasso, 2005). Plant invasions may locally depress diversity, inhibiting regeneration within areas of heavy invasion (Hutchinson and Vankat, 1997, Walker et al., 1997, Christian and Wilson, 1999, Martin, 1999, Miller and Gorchov, 2004, Vilà et al., 2011). At the same time, diversity at the scale of species interactions is often linked with species colonization, with less diverse areas more susceptible to invasion (Knops et al., 1999; Levine, 2000; Naeem et al., 2000; Troumbis et al., 2002; Wilsey and Polley, 2002).

We also see evidence of diversity as both a cause and consequence of species invasions in the BSS (Meiners and Cadenasso, 2005). Of 26 species abundant enough to statistically test for direct impacts of invasion on local species richness, only five species exhibited significant impacts (Meiners et al., 2001). Of these five species, the impacts of the only native species, *Solidago canadensis*, were dependent on the exact timing of the analysis and were much weaker than the non-native species tested (Meiners et al., 2001, Yurkonis et al., 2005). In fact, at short temporal scales, this species had a positive association with local species richness and no effect on colonization rates (Pisula and Meiners, 2010a). A more recent invader to the site, *Microstegium vimineum*, was also associated with a depression in forest understory richness and native cover, though it increased the total cover of the herbaceous understory (Cadenasso, unpublished data). Impacts of invasion in this system were largely driven by reductions in community colonization rates, which were mechanistically linked with the reduced probability of colonization by individual plant species (Yurkonis and Meiners, 2004; Yurkonis et al., 2005). The lack of an increase in local extinction rates associated with invasion suggests that the key interaction in generating invasions impacts may be between established invaders and regenerating species. This pattern is in contrast to many experimental studies, which focus on competitive interactions between established native and non-native species.

While species invasion could impact local richness, the BSS data also showed evidence of local richness regulating invasibility. Analyses of the same 26 species used to determine invasion impacts revealed that plot colonization of 15 species was influenced by species richness (Meiners et al., 2004). Species exhibited a range of diversity responses, with colonization positively, negatively or unimodally related to richness. Colonization in most invading species was related to the abundance of one or more resident species, indicating that an ecological sampling effect was contributing to the association (e.g. Huston, 1997; Tilman et al., 1997a; Crawley et al., 1999). Most important to the discussion here was that there were no fundamental differences in the controls on invasion of native or non-native species. While analyses of species invasion and impacts show overall equivalence between native and non-native species, the one glaring exception is that clear and consistent impacts have not been documented for any native species.

Implications of ecological equivalence

The preponderance of data from the BSS supports the equivalence of native and non-native species. These data strongly argue that the separation of invasion biology as a subdiscipline of ecology is an artificial one (Davis et al., 2001; Valéry et al., 2013). The implication of acknowledging equivalence is that invasion biology can inform ecology as a whole, just as information gathered from non-invasion systems can inform invasion biology. Colonization by native species and the invasion of non-native species are similarly equivalent processes. What makes one species invasive is a remarkably similar question to what makes another species successful, the distinction being the origin of the species and the perception of the person asking the question.

Non-native species invasions provide unique opportunities to examine the processes that limit the spread and dominance of species in communities. At one time, all species must have been colonizers, whether moving northward after the retreat of a glacier, expanding following a rare long-distance dispersal event or spreading following a speciation event. Investigating the spread of a species new to a system gives insight into the processes that historically and currently operate to maintain the structure of the community. The dynamic perspective provided by studying new species invasions in relation to existing populations is analogous to that provided by studying successional dynamics in relation to modern forests. Both provide a basic understanding of the range of mechanisms that may operate in concert to generate the structure and composition exhibited by a system.

This is not to say that all species are equivalent. Some species will possess a characteristic or suite of characteristics that allow them to be successful relative to most other species. A suite of characteristics or a single one may determine a single species' success, but this is a quite different question from whether the trait is key in determining invasion in a broader sense. Individual non-native species also represent serious management concerns for a variety of habitats. Our argument here does not diminish the need for active study and remediation of these species. Rather it is a call to take a broader view of invasion biology and draw from the breadth of information and theory that ecology can provide. Understanding the characteristics that contribute to the success or impacts of an invader will certainly provide information critical to effective management.

If indeed native and non-native species are ecologically equivalent, should we have expected it? Over the last few decades, many mechanistic explanations have been proposed to explain why invasions occur, which species should be invasive, and what the impacts of those invasions should be. In isolation, these studies provide case histories that can provide insight into the invasions of species under similar conditions and provide information for the remediation of the individual species. Taken together, they provide a vision of a diverse assemblage of invasion controllers. The challenge, therefore, is to integrate these *individual* mechanisms into a conceptually unified view.

In looking across the diverse list of mechanisms generated in invasion biology, we are struck by the similarity to the long list of potential drivers generated in successional biology. If a list of invasion mechanisms is merged with the conceptual structure used to guide this book, we see that invasion mechanisms fall nicely into the three general classes of successional drivers (Figure 10.10). Differential site conditions and history encompass the influences of cultivation (Mack, 2000), fertility (Hobbs and Drake, 1989; Burke and Grime, 1996; Stapanian et al., 1998) and disturbance (Burke and Grime, 1996; Hobbs and Drake, 1989; Stapanian et al., 1998; Moles et al., 2011; Catford et al., 2012). Differential species availability contains issues dealing with the vagility of species (Bazzaz and Mooney, 1986; Myers et al., 2004; Aronson et al., 2007; Lenda et al., 2012) and the importance of propagule pressure in determining local invasion

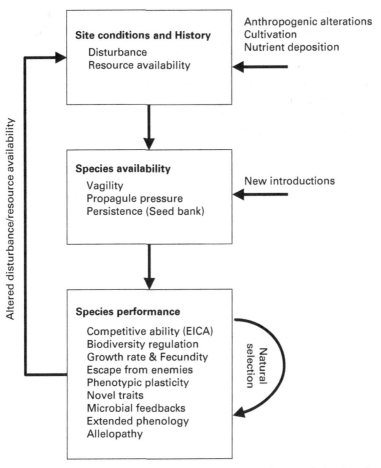

Figure 10.10 Model template placing determinants of invasion in the framework developed to explain community dynamics. Drivers external to the local system are indicated by arrows entering from the right.

success (Levine, 2000; Brown and Peet, 2003; Lockwood *et al.*, 2005, Pyšek *et al.*, 2009). At a coarser scale, the historical introduction of species sets the regional species pool (Pyšek *et al.*, 2003; Pemberton and Liu, 2009) and landscape position determines availability of species to a site (Moran, 1984; Benninger-Truax *et al.*, 1992; Hutchinson and Vankat, 1998; Cadenasso and Pickett, 2001; Pauchard and Alaback, 2004). Finally, the vast majority of invasion mechanisms fall under differential species performance. Escape from natural enemies (Wolfe, 2002; Parker and Gilbert, 2007; van Kleunen and Fischer, 2009), competitive ability (Daehler, 2003; Vilà and Weiner, 2004; Leicht *et al.*, 2005; Dawson *et al.*, 2012), allelopathy (Gómez-Aparicio and Canham, 2008; Thorpe *et al.*, 2009; Pisula and Meiners, 2010b; Uesugi and Kessler, 2013), soil-mediated biofeedbacks (Klironomos, 2002; Callaway *et al.*, 2003; Bever *et al.*, 2010; Lankau, 2010; Kim and Lee, 2011) and greater allocation to reproduction (Baker, 1965; Bazzaz and

Mooney, 1986), among others, all determine the performance of non-native species relative to the recipient community. As a whole, invasion biology has gradually replicated the diversity of ecological drivers that we see controlling the success of species in plant communities. Integration is needed to fully explore the linkages among these drivers and to generate a conceptual basis for determining the conditions under which individual drivers may be favored.

11 Heterogeneity in dynamic systems

"No part of the triangle of ecology has yet really faced the fact that to an organism and its progeny the environment is a richly dappled, minutely speckled, thing."
John Lander Harper, The contributions of terrestrial plant studies to the development of the theory of ecology (1977)

"Heterogeneity in the structure of an association may be explained by the accidents of seed dispersal and by the lack of time for complete establishment."
Henry Allan Gleason, The individualistic concept of the plant association (1926)

It took a long time for the study of succession to emerge from the focus on uniformity that characterized its early history, and from its standard textbook presentation as a smooth sequence of distinct phases. There were several reasons why the potential role of spatial heterogeneity within succession theory was ignored for so long. One was the tendency of ecologists, like other biologists of the early twentieth century, to view their systems as expressions of an ideal form. Idealization included uniform structure and composition, or a high degree of spatial mixture. The second reason was the assumption that the really important drivers of system structure and dynamics occurred within the spatial boundaries of the system. Consequently, there was no need to look outside of successional communities – aside from a well-mixed rain of seeds or other propagules – for drivers of successional change. There was thus no reason to ask about the heterogeneity of neighboring sites and influences, or the contingencies they may impose. Indeed, for much of the history of succession research, very little or no information was provided about the variety and spatial arrangement of neighboring or nearby systems that might influence succession in a particular place. Finally, spatial heterogeneity itself turns out to be a complex issue. This is because spatial heterogeneity can affect succession, as well as be an outcome of the processes of succession. In other words, heterogeneity can be both cause and consequence of successional patterns and processes.

The two quotes that begin this chapter reflect two very different perspectives on the potential causes of heterogeneity in plant communities. Harper laments the little ecologists know of how environmental variation impacts ecology – an argument for the abiotic generation of heterogeneity. Gleason, in contrast, invokes dispersal limitation as the driver of heterogeneity and even suggests that heterogeneity should dissipate in time – a biotically generated process. Both of these perspectives will be addressed in the following sections. What are not incorporated into these quotes are the subsequent constraints that either source of heterogeneity can impose on species interactions and performance. Though this is a much more complex task, we will also address this through various examples from HMFC and the BSS. Contingencies, the heart of modern ecology, provide a rich interaction landscape that generates heterogeneity at all scales.

The turn toward heterogeneity

The science of ecology has, over its long history, become much more attuned to and aware of the role of spatial heterogeneity. Once the basics of the process of succession were theorized, it became possible to accommodate variation in composition. Heterogeneity was one of the beneficiaries of this conceptual growth. There are several milestones along the way. One of the earliest leaders in American ecology to emphasize the role of spatial heterogeneity was William S. Cooper, of the University of Minnesota. He explained the dynamics of boreal forests as a phenomenon that depended upon the creation and filling of canopy gaps (Cooper, 1913). British ecologist Alex Watt (Watt, 1947) similarly explained the dynamics and coexistence of species in calcareous grassland as a cycle of patch occupancy, maturation, senescence and the opening of the area to colonization by species that had been excluded by the previous dominants. The theory of island biogeography, still relatively new in the 1970s, stimulated a spatial and temporal conception of succession (Pickett, 1976). Generalizing on these sorts of heterogeneous and dynamic phenomena, Simon Levin (1976) presented a mathematical theory of such changes, which he articulated as patch dynamics. Levin's work was instrumental in promoting the hypothesis that dynamic spatial heterogeneity is a key to understanding ecological communities, and to developing effective conservation and management strategies (Pickett and Thompson, 1978). Ultimately, a patch dynamics perspective facilitated the recognition of a non-equilibrium paradigm of ecology, in which species coexistence and the structure of communities and ecosystems was seen to depend in no small part on the recurrence of disturbance and a variety of subsequent pathways of succession (Simberloff, 1980). This perspective stands in marked contrast to the much more deterministic paradigm of the foundational succession work. The long-term study of succession in permanent plots would seem an ideal venue in which to test the role of heterogeneity in dynamic systems.

For several decades, succession theory has asserted that spatial heterogeneity in composition increases through succession. Heterogeneity has in fact been empirically documented to generally increase with succession (Bazzaz, 1975; Finegan and Delgado, 2000). In some forest successions, heterogeneity peaks early, then declines. For example, in conifer stands in the western Cascade Mountains, heterogeneity peaked at 15 years (Schoonmaker and McKee, 1988). Species richness and diversity follow separate patterns from heterogeneity there. The plant cover of early successional areas may be quite sparse or even absent in some cases. By definition, such unoccupied sites exhibit low vegetation heterogeneity. However, increasing evidence has shown that even newly disturbed sites can possess high spatial heterogeneity (Finegan and Delgado, 2000; Walker and del Moral, 2003). The expected increase in spatial heterogeneity of the plant community is largely generated by the increased size or clonal extent of plants as they age, producing patchiness. Clones of shrubs and even some trees increase the number of above-ground stems attached via root sprouts or rhizomes. Trees extend their crowns over larger extents. Individuals of clonal graminoids and forbs become larger and larger with time, displacing inferior genotypes via intraspecific competition, while maintaining dominance in the community (Hartnett and Bazzaz, 1985). Furthermore, the infilling of the open sites of both primary and secondary succession by additional species generates contrasts in composition and plant architecture. If there is a dominant competitor amongst the long-lived plants of the canopy in mesic systems, or of the root mat in grassland or arid systems, it may ultimately reduce compositional heterogeneity. While heterogeneity initially increases in many successions, it may in fact peak at early or in intermediate ages, before competitive species (*sensu* Grime, 1977) dominate.

Whether the early and mid-successional increase in heterogeneity continues indefinitely depends not only on the growth habits and interaction of the members of the community, but also on the occurrences of disturbances that disrupt the growth of dominant clones or canopies. Disturbances, as Cooper (1913) and Watt (1947) indicated, often introduce heterogeneity at new, gap-based scales. For example, in mesic, broad-leaved forest of temperate and tropical zones, later successional dominance of a few species is often disrupted by wind damage or uprooting of canopy trees. Lightning, fire and herbivore irruptions can have similar effects. Thus, succession shifts from an accumulation of heterogeneity based on the life history and architecture of individual plants or clones, to a process involving external disturbance agents like wind, flood or fire. The changes in spatial heterogeneity are an important part of succession nonetheless (Getzin *et al.*, 2008).

Heterogeneity can also influence succession through the effects of contrasting neighboring systems or boundaries between a successional system and neighboring patches (Cook *et al.*, 2005). If an open site is adjacent to older successional communities, those other communities can serve as sources of various influences and fluxes that drive succession in a target field (Lawson *et al.*, 1999). Most obvious is the source of seeds or vegetative fragments, or of animals that can carry those propagules by different means (Meiners and Pickett, 1999). Fruit-eating birds and scatter-hoarding mammals from an adjacent patch can affect the composition and success of the propagule pool. In addition, they may determine the specific locations in which colonizers of the successional patch establish. Wind-dispersed seeds will have a different spatial distribution from those dispersed by frugivorous birds. For example, bird dispersal into open sites is typically clustered around perch sites (McDonnell and Stiles, 1983), while wind-dispersed species typically exhibit leptokurtic distributions (Greene and Johnson, 1996). Successional influences of neighboring patches extend beyond establishment as well (Meiners *et al.*, 2000; Meiners and LoGiudice, 2003). Seed predators and herbivores can shelter in an adjacent patch and affect the survival and performance of plants in a target patch. Deer may shelter in forest, feed in yet another adjacent farm field or suburban garden, and regularly visit a successional field to feed as well, impacting the composition and architecture of plants there. Interactions such as these have permitted deer, *Odocoileus virginianus*, to become a virtually new dominant factor for understanding and managing forest restoration in eastern North America (Inouye et al., 1994; Baiser et al., 2008; Nuttle et al. 2013). Adjacent forest patches can influence local wind patterns, moisture and shade in portions of a focal successional patch, influencing plant performance. All of these biotic and abiotic drivers exemplify the role of neighborhood heterogeneity as a cause of succession.

The initial conditions at a site may exhibit heterogeneity based on two main effects. One is the underlying template of resource availability or capacity to retain resources in the soil and substrate (Wardle et al., 2004). Such templates are very often the result of long-term soil formation processes (Vitousek, 2004), or long histories of human use of the site (Baeten et al., 2010). If those conditions vary within the boundaries of the successional patch being studied, they can have an initial and potentially persistent effect on the identity and architecture of the communities that later come to occupy the site (Walker et al., 1996). The expectation is that both coarse- and fine-scale patches having greater initial availability of resources will experience faster rates of succession due to the exacerbation of competitive interactions (Inouye et al., 1987; Collins and Wein, 1998). Heterogeneity has been shown to influence succession in various situations. For example, within forest canopy gaps, environmental heterogeneity is a driver of the complexity of succession (Veblen, 1992; Sipe and Bazzaz, 1994, 1995). Echoing empirical findings, theory has demonstrated the power of heterogeneity in plant communities to permit species' coexistence (Pacala and Tilman, 1994). Such ideas have been generalized as non-equilibrium coexistence of plant species (Pickett, 1980).

The second major determinant of heterogeneity in site conditions is the nature of the disturbance event that initiates the succession. For example, hurricanes as events generate great heterogeneity, depending not only on the force of the wind, but also on the associated rainfall and soil saturation, and interaction with slope and aspect (Walker et al. 1996). Small-mammal burrowing and foraging create heterogeneous disturbance conditions within plant communities (Tilman, 1983). The heterogeneity created by fires, and their effect on succession, is legendary (Halpern et al., 1990; Turner et al., 1994). Together, these two sources of heterogeneity provide a dynamic environmental template that influences and responds to succession.

The BSS and heterogeneity

Was the BSS set up to examine heterogeneity? In a sense, no, because the design of 48 plots per field was said by Helen F. Buell (personal communication), one of the founders of the project, to have been recommended by a statistician to allow rigorous comparison among fields. The implication was that such a number would be adequate to represent each field as a population of plots. Clearly, this was not an "essentialist" view of fields as uniform units, but it was one not primarily focused on spatial variation per se within fields. On the other hand, the plots are arranged in a grid in each field (Chapter 2), and 8 of the 10 fields abut the old-growth forest, resulting in an edge of strong contrast. Hence, each row of plots is a different distance from the old forest on the one hand, and from younger or older fields on the other, allowing an opportunity to examine important aspects of within-field heterogeneity. The fields were intentionally placed as they were, in order to grow a buffer for the 64 acre (26 ha) remnant primary forest, which was the centerpiece and the *raison d'etre* of the Hutcheson Memorial Forest Center at Rutgers. Already in the mid-1950s, the old-growth forest was the last remaining upland example of ancient forest in New Jersey (Buell, 1957). The desirability of a buffer for the old-growth forest exemplifies the early landscape perspective that adjacent land use could affect biotic and abiotic conditions (Wales, 1972).

This chapter explores heterogeneity in the BSS from the perspectives of both cause and consequence. Several questions motivate the assessment of heterogeneity in the Buell–Small Succession Study: (1) What is the spatial heterogeneity of soil conditions across the array of fields, and does this reflect underlying or emergent patterns? (2) What is the nature of spatial heterogeneity within fields, and does this reflect neighboring conditions or biotic processes, such as dispersal, invasion and decline, and fine-scale interaction among plant species? In short, are there uniformities in the temporal and spatial patterns of heterogeneity and are there biologically interpretable aspects of any such patterns?

Mechanisms of heterogeneity in the BSS

Questions about the pattern of heterogeneity through succession are important because the spatial pattern may influence the function of successional communities and hence alter subsequent dynamics. This section introduces the concepts and some key examples from mechanistic work conducted at the Hutcheson Memorial Forest Center. Examples from novel long-term analysis of relevant patterns from the Buell–Small Succession Study will be presented later in this chapter. A number of experiments conducted at the HMFC have examined the importance of a variety of factors and processes that create or respond to heterogeneity. We highlight several of these in the context of the conceptual model of successional drivers used throughout earlier chapters (Chapter 4). The conceptual model of succession based on three determinants of vegetation change identifies potential impacts of spatial heterogeneity in succession (Figure 4.4). The three realms of factors are: (1) site conditions and history, (2) differential species availability and (3) differential species performance. These three groups of factors can be considered to be filters on the identity and architecture of the members of communities through succession.

Site conditions and history – The role of site history and initial conditions shows up clearly in earlier analyses of the BSS data. Myster and Pickett (1990) assessed successional heterogeneity over 22 years of the data set. Principal components analysis (PCA) revealed that during the first eight years of the succession, the trajectories of hay fields abandoned with a cover of *Dactylis glomerata* (orchard grass) differed from all other treatments, including row crops with spring or fall plowing, or no plowing. The species that defined the extremes of the first three ordination axes were all clonal perennials. Earlier chapters in this volume have extended the analyses of starting condition, confirming the contrasts in abandonment as a filter on succession. Typically the initial contrasts detectable in the BSS abandonment disappear by year 10.

An important theoretical contribution of work that emerged from the BSS is a critique of the Markovian approach to successional modeling (Facelli and Pickett, 1990). The Markovian approach assumes that history, such as the abandonment conditions or the order of invasion of species, does not affect the transitions over time. The historical patterns in the BSS clearly deny the validity of assuming the initial history to be irrelevant. Once again, the complexity of ecological systems defies the simplifications necessary to develop specific prediction of composition.

Differential species availability – Species characteristics per se are an important determinant of heterogeneity in species availability at any given time. Similarly, the distribution and movement of dispersal vectors, whether biotic or abiotic, can affect species availability. One of the most compelling demonstrations of the role of species availability at the HMFC experimentally examined the role of perches for birds by adding artificial perches at distance, within large abandoned fields dominated by low herbaceous canopy. Although bird dispersal differed in unmanipulated fields based on the complexity and height of their dominant canopies (McDonnell and Stiles, 1983; McDonnell, 1986), the addition of artificial perches was a powerful influence on input of bird-dispersed seeds into successional old fields. The input of seed depended upon the relative nutritional attractiveness of the fruit to birds, and on the seasonal dynamics of both the birds and the presentation of ripe fruit. Notably, later in succession, bird-dispersed woody stems are more highly clumped than are the stems of wind-dispersed species (Myster and Pickett, 1992b). However, there are several critical transitions between dispersal and the pattern of surviving stems (Houle, 1992), as explored below.

Many of the dominant, earliest invaders in the post-row-crop fields possess persistent seed banks. *Ambrosia artemisiifolia*, for example, depends upon large and long-lasting banks of dormant seeds in the soil (Willemsen, 1975). *Setaria faberii*, now the predominant annual first-year colonizer at the HMFC, exhibits physiological dormancy and forms a short-term seed bank (Nurse *et al.* 2009).

Of course, the characteristics of the seeds and architecture of seed parents also determines the availability of species in succession. Traits such as seed weight are analyzed elsewhere in this volume (Chapter 12). However, seeds of species contributing to the successional communities at HMFC range from the small, wind-dispersed seeds of many composites, such as *Heiracium* spp. and *Erigeron annuus*, through the larger-seeded, wind-dispersed *Acer* spp. and *Fraxinus* spp., to the animal-dispersed *Quercus* and *Carya*, which possess the largest seeds in the BSS species pool. Bird- and mammal-dispersed seeds, such as those of *Cornus florida*, and *Rubus* spp., respectively, are less limited by size, and more by their attractiveness to dispersal agents.

Differential species performance – a species' populations and individuals can perform differently in a successional community, depending upon their location within it. Stem density, as an aggregate measure of performance of woody species, is strongly affected by distance to older vegetation, such as the primary forest or older successional woodlands and hedgerows. For example, stem density decreased exponentially in space from each field edge early in succession (Myster and Pickett, 1992a). The patterns weaken through the 31 years of succession examined. The filters of both availability and performance can drive these spatial patterns.

Not only woody species, but also herbaceous species exhibit complex spatial patterns relative to forest-field edges. A sampling grid of 10 × 10 cells was established as an experimental platform at the HMFC. The grid was 100 m on a side, and extended from 30 m within a secondary forest, across the forest–field edge and ended 60 m into an adjacent, nine-year-old, perennial-herb-dominated field. All herbaceous population and community variables changed over the cross-edge gradient. Species richness, species diversity and total community cover increased from within the forest to the edge, with slight declines to 60 m in field (Meiners and Pickett, 1999). Compositional heterogeneity among plots was greater at the edge than elsewhere in the grid. Furthermore, exotic and native species behaved differently in space. Exotics peaked within the 20 m band inside the forest and were concentrated along the edge (Meiners and Pickett, 1999). An important aspect of this study was the sampling methodology, which incorporated both sides of the edge. Most forest–field edge studies previously had neglected the forest side of the pattern (Willson and Crome, 1989). This sampling grid was used for a number of the other analyses and experiments, to be described below.

The patterns of both woody and herbaceous heterogeneity across the forest–field grid suggests questions about the mechanisms and specific interactions that might generate the patterns. Consequently, several experiments were conducted to parse out the factors that may cause such distance decay patterns. The successional framework (Figure 4.3) and research and theory summarized elsewhere (Chapter 4) points to the potential for seed predation, seed germination requirements, stress tolerance, resource demand and competitive ability, allelopathy, interaction with browsers and herbivores, and the response to fine-scale disturbance and disease to be likely causes of spatial heterogeneity within the fields.

Consumer activity and heterogeneity

An example of the processes that could influence spatial patterns is predation of seeds. Meiners *et al.* (2002) planted seeds of *Acer rubrum*, *Acer saccharum* and *Quercus palustris* in the experimental grid described above. The emergence of *A. saccharum* and *Q. palustris* increased with distance into the old field compared to the forest and edge. *A. rubrum* emergence peaked in the +20 m band in the field, declining to +60 m. Survival of *A. rubrum* was virtually nil at all distances, while *A. saccharum* survival increased with distance into the field and *Q. palustris* showed no change in survivorship. Another experiment examined the predation by white-footed mouse (*Peromyscus leucopus*) on *A. rubrum* seeds over the grid. *Peromyscus* is documented to be the most common mammalian seed predator in the area of the study (Manson and Stiles, 1998; Meiners and LoGiudice, 2003). Within 30 days after the initiation of the experiment, only 28% of seeds survived during the fall trial, whereas in the winter trial, only 4% survived. The rate of removal was greatest near the edge and within the 30 m distance band in the field (McCormick and Meiners, 2000). The pattern of survival also varied between years of high and low population density of *Peromyscus*, with years of high population density having very little seed survival, regardless of distance (Meiners and LoGiudice, 2003).

Herbivory and browsing on established old-field species, both herbaceous and woody, are also potential drivers of succession. An experimental cohort of *Quercus rubra* seedlings was planted across the grid from 30 m within forest to +60 m in the field. Seedling survival was lowest in the forest, increasing to 49% far into the field. Herbivory by deer decreased with distance into the field, leading to greater survival at those distances. Local vegetation patchiness was also important along the transect, with plots containing patches of *Rosa multiflora* having greater survival of *Q. rubra* (Meiners and Martinkovic, 2002).

A second study at HMFC experimentally tested the role of mammalian herbivores in successional communities. Large fences, measuring 3 m high and surrounding 5 × 5 m plots, were built atop sheet metal skirts that extended into the soil to excluded all mammals in a 17-year-old successional field. Mammalian exclusion for over 10 years mainly affected community architecture, but not community composition (Cadenasso *et al.*, 2002). Herbaceous canopy was taller in the absence of herbivores. Saplings of *A. rubrum* and *Cornus florida* planted into the exclosures survived better than control individuals in the open. *Juniperus virginiana* demonstrated no difference. *Acer* and *Cornus* are classified as more palatable at HMFC than *Juniperus*, mirroring the observation that *Juniperus* is usually an earlier woody invader than *Acer* and *Cornus*. Therefore, mammalian herbivory has the potential to slow succession, and heterogeneity in herbivore pressure may generate variation in successional turnover. This experiment, though a compelling documentation of the general and specific effects of herbivory and browsing together, does not address any *spatial* filter for the process. The spatial heterogeneity documented by other experiments suggests that the selective damage seen across edge gradients will generate differentials in species performance and therefore compositional heterogeneity.

Observations of the differential distribution of species across space in fields based on their palatability to mammalian herbivores are highly suggestive of heterogeneity in foraging activity. Palatable woody species were more common close to forest-field edges than in the fields later in succession (Myster and Pickett, 1992b). This result may seem counter-intuitive, based on the assumption that herbivorous mammals would be more abundant at the edge. However, observations of the nature of damage on woody stems in another study suggested that field-dwelling mammals that preferred woody patches as habitats were the agents of damage (Myster and Pickett, 1993). The spatial association of woody stems with the forest-field edge decreased with time in the 31-year data set from the BSS.

Invertebrate damage can also be important to tree seedlings, particularly early in establishment. Meiners *et al.* (2000) tested the impact of reduced insect herbivory on the establishment of *A. rubrum*, *Fraxinus americana*, *Q. palustris* and *Q. rubra* seeds in plots positioned across a forest–old-field edge. Distance to edge was most important for seedling emergence and mortality (Meiners *et al.*, 2000), with performance of the surviving seedlings increasing from field to forest. Drought in the first year of the experiment produced a greater effect of insect removal on emergence and survival of seedlings. *Acer rubrum* was most susceptible to insect damage, while *F. americana* was least impacted. This experiment demonstrates an interaction of spatial heterogeneity, relative to the edge gradient, and temporal heterogeneity, based on drier vs. moister years. By altering initial regeneration patterns in trees, these effects generate spatial heterogeneity that may persist for decades.

The distributions of many woody species are simultaneously impacted by vertebrate and invertebrate consumers, as well as competition. For example, experiments assessed the response of the large-seeded *Carya tomentosa* to mammal browsing, insect herbivores and above-ground competition from old-field herbaceous canopy (Myster and McCarthy, 1989). Mammal effects were by far the most influential, with no effect of invertebrate herbivores on mortality. Damage by mammals in this case decreased with distance from the forest edge. Glasshouse experiments in which simulated seedling herbivory was tested in the context of competition with old-field sward, showed mortality to increase under the interaction of herbivory and competition. Growth was not affected by the interaction (Meiners and Handel, 2000).

Competition is a long-expected driver of succession. The contrast between competitive abilities of native and exotic species is of growing concern in ecology and the practice of conservation and restoration. If species of different origin are differentially distributed in successional space, such heterogeneity might affect the trajectories of compositional change. Assessment of the first 40 years of data from the BSS (Meiners *et al.*, 2001) revealed field-level differences to have a greater effect on species richness than the identity of native or exotic species within plots. Most species, regardless of origin, had little effect on species richness. However, those places in which the exotics, *Elytrigia repens*, *Lonicera japonica*, *Rosa multiflora*, *Trifolium pratense*, and the native, *Solidago canadensis*, came to dominate did exhibit reductions in species richness over time. This result echoes an association analysis conducted on 31 years of the BSS data (Myster and Pickett, 1992a). In that analysis the largest number of negative associations involved the exotic species *L. japonica* and *R. multiflora*. The longer analysis indicated that there was little difference in the effect of exotic vs. native species on community richness. However, the effect of dominance by exotics as a group was greater than that of natives. Both the Meiners *et al.* (2001) and the Myster and Pickett (1992a) analyses together suggest that either cover or architecture are the relevant features to determine impact of species on successional community structure rather than geographic origin. Field-by-field differences existed in the effect of exotics vs. natives, exhibiting heterogeneity in the strength of species interactions (Meiners *et al.*, 2001).

Together these experiments and analyses of the BSS data highlight the contingent nature of interactions during succession. Combined forces of environmental tolerances, animal foraging activities and plant–plant interactions result in complex patterns of regeneration that create heterogeneity in community composition.

Disturbance and plant–plant interactions

To this point, we have reviewed research that emphasizes the detection of existing spatial patterns in fields and the experimental discovery of mechanisms of dispersal, seed predation and seedling predation. Some of the work so far has pointed also to the role of competition. In the rest of this chapter, we collect research that has emphasized the role of nutrient and water resources, litter, competition and disturbance in structuring old-field communities at HMFC.

One of the first experiments to examine the role of disturbance was conducted by Armesto and Pickett (1985). This experiment examined fine-scale disturbance to the canopy. A seven-year-old field dominated by the perennial *Solidago canadensis* and a two-year-old field dominated by the annual *Ambrosia artemisiifolia* were the comparative systems. During the growing season, the richness of species in control plots declined in the older field, but this was mitigated by increasing intensity of disturbance. However, in the younger field, which was, as is typical for the age, undergoing rapid successional changes, the disturbance treatments generated no differences in richness. The primary difference was that the dominant plant canopy was denser in the older field than in the younger field. Disturbance in the older field increased the amount of light in the understory and moisture content of the first 10 cm of soil depth. Not only was species richness and frequency altered by the disturbance treatments, but species cover was also altered (Armesto and Pickett, 1986). Successional advance was enhanced by canopy disturbance in the older field, suggesting an inhibitory rather than facilitative effect of the seven-year-field dominants.

Experimental assessment of the role of specific resources was next investigated along with manipulation of soil and canopy integrity (Carson and Pickett, 1990). The treatments were adding macronutrients (N, P, K), supplementing soil water, tying back the *Solidago* canopy to increase subcanopy light and disturbing the soil early in the growing season. These treatments were designed to isolate the potential mechanisms that generate disturbance effects. Soil disturbance alone had little effect on community structure. However, the dominant canopy itself was increased by the addition of nutrients, and in the second, drier year of the experiment, by water addition. Creating canopy gaps released understory species, especially *Fragaria virginiana*. Among the understory species, there was variation in the identity of the predominant limiting resource. *Rumex acetosella* responded to nutrients, while *Hieracium pratense* was released by the water supplementation. The limiting power of different resources depended on timing within the growing season, and contrast between the rainfall of the two years of the experiment. The results suggest shifting, non-equilibrium coexistence in this successional plant assemblage and a strong role for heterogeneity.

The experiments summarized so far have addressed direct biological interactions such as predation and competition, along with the complex gradients implied by forest-edge–field transects. However, the drivers along this transect can be still more varied. One complication is the spatially patchy nature of leaf and stem litter. Predation on seeds of six tree species was tested against additions of *Solidago* (herbaceous) and *Quercus* (woody) litter along a field transect (Myster and Pickett, 1993). Few seeds of *Juniperus* were killed by predators regardless of litter addition. However, seed predation on the other five species was reduced by both kinds of litter. *Carya tomentosa* predation also increased with distance from forest edge. Overall, this experiment exposed effects of litter, distance, density and patch type on predation of successional tree seeds and reflects the complex nature of safe sites for regeneration in succession (Grubb, 1977).

Litter effects also represent a potential mediator of resource and predator impact on the heterogeneity of the community. Facelli and Pickett (1991c) conducted a field experiment to expose the role of litter on community structure. Three species of litter were chosen to represent a measured gradient of light attenuation for *Solidago* (least opaque), to *Setaria faberii*, to *Quercus alba* (most opaque). The three litter types produced different light mosaics at the 0.8 cm scale, and all affected community structure. Litter addition decreased the density of the dominant grasses *Setaria* and *Panicum dichotomiflorum*. Litter also affected biomass, such that litter of *Quercus* and *Solidago* resulted in fewer, larger *Setaria* individuals, but did not affect *Panicum* biomass, likely because it was outcompeted by *Setaria*. *Setaria* and *Solidago* litter increased the establishment of the early successional *Erigeron annuus*, while *Quercus* litter inhibited this species. Because these effects are strong and fine scaled, litter has great potential as an agent of heterogeneity and successional influence in old fields (Facelli and Pickett, 1991b).

Importantly, litter not only has direct effects, as shown above, but it also has indirect effects on old-field community structure (Facelli and Pickett, 1991a). Field addition of *Solidago* and *Quercus alba* litter with and without intact herbaceous cover, showed a strong negative effect of herb competition on establishment of woody species, while litter led to an indirect positive effect on woody establishment. Litter addition experiments decreased the intensity of herbaceous species competition with a test tree species, *Ailanthus altissima*, in the glasshouse. The complex interactions between woody establishment, herb competition and litter accumulation may have heterogeneous effects in successional communities dominated by perennial herbaceous canopies.

Not only does litter have an effect in middle-aged old fields in our 50-year scope, it is a significant factor in first-year fields as well (Facelli and Facelli, 1993). A field experiment was used to test whether litter of the dominant species of year 1 affects the plant community structure the next year. The removal of litter of the dominant *Setaria faberii* in the fall and in the early spring allowed *Erigeron annuus* to colonize the site prior to other species and to gain dominance in the community. Intact litter strongly inhibited *E. annuus*, to the benefit of *Setaria faberii*. Litter can thus act as a strong historical factor, acting across years and shaping the composition of subsequent successional communities. There is the potential for such effects to be patchy, based on the spatial pattern of dominance in fields, and the differential lodging of grass litter at the end of the growing season or over the winter.

Allelopathy, the inhibition of plants by the chemical products of other plants of the same or different species, can play a role in succession (Kaligarič et al., 2011). For example, the failure of *Ambrosia artemisiifolia* to re-colonize bare soil in the second year of succession at HMFC was experimentally attributed to chemical effects (Jackson and Willemsen, 1976). Recent research at HMFC with *Solidago*, a genus widely recognized to have chemical effects, has confirmed those effects with test species in the lab, but the lab results did not mirror field patterns of dominance (Pisula and Meiners, 2010a). The environmental heterogeneity experienced by plants may also alter their chemical interactions during succession. Early successional lianas from shaded environments of HMFC had a greater allelopathic potential than those growing in full sun (Ladwig et al., 2012). In contrast, the lianas characteristic of more mature forests, *Vitis* spp. and *Celastrus orbiculatus*, were unresponsive to light environment. It appears that early successional lianas are able to respond to environmental heterogeneity in a way that may increase their persistence in the community.

The joint effects of growth forms and species origin show another mode of complexity that may have spatial ramifications. Lianas are a specialized growth form that depends upon the architecture of other plants for its own success. We present their distribution as the final example of spatial heterogeneity derived from prior work on the BSS data. In the fields over 50 years of succession, the native *Vitis* spp., *Toxicodendron radicans* and *Parthenocissus quinquefolia*, compared to the introduced *Lonicera japonica*, exhibit contrasting and changing patterns (Ladwig and Meiners, 2010b). Native liana species were more common on the forest–field-edge, while exotic lianas were more common farther out in the successional fields. Although the lianas were widespread in time and space, they were most abundant from 20–30 years of succession, when herbaceous perennials were still the predominant cover in the fields. Liana cover decreased later in succession, with their decline accompanying the closure of the successional woodland canopy. The interaction of lianas with trees in the succession was frequent, but heterogeneous (Ladwig and Meiners, 2010a). Of the 798 trees greater than 5 cm dbh in the 10 BSS fields, 64% were colonized by lianas. Lianas preferred early successional trees, especially *Juniperus virginiana*. *Vitis* spp. were more common on hardwood canopy trees.

Liana canopy cover decreased the growth of colonized trees (Ladwig and Meiners, 2009). Diameter growth of 606 trees ≥8 cm dbh was examined over nine years in the 50-year-old secondary woodlands of the BSS fields. Five species, *Celastrus orbiculatus*, *Lonicera japonica*, *Parthenocissus quinquefolia*, *Toxicodendron radicans* and *Vitis* spp., occurred throughout the forest. On average, each tree supported 9.7 cm^2 of liana basal area, which covered a mean of 23% of the tree canopies. Canopy cover by lianas reduced growth of dominant and codominant trees only, not affecting the performance of subcanopy individuals. Lianas had most impact on host tree growth in those relatively rare, unsuppressed trees in which a majority of the canopy was occupied by the lianas. This suppression was not related to differential liana colonization of canopy trees, as all canopy classes supported equivalent liana burdens. The infrequent dominance of a tree canopy by lianas means that liana impacts will be heterogeneously spread across the forests. Notably *C. orbiculatus* and *Vitis* spp. are still increasing in the BSS; hence they may influence future forest growth and development.

The prior experimental data and spatial analyses for the BSS suggest several roles for spatial heterogeneity. At this point we turn to an analysis of the patterns of spatial heterogeneity over time within and across the BSS fields. This constitutes an unusually dense and complete record of the potential patterns of spatial heterogeneity in a continuously measured succession.

Heterogeneities in the Buell–Small Succession Study

The previous sections have dealt with heterogeneity, its sources and its impacts on the old-field ecosystems of HMFC. While these studies illustrate the potential for heterogeneity to shape succession, they have not directly come from the data set that forms the core of this book. We now turn to the BSS data to address patterns of heterogeneity within the BSS. We group the presentation into five subsections. The first three focus on the three causal processes of succession that can be considered to operate as filters on the assembly of successional communities: (1) site conditions and history; (2) species availability and (3) species performance. We then examine: (4) patterns of life form as integrative phenomena and (5) end by answering the question, are there cycles of heterogeneity in the first 50 years of succession, as was proposed by Armesto *et al.* (1991)?

Site and history – The BSS was established on land that had been farmed since 1711 (Chapter 2). The farm was said to be well managed and was owned by descendants of the original Dutch colonists, suggesting a relatively homogeneous past. Furthermore, the location of the fields adjacent to the old-growth forest was a conservation, rather than an experimental design, consideration. However, the soils of the HMFC are relatively uniform in physical structure and origin (Ugolini, 1964). The fine-scale heterogeneity of soils in the 10 fields was assessed in 2006, during the later phase of the 50-year data run reported in this book. There was heterogeneity among the 10 fields (Figure 11.1); however, examining an east (Fields E1-D1)–west (Field C7) gradient of fields, showed no consistent increase or decrease in percentage N and C, or pH. Nor can we discern any patterns of these three soil parameters that relate to: (1) year of abandonment, (2) fall vs. spring abandonment or (3) abandoned bare or with crop residue. Even the fields that had been dominated by *Dactylis glomerata*, which had a strong vegetation and successional structure signal (Chapter 9), did not have a consistent soil signature compared to other kinds of fields. Consequently, the template of soil variability seems neither to explain any initial differences among fields, nor to be an outcome of 50 years of successional change.

Light at ground level, measured in 2006 and 2007, did exhibit some weak trends reflecting abandonment treatments (Figure 11.2). As expected, however, there is great spatial variation in light levels measured in the fields. Fields abandoned with a crop of hay (*Dactylis*) had less light at the soil surface than fields abandoned as bare soil. Hay fields abandoned bare had higher light than the fields with intact *Dactylis* canopy. Generally, fields that had supported row crops and were abandoned as if being prepared for another crop, exhibited greatest light levels at the soil surface many years into succession. Row crops abandoned with crop residue intact had levels of light that were similar to some of the hay fields.

Species availability – We expect contrasting life forms – annuals, biennials, herbaceous perennials, shrubs, lianas and trees – to reflect different strategies of interacting with the physical environment and with other producers and consumer species. To relate variation among life forms to heterogeneity, we analyzed the temporal pattern of spatial variability over 40 years, presenting the coefficient of variation (CV) among fields for each life form separately (Figure 11.3). Notably, CV of life forms was less variable than for individual species (Chapter 9). Annuals showed no clear statistical pattern of spatial variation among fields. However, it is intriguing that their highest CVs appear at phases of succession in which dominance has shifted to a combination of perennial herbs and

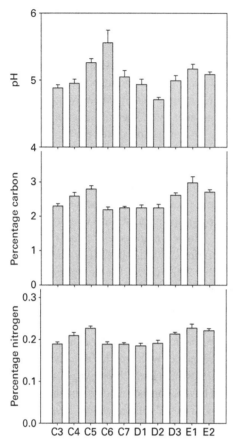

Figure 11.1 Heterogeneity in soil conditions across the 10 fields of the Buell–Small Succession Study. See Chapter 2 for details of the abandonment treatments. Fields are arrayed from east (E1 and D1) to west (C7).

short-lived or shrubby woody species. The later increase in CV may reflect another phase of community re-organization, during which invasible space opens up patchily. Certainly the lower variation of annuals among fields less than 10 years old is due to their relatively uniform dominance during that phase, and the ready availability of annual propagules throughout the fields. At no point during succession do annuals not exist in any of the fields, suggesting *in situ* seed banks or seed sources are available to take advantage of small, patchy disturbances in later years.

Biennials likewise show no statistically detectable trend. Their variation among fields was generally lower than those of annuals, perhaps reflecting the longer and more flexible life cycles of biennials and hence some buffering of availability throughout the fields relative to the shorter-lived annuals. The vagaries of biennial variation, like those of the annuals may reflect their opportunistic dependence on local disturbances and on dispersal. Both of these phenomena are expected to be unpredictable at fine spatial scales.

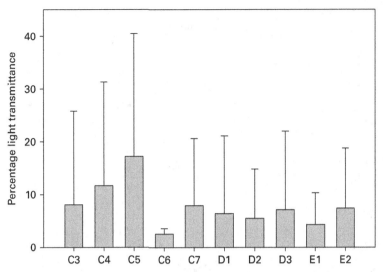

Figure 11.2 Heterogeneity in light transmittance as a measure of community-generated heterogeneity in abiotic conditions. Measurements taken mid-July 2006 and 2007.

Herbaceous perennials show a shallow concave trend in CVs through the 40 years analyzed. Their highest variation was in the first year of succession, when their availability is expected to be quite limited and localized. Furthermore, the freeze–thaw cycles during the winter in the earliest years of succession in sites where the soil surface is not insulated by plant litter are expected to reduce establishment and survival of later successional dominants (Buell *et al.*, 1971). Variation in perennial abundance declines during the phase of succession where their dominance consolidates, again suggesting some homogenization of propagule sources, along with the increasing size of clones or plant canopies. The increase in perennial CV toward the end of the 40-year span likely reflects the patchy and gradual decline of perennials adapted to mid-successional environments and a re-organization of the ground layer toward dominance by forest-adapted species. Again, during such phases, the heterogeneity of establishment safe sites and the propagule rain would both be expected to increase.

Shrubs and lianas exhibit similar, concave temporal patterns of spatial heterogeneity (Figure 11.3). Both of these life forms have greatest heterogeneity, as indexed by CV among fields, in the beginning of succession. There is also a slight uptick of the trend line for both toward the end of the 40-year window of observation. The high CV early in succession is likely due to the initial rarity of colonization across plots, as these groups typically do not survive long agricultural practice and must invade anew. As individuals of these life forms occupy increasing proportions of the fields, CV declines. It may be that the slight increase toward 40-year-old fields represents the establishment of a forest complement of shrubs and lianas, compared to the assemblages characteristic of the teens through thirties.

Trees, as the largest seeded invaders, the least effective in forming seed banks and the slowest species in this succession to mature, show the most intense spatial variation in

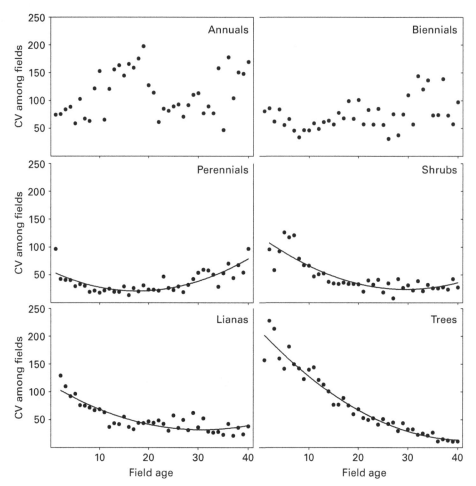

Figure 11.3 Temporal patterns of variability among fields, as indicated in coefficient of variation of cover for each life form (all species pooled) in the BSS.

early succession (Figure 11.3). The steady and steep decline in the spatial heterogeneity of tree abundance is notable. Like the other life forms, the greatest heterogeneity seems to be associated with regeneration patterns. Persistence and spread within the communities is represented by lower CVs. In trees, there is no evidence within the 40 years analyzed that new opportunities for tree establishment open up unpredictably across the fields. Indeed, it may be that the 2-m long sample plots are simply beginning to record the dominance of tree canopies that are coming to occupy a length scale of tens of meters. We would not expect an increase in CV for trees until some distant future period when gap-phase regeneration became the predominant mode of canopy replacement.

Species performance – The aggregate patterns of spatial variation over time for the different lifeform groups can be understood better based on the performance of exemplary species from each group (Figure 11.4). Plotting the total abundance of a species over time in all fields against its variation in abundance across plots adds interpretive power to

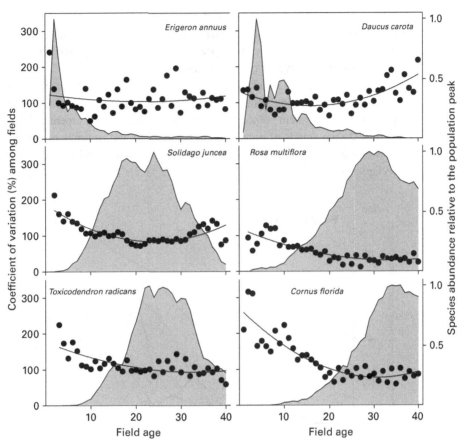

Figure 11.4 Examples of the relationship of species cover trends through succession and the CV (%) among fields of the cover of those species.

the patterns of spatial heterogeneity seen in life forms. In general, when a species is most abundant over all fields, its variability among plots is least. Even the annual *Erigeron annuus* supports this generalization. Because annual performance is most contingent upon spatially unpredictable disturbances, its first entry into the fields is the most spatially heterogeneous part of its successional trajectory. Once it becomes dominant or highly important in the communities, the CV settles down into the range from 40% to 100%. When its cover begins to decline precipitously, at roughly year 8, the CV becomes quite unstable throughout the remainder of the 40-year window. The persistence of *Erigeron*, although at very low total cover values, throughout this succession is typical of dominant early succession annuals (Chapter 6). The long tail is associated with continued, but diverse patterns of spatial heterogeneity, reflecting the opportunistic nature of its life history strategy.

In the case of the biennial *Daucus carota*, a similar trend is evident. Years with similar levels of heterogeneity only appear during the period of its greatest abundance. It may be that the slightly longer life cycles of biennials, and the potential for different cohorts to be present in the same year, explains the somewhat more constrained variation of this life form compared to that of the annual. The longer-lived perennial and woody species, especially those that have a strong peak within the 40 year span of succession analyzed, present the best examples of the congruence between dominance and reduced heterogeneity across fields. Colonization and population collapse generate more spatial heterogeneity than population dominance. Such trends may support the use of variation as a predictor of ecosystem vulnerability to change (Brock and Carpenter, 2006).

How do the details of population dynamics play out in terms of heterogeneity? We answer this question using *Solidago juncea*, a dominant species that both increased and collapsed within the 40-year window (Figure 11.5). The temporal dynamics by which

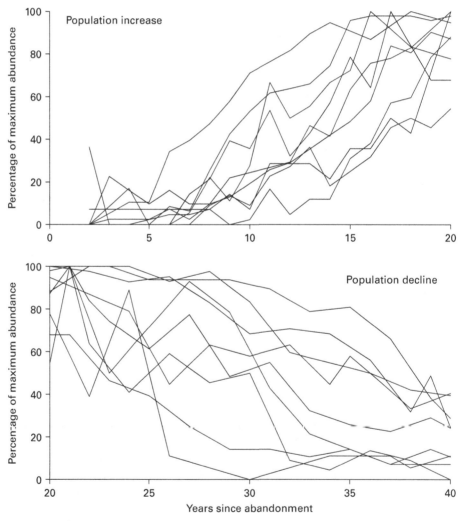

Figure 11.5 Variation in population trajectories among fields during population increase and population decrease. The abundance of *Solidago juncea* (as a percentage of maximum cover) across time is shown for the first 20 and the last 20 years of its successional trajectory.

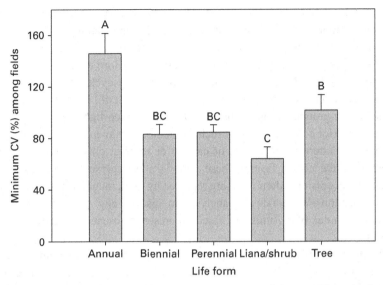

Figure 11.6 Variation among life forms in heterogeneity. For each of the core 84 species in the BSS the minimum coefficient of variation among fields was calculated. Data are means of the species in each life form.

maximum abundance in each field is achieved and then lost are described by splitting the 40-year window into halves. The period of population increase showed a remarkable variety of patterns – some convex, some concave and some saw-toothed. However, following an initial lag time, all increased fairly consistently. Population decline was also a collection of diverse patterns, but they were more diverse than the crescendos of increase. Populations of this common species collapsed rapidly in some fields but very slowly in others. The implication is that there are perhaps more different causes of species demise than of species ascendancy.

Life-history integration – When we examine the population dynamics of the 84 core species that the BSS contains, we can use the minimum CV achieved across fields as the base measure of spatial heterogeneity for each species (Figure 11.6). We note that individual species are far more variable across fields than life-form aggregations where species are pooled. Annuals exhibit the greatest heterogeneity across fields, followed by trees. As mentioned earlier, both of these life forms may be particularly susceptible to limitation by dispersal and the existence of fleeting recruitment safe sites. Lianas may be least heterogeneous due to their dependence on what ultimately is a widespread resource of tree stems in the developing successional woodlands. Both lianas and shrubs are predominately bird-dispersed in the BSS, also allowing widespread colonization across the site. The variety of biennial and perennial colonization and persistence strategies may result in their intermediate status of heterogeneity. Overall, these data suggest that complementarity among species leads to less variation at the life-form scale – an increase in predictability. However, the "patchiness" of an individual species is somewhat predictable based on life form. In other words, the unpredictability of a species across a landscape is...predictable.

A temporal heterogeneity cycle? – In forest succession, a cycle of heterogeneity, driven by canopy closure, gap creation and gap filling, is common. Grasslands, deserts and other ecosystem types also exhibit such patterns. Indeed, such cycles are an important part of contemporary successional theory (Horn et al., 1989; van der Maarel and Sykes, 1993; Pickett et al., 2011). Does such a cycle appear in the "old-field" part of mesic, post-agricultural succession (Armesto et al., 1991)? If old fields possess dense above-ground canopies that significantly reduce light, or root arrays that can exclude inferior competitors, then such communities may indeed have cycles of heterogeneity that alternate in periods of canopy or rootmat closure with periodic, patchy opening of the closed layers. These periods may represent successional transitions as earlier successional dominants are replaced by later, inducing a period of community re-organization. We conducted analyses to discern any such patterns in the BSS. Sorensen's index of dissimilarity based on presence/absence was used across the plots in each field.

We show results for fields representing abandonment in the fall with litter present, but contrast two row-crop fields with one hay field (Figure 11.7). All three fields show fluctuations of heterogeneity very early in the succession. In field C3 the early cycles of heterogeneity include one linked to the severe 1965 drought. However, field D1, which shared a history of fall abandonment from a row crop with intact crop residue, showed much less clear cycles, and responded to the same drought only modestly. After approximately 18 years, there is virtually no fluctuation in plotwise dissimilarity in field D1. C3 shows some long cycles after 10 years. Field C6, abandoned from hay, shows some early cycling, but none after the mid-teens. However, its marked homogenization after the late 30s is likely due to the explosive spread of the recently introduced *Microstegium vimenium*.

All in all, there is little evidence that the old fields of the BSS beyond their earliest years exhibit cycles of heterogeneity. This may reflect the diffuse canopies, and the relatively poor soil resource base of these fields. The possibility remains that cycles may appear more distinctly in the herbaceous phases of more resource-rich situations. However, there is clearly no universality to cyclic heterogeneity in the largely herbaceous phases of succession (Rejmanek and Rosen, 1992).

Conclusions

Although we have shown the manifold importance of heterogeneity, through experiments and through both within-field and across-field analyses, the patterns through time are complex. There are many powerful drivers that can and do influence heterogeneity in successional communities. Most of these relate to edge gradients of fields that border older successional or forest patches. The patterns of heterogeneity in old fields confirm and contribute to the theory of edge dynamics (Cadenasso et al., 2003a). Some of the heterogeneity relates to the spatial patchiness within fields, such as those provided by perches for fruit-dispersing birds, woody patches sheltering small-mammal seed and seedling predators, or patches of greater nutrient availability that alters competitive

Conclusions

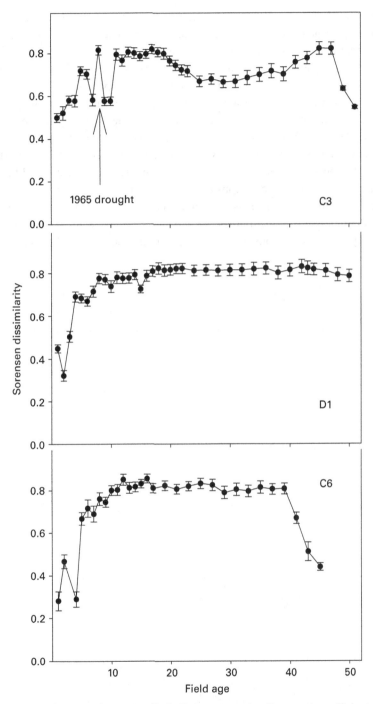

Figure 11.7 Temporal patterns in average dissimilarity, assessed as Sorensen's coefficient, for three fields over from 46 (C6), 48 (D1) or 50 years (C3). All fields were abandoned in autumn with litter intact. C3 and D1 originated from row crops, while C6 originated from hay.

interactions. With these ubiquitous sources, succession cannot be made sense of without accounting for heterogeneity. Scale of interactions sometimes makes heterogeneity a background, slow variable, and at other times makes it an active part of the local dynamic model. Finally, the power of examining successional patterns through the lens of growth forms is confirmed. While individual species and their performances are the nuts and bolts of successional turnover and persistence, aggregating species into groups that share broad outlines of architecture, reproduction and life history simplifies the understanding of the resultant patterns. This is, not surprisingly, the place from which Frederic Clements started his theory of succession roughly 100 years ago (Clements, 1916). The rough sequence of life forms from smaller and shorter-lived to larger and longer-lived is a coarse-scale constant in mesic environments with relatively secure supplies of soil nutrients. This of course reflects the fundamentals of physiology, allocation of assimilated resources and strategies to deal with whether resources are externally available or conserved within persistent biomass of plants (Pickett *et al.*, 2011).

12 Functional ecology of community dynamics

> "Strategies may be defined as groupings of similar or analagous genetic characteristics which recur widely among species or populations and cause them to exhibit similarities in ecology. The potential value of the concept of strategies as a unifying approach to ecology depends on the extent to which it may be applied to all living organisms."
> John Philip Grime, *Plant Strategies and Vegetation Processes* (1979)

One of the major shifts in focus for ecology over the last decade or so has been the increased interest in functional ecology, or as it is commonly practiced, the ecology of traits. This paradigm shift has been a slow one to develop as its early champion, Phil Grime, was examining functional species differences and discussing plant strategies throughout his career in what he described as comparative ecology. The British Ecological Societies' journal *Functional Ecology* began in 1987 with a focus on the roles that species play in communities. The surge of functional work, which continues as of the writing of this text, began in the late 1990s. As with any paradigm shift, some of the work done under functional ecology is really a re-phrasing of the same sort of work that has been done for years, such as was done in placing species along the r- and K-selection gradient or in testing the ecological determinants of rarity among many others. At its best, however, the focus on functional ecology can and has given us real and new insight into the forces that structure communities.

In a very simple way, functional ecology differs from other sub-branches of the ecological shrub in that it focuses on the ecological attributes of species. The functional aspect of functional ecology directly focuses on mechanisms – what an individual species does in a community or what a particular trait does for a species. The functional approach is in contrast to the more traditional approach to community ecology where the species is the ecological unit of interest. This is not to say that researchers have ignored the mechanisms of community structure. Nothing could be further from the truth. Functional ecology merely starts with those mechanistic relationships as the core interest. The shift in focus is analogous to what we have seen develop with ecosystem ecology. In conceptual views of ecosystem dynamics, the critical units are pools and fluxes, with the species composition within individual pools represented as black boxes. Researchers differ widely in how opaque the black box of species composition is. No one (largely) would ever argue that species do not matter to ecosystems; they are just not the main focus of the research program. Instead, the mineral and energetic contents of the constituent species are followed. In a similar way, species composition takes a back seat in functional ecology, while the traits of those species are driving the research car.

Ultimately the functional approach, like others, is focused on understanding ecological patterns in communities.

Trendy research topics are sometimes problematic and short lived. We must ask whether there is an inherent value to the functional ecology approach, or will we look back on this period with the conceptual equivalent of a hangover and wonder "what were we thinking?" The real, and likely lasting, value of functional ecology is that it focuses on the characteristics of species that determines their role in communities (Shipley, 2010). These traits should be linked with how a species disperses, establishes, grows, captures resources, and if successful, reproduces (Lavorel et al., 1997; Weiher et al., 1999; McGill et al., 2006; Westoby and Wright, 2006; Violle et al., 2007). These are the mechanisms that drive community composition and structure. The difficult part of this approach is in determining whether the traits we measure are really the mechanistic drivers or closely related to the critical, and perhaps unmeasured, traits.

Functional ecology is a tool for integration across systems and across scales. From a taxonomic view, we can say that *Aster pilosus* is replaced by *Solidago canadensis* within the BSS. This research statement is perhaps interesting to researchers in other systems with the same two species and could be perhaps studied experimentally and linked mechanistically to competition for light (Banta et al., 2008). This statement may also be useful to researchers working with other *Solidago* and *Aster* species as a starting point. However, systems that do not contain those two species gain little from such statements. For those researchers interested in understanding the impacts of *S. canadensis* on invaded systems in Europe this information provides little predictive ability. If, however, we describe the species transition from a functional standpoint it gives a very different view. Within the BSS, non-clonal herbaceous species are replaced by aggressive clonal herbs. The functional statement may be applied to any successional system to examine its generality. When the dynamics of a system do not follow the same functional trajectory, we have identified an opportunity to further our basic understanding of how succession, or any ecological process, operates.

Another main avenue of research in functional ecology is in the identification of the relationships among traits. Relationships among traits allow researchers to measure simpler, but correlated, traits and still capture the same bit of ecological information (Weiher et al., 1999). For example, while plant ecologists are really interested in species growth rates, it is difficult to repeatedly measure plants in the field or to cultivate all species within a community to assess growth. Instead, researchers focus on a structural characteristic, specific leaf area, as a surrogate. The relationships among traits also helps researchers to identify suites of correlated traits or plant strategies (e.g. Grime, 2001; Westoby et al., 2002). These strategies allow species to be successful under a range of biotic and abiotic conditions, but are also constrained by tradeoffs among traits.

The conceptual framework around which we have structured this book provides some guidance in understanding the role of traits in ecology. Functional ecology largely lies within the coarse mechanisms of differential species availability and differential species performance. Most traits map out fairly cleanly on these two groups of successional drivers. However, this approach also allows us to separate out more subtle effects between groups. The species pool necessarily constrains the range of species available to sort out during succession. Therefore the species pool also determines the range of traits available for differential performance to act upon. When the traits of the available species pool are broader than the patterns expressed by the community, this is evidence of differential performance. Trait dynamics that follow the available species pool are more ambiguous. Differential species availability may constrain the dynamics of the system, or may reflect selection imposed by differential performance, even before entry to the regional species pool. In other words, only species that are successful in these habitats are available in this landscape. Prior selection may be strong in studies of introduced species or in largely agricultural landscapes, like those surrounding the BSS when it was started (Pyšek *et al.*, 2003, 2009; Gavier-Pizarro *et al.*, 2010).

In this chapter we will present information on the pool of available traits – species availability – and on the community-scale changes in traits that occur during succession – species performance. To place the trait pool into an ecological context, we have split the species pool into life forms. To generate community trait dynamics, we have weighted each species trait by its relative abundance in the community. This weighting is commonly used to reflect the contribution of each species to the community (e.g. Kahmen *et al.*, 2002; Garnier *et al.*, 2004; Pakeman *et al.* 2008), as the influence of a species should scale with its abundance (Grime, 1998). These data, in combination with the trait information on when species peak in succession (Chapter 6), provide a comprehensive view of trait dynamics in this system.

We will start each suite of traits with an exploration of variation among life forms to understand the components of the species pool and then move on to an examination of trait dynamics for the entire community. One of the unique aspects of applying a trait-based approach to succession in the BSS is that it encompasses the full structural transition from an herbaceous to a woody-dominated system. Many studies of traits in communities focus on either herbaceous or woody species, which simplifies the transitions and tradeoffs that are examined. Shifts in life form may represent positions along a continuum of lifespan, or may represent fundamental shifts in strategy (Tecco *et al.*, 2010). For example, a large seed from an herbaceous plant would typically be similar in mass to a small seed in a woody plant. The herb would be interpreted as being able to regenerate well under competition; the woody plant would not, at least in comparison to structurally similar species. On the other hand, woody species largely replace herbaceous species in succession to forest, confirming the shift in traits over time. We will also return to the utility of life forms as a functional group at the end of this chapter.

Plant reproductive traits

The changes in plant reproductive strategies during succession have by far the longest history of study, so we will begin our exploration of traits in succession with them. Early views of reproductive characteristics were largely based on comparisons between early and late successional species and so do not reflect the continuous nature of successional changes or of species' strategies – our approach will be quite different. Reproductive traits, unlike most of the others that will be discussed, mechanistically occupy the border between differential species availability and differential performance. For this reason it is difficult to cleanly integrate these into the hierarchical view of succession. This is indicative of the linkages and feedbacks among the main classes of drivers. For example, seed mass and dispersal mode are linked. Larger-seeded species are more likely to be dispersed by vertebrates and would have the energy reserves to become established under a wide range of local conditions. As a species' success, and therefore fitness, would be affected by both its ability to disperse into spatially and temporally unpredictable habitats and to become established, natural selection should focus simultaneously on traits that control differential availability and performance. The ecological strategies that we will explore later in this chapter also integrate across diverse community drivers and a similarly broad suite of traits.

Pollination – As would be expected, there was large variation among life forms in their reproductive traits (Figure 12.1). A dichotomy between woody and herbaceous species was also evident in some traits. Overall, most species present in the BSS (70.5%) were biotically pollinated – almost entirely by insects. Tree species differed from other life forms in that only 32% were biotically pollinated. The dominant tree species in the BSS, *Acer* spp., *Carya* spp., *Quercus* spp. and *Juniperus virginiana* are

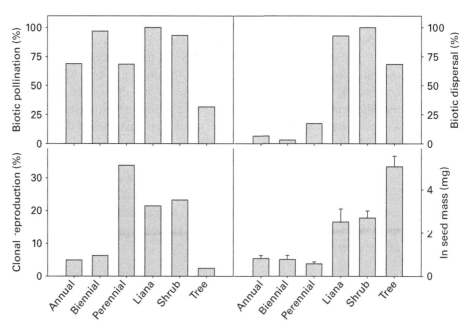

Figure 12.1 Variation among life forms in reproductive characteristics. Data plotted are the proportion of species with a trait (categorical traits) or means ± SE (seed mass). Seed mass data were ln +1 transformed.

all wind pollinated. Annuals and perennials also had a slightly lower proportion of biotic pollination than biennials. Both of these life forms contained many grass species, which are only wind pollinated. Several of the dominant annuals in early succession such as *Ambrosia artemissiifolia* and *Chenopodium album* are also wind pollinated. The remaining life forms, biennials, lianas and shrubs were nearly all pollinated by insects. In fact, all of the liana species recorded in the BSS were insect pollinated. The dominance of the species pool by insect-pollinated species is somewhat expected based on the prevalence of insect pollination in angiosperms (Pellmyr, 2002). What is perhaps more informative are the life forms that deviate from that dominance.

While species pools set the range of traits available for sorting through succession, these pools do not necessarily generate successional patterns. Differential performance, whether mediated through plant–environment interactions or plant–plant interactions exhibits a strong selective force on the species pools, which may generate patterns quite different than those set by the initial pool of traits. Though the majority (70%) of annuals and nearly all biennials were biotically pollinated, succession began with a much lower representation of biotic pollination (Figure 12.2). Prevalence of wind pollination in the hay fields would be expected, however there did not seem to be large differences based on site history as the pattern was consistent across hay and row-crop fields. Fields with large representation of annual mustards, which are all pollinated by insects, started out somewhat higher, but then also decreased. The proportional abundance of biotically pollinated species increased dramatically during the first 8–12 years of succession for most fields. Following this initial peak, there was a long and slow decline in the prevalence of biotic pollination as wind-pollinated trees slowly increased in cover. While the initial increase in biotic pollination and its peak abundance varied dramatically among fields, the decrease in biotic pollination was much stronger and the fields appeared to be functionally converging, despite marked variation in tree composition.

Biotically pollinated species would be expected to have more efficient pollen transfer among widely spaced individuals than wind-pollinated species, perhaps explaining their abundance in mid-successional communities (Pellmyr, 2002). At that point in time, there would be adequate pollinator resources across the site to attract and maintain a suitable suite of pollinating insects. Early in succession, these resources may not be sufficiently available within the BSS fields or in the surrounding landscape (Steffan-Dewenter and Tscharntke, 2001; Steffan-Dewenter *et al*. 2002). If pollinators also have to colonize a site following abandonment of agriculture, then there may be a time lag in pollinator services as well. The dominance of early successional communities by wind-pollinated plants fits with some of the successional generalizations that have been proposed (e.g. Odum, 1969; Pianka, 1970; Prach and Pyšek, 1999), though this has not been seen in all systems (Prach *et al*. 1997). Dependence on mutualists is risky early in succession, but may become relatively safe as the plant and insect communities assemble following abandonment. The shift in forest trees towards wind-pollinated species is probably facilitated by the relatively species-poor nature of tree canopies in eastern deciduous forests. The prevalence of wind pollination at the forest stage may also be related to a

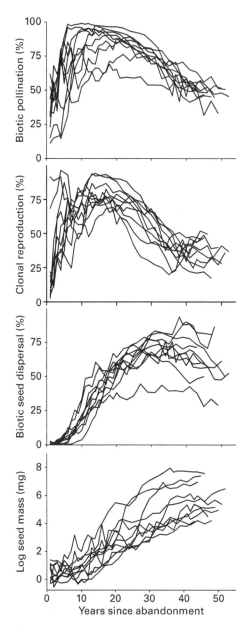

Figure 12.2 Changes in reproductive traits during succession. Each line represents the abundance weighted trait values from an individual BSS field.

shift towards earlier bloom times before the plant canopy develops and forms barriers to pollen movement (Rye and Meiners, unpublished data).

Clonal reproduction – Clonal reproduction is an important plant trait that allows the vegetative expansion of an individual without the need for regeneration via seeds. When growing in good conditions, this trait would allow a plant to spread and capitalize on local resource pools and may even support some level of resource foraging (Hartnett and Bazzaz, 1983; Hutchings and Wijesinghe, 1997). Life forms varied dramatically in their ability to spread clonally (Figure 12.1). Annuals, biennials and trees had very low incidence of clonality in the species pool. As these life forms each capitalize on local and ephemeral opportunities for establishment, their reproductive cycles are dominated by seed regeneration. In contrast, perennials, lianas and shrubs had much greater incidence of clonality. One-third of perennial species had clonal reproduction and a little over 20% of both shrubs and lianas had clonal reproduction. These three life forms are in general more competitive and may form extensive patches within plant communities. Growing in competitive environments may also reduce the chance of successful regeneration from seed. Seeds in this situation would predominately be involved in initially colonizing new habitats, whereas clonal expansion would take precedence following colonization. We would therefore expect clonal species to produce smaller seeds than their non-clonal counterparts, as the necessity for larger seed reserves would be less. To test this prediction we can examine the influence of clonality on seed mass in perennial herbs and shrubs, which had a sufficient sample size. In both life forms, seed mass was lower in clonal taxa, though only significant in shrubs (perennials – Kruskal–Wallis $\chi^2 = 2.80$, P = 0.09; shrubs – Kruskal–Wallis $\chi^2 = 4.71$, P = 0.03). Again, this relationship emphasizes the interdependence of species traits.

Based on the composition of the species pool, we may make a few projections about clonality in succession. In general, clonality should represent a minority of the community, as only 20% of species in the entire pool are clonal. Second, as perennials, lianas and shrubs have the greatest representation of clonal species, we should expect clonality to peak in mid-successional communities, when these life forms were dominant. When we examine the temporal patterns in clonality exhibited by the community, we initially see dramatic variation among fields (Figure 12.2). Most fields had very low cover of clonal species at less than 25% of the community, while two fields, those which were abandoned hay fields that were left unplowed, had much greater cover of clonal species (grasses). These initial trait dynamics resulted from pre-abandonment conditions, but were short-lived. All fields rapidly increased in the dominance of clonal species, with clonal species reaching 65% or more of the community. Dominance of clonality in the communtiy was much greater than would be expected if the community were composed of random draws from the species pool. While a minority of species was clonal in the species pool, they represented a majority of the cover in mid-successional communities. As clonality is considered a major competitive trait in Grime's c-s-r scheme, the peak in clonality at intermediate successional ages fits with his predictions for successional transitions in productive environments (Grime, 2001).

Seed dispersal – Dispersal mode varied widely among life forms within the species pool, with a major dichotomy between woody and herbaceous taxa. Herbaceous species had a quite low prevalence of biotic dispersal (Figure 12.1). Most of the biotically dispersed herbaceous species were adhesively dispersed, such as *Desmodium*, *Galium* and *Polygonum*. A much smaller group of herbaceous species were fleshy fruited, with

Fragaria virginiana and *Solanum carolinense* the most abundant. In contrast to the herbaceous species, over 80% of all woody species were vertebrate dispersed, including all of the shrub species and all but one liana species. The majority of these were bird-dispersed, fleshy-fruited species, though late successional trees tended to be dispersed by small mammals that function as both seed predators and dispersers (e.g. *Carya* and *Quercus* species). The shift from herbaceous, abiotically dispersed species to woody species with primarily bird dispersal in mid-successional communities would also lead to changes in the spatial patterning of dispersal and recruitment. Specifically, the establishment of early perch sites would lead to localized recruitment of bird-dispersed species (McDonnell and Stiles, 1983; McDonnell, 1986; Robinson and Handel, 1993). Within the species pool we also see the interdependence of reproductive traits. Seed dispersal and pollination modes were somewhat linked, as species tended to utilize the same mode for both ($\chi^2 = 3.90$, P = 0.048). Similarly, biotically dispersed species tended to have larger seeds than species with abiotic dispersal (median seed mass: biotic – 3.46 mg, abiotic – 0.45; Kruskal–Wallis $\chi^2 = 14.8$, P = 0.0001). Linkages such as these limit our ability to fully isolate the role of individual traits in succession, reflecting the importance of suites of traits in forming plant strategies.

Patterns in seed mass seen at the community level largely followed the constraints of the species pool during the transition from herbaceous to woody species (Figure 12.2). Biotic seed dispersal was consistently low early in succession. The proportion rapidly increased after five years of succession as bird-dispersed shrubs and lianas expanded and came to dominance. The increase continued until 40 or so years after abandonment, when some fields began to decrease. The decrease late in succession is linked with the dominance of the wind-dispersed trees *Acer rubrum* and *Fraxinus americana* in some fields, and in the development of a understory flora that is primarily abiotically dispersed. Late in succession, we also see evidence for increased variance among fields in dispersal mode. Though we may expect biotically dispersed species to have greater potential dispersal ability, the data suggest that dispersal limitation was important in limiting the functional responses of the community. Similarly, the development of a forest understory flora was incomplete in the BSS and species distributions were still very patchy among fields. As canopy and understory species become available across the site, we would expect variation in dispersal mode to also decrease across the site.

Seed mass – Seed mass is perhaps one of the best measures of a plant's regeneration capabilities (Westoby, 1998) and typically exhibits large variation within plant communities (Baker, 1972; Mazer, 1989; Lord *et al.*, 1995; Leishman *et al.*, 2000; Westoby *et al.* 2002). Seed mass is linked with seed longevity, seed dispersal and with the ability to regenerate under limiting conditions such as low light, water stress or from under deep litter layers (Jurado and Westoby, 1992; Saverimuttu and Westoby, 1996; Norden *et al.*, 2009). As such, it is not surprising that seed mass was strongly associated with life form in the BSS (Figure 12.1). Herbaceous species were all remarkably consistent in seed mass with an average of 1.7 mg. In contrast, the seeds of woody species were on average more than two orders of magnitude larger with a seed mass of 735 mg. Within woody taxa, seed mass increased from lianas to shrubs to trees, with tree seeds being much larger than any other life form. These patterns are consistent with the weedy nature of lianas and with the dependence of many late successional trees on their seed reserves for successful regeneration.

Succession led to a gradual and consistent increase in seed mass at the whole-community scale (Figure 12.2), as has been documented in other successional communities (e.g. Navas *et al.*, 2010; Douma *et al.*, 2012). The pattern in the BSS largely followed the constraints set by the species pool and transitions among life forms. During the first 8–10 years, when herbaceous species dominated the community, seed mass was relatively consistent. As woody species increased in dominance following that period, seed mass also increased steadily. There were large differences among fields, even late in succession, showing the influence of canopy composition on trait patterns. As has been noted previously (Chapter 9), dispersal limitation in woody species, particularly large seeded ones, functions to constrain composition and dynamics during succession.

Leaf structural characteristics

The measurement of leaf characteristics should provide the basis for a good assessment of the physiological behavior of plant species so that we may place them in a successional context. As with plant reproductive traits, we will address the species pool separately from changes in the community scaled by the relative abundance of species. In this way we will structure our examination of leaf traits to follow the guiding conceptual model for this book.

Leaf dry matter content (LDMC) – The proportion of a leaf's biomass that is not composed of water is a measure of physical and chemical allocation to the leaf. These components may be structural support, chemical defenses or physical structures, such as hairs and glands that may serve a variety of functions. As such, LDMC is a measure of the plant's investment in, and potentially the value of, a leaf to the metabolism of the plant (Navas *et al.*, 2010). The species pool showed a relatively large range of variation in LDMC. Unlike the reproductive characters, which showed a clear dichotomy between woody and herbaceous species, LDMC revealed more of a continuous gradient from annual/biennial species through trees (Figure 12.3). As would be expected, annuals and biennials allocated the least to their photosynthetic tissues, as these species typically do not allocate to chemical defenses and leaf turnover is high (Coley *et al.*, 1985; de Jong, 1995). As plant life span increased, so did the investment in leaf structures. Part of this is likely a physical requirement to support larger leaves, but there is also likely an increase in allocation to chemical and physical defenses in plants with more persistent tissues. For example, oak (*Quercus* spp.) leaves contain a large amount of tannins, a digestibility reducer (Forkner *et al.*, 2004). These also tend to be physically tough leaves with a large investment in structural compounds. Similarly, other common trees such as *Carya* spp. and *Juglans nigra* (both Juglandaceae) produce large amounts of aromatic chemicals.

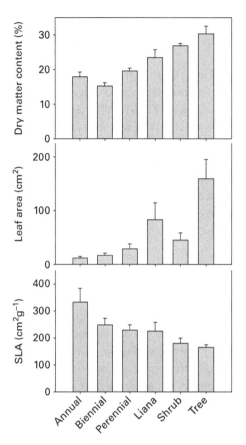

Figure 12.3 Variation among life forms in leaf structural characteristics. Data plotted are the means ± SE for each trait.

When we look at the changes in community level LDMC, there was a clear and strongly convergent trend towards increasing LDMC (Figure 12.4). Early successional fields varied dramatically in LDMC, though most started very low, as would be expected based on the pattern seen in life forms. However, fields that were planted in hay started out with a much higher LDMC, due to the abundance of grasses that contain silica as a structural component. All fields became much more consistent in LDMC as succession proceeded, with the highest values and least variation 35 years after abandonment. Following this peak there has been some divergence among fields, likely driven by heterogeneity in canopy and forest understory composition. Of all of the leaf characteristics, LDMC generated the clearest successional dynamics with the least variability across sites.

Leaf area – The size of leaves varied dramatically across life forms with another clear dichotomy between herbaceous and woody plants (Figure 12.3). Within herbaceous plants, there was a slight increase in leaf area from annual to biennial to perennial species. In woody plants, shrubs had the smallest leaves, followed by lianas and trees. Tree leaves were on average twice the size of other woody life forms. Physiological limitations occur with larger leaves as they may more easily overheat, leading the plant to evapotranspirate

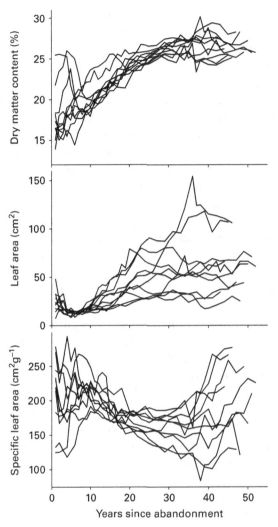

Figure 12.4 Changes in leaf structural traits during succession. Each line represents the abundance-weighted trait values from an individual BSS field.

more water to maintain functional leaf temperatures. Early successional species would be exposed to full sun and may have poor ability to retain water (Bazzaz, 1979). This may constrain them to having smaller leaves that disperse heat more readily. Changes in water use efficiency may explain the general increase in leaf area with plant longevity (and presumably allocation to roots) in herbaceous species. The linkage between leaf size and water loss may also explain the patterns seen in woody species. Shrub species within the BSS were primarily shade-intolerant species that should be influenced by the need to dissipate heat and conserve water. In contrast, lianas tended to have larger leaves than shrubs, despite their similar period of dominance in succession. Most lianas would be associated with other woody species that would not only provide physical support, but may provide some shading. Lianas should therefore have leaf characteristics that would maximize light interception rather than minimize heat gain.

When we examine the whole plant community during succession, we see a much less clear pattern for leaf area (Figure 12.4). Leaf area was initially low, though variable across fields. Years 4–8 were characterized by very low leaf area and very little variation among fields. Following this minima, leaf area increased for the next 30 years, as did variation among fields. At its greatest, there was over a four-fold difference in community-weighted leaf area across fields. Though successional transitions among life forms would suggest a relatively predictable increase in leaf area with succession, functional dynamics at the community scale followed this pattern only loosely. We may invoke dispersal limitation in trees as generating late successional variation in leaf area among fields, but the differences began well before the increase in dominance of trees in the system. Though leaf area was associated with the successional status of species (Chapter 6), the relationship was not strong enough to generate clear successional trends when compared to other traits. The weakness of this relationship argues that leaf area by itself is not a strong driver, nor is it associated with a strong driver of successional dynamics.

Specific leaf area (SLA) – The strong experimental linkage between SLA and growth rates as well as its role in global patterns of leaf economics (Reich *et al.*, 1998; Westoby *et al.*, 2002; Garnier *et al.*, 2004; Wright *et al.* 2004) would lead to the a-priori prediction that SLA should track well with life history and succession. Data from the species pool bears this prediction out (Figure 12.3). As would be expected in rapidly growing plants with high leaf turnover, annuals had distinctly higher SLA than other species. Trees, to the other extreme, grew the slowest because of their allocation to woody tissues and allocation to physical and chemical structures in the leaves (see above) and had the lowest SLA. Between these functional endpoints, all life forms fell out along a gradient of SLA. The species pool therefore would suggest, based on transitions in life forms and the small differences among most forms, that succession should generate a relatively predictable and consistent decrease in SLA.

While analysis of the species pool revealed very clear differences among life forms, these differences did not translate into a simple or predictable successional trajectory. The earliest successional communities exhibited wide variation in SLA, though they were, on average, relatively high, as would be expected (Figure 12.4). Part of this early variation would clearly be generated by the differences between row crop and hay fields, as grasses typically have low SLA values. As grass cover began to collapse, SLA increased temporarily, followed by a 20-year decline in SLA that was very consistent among fields. However, fields again began to diverge as SLA increased late in succession. A general increase in SLA may be expected with the development of an herbaceous forest understory, but the magnitude of this change would not be expected to be as large, based on the species pool. It appears that the dynamics of individual species may be largely responsible for temporal patterns in SLA. The consistent decrease in SLA in mid-successional communities coincides with the dominance of *Rosa multiflora* across all fields (SLA 125 cm^2g^{-1}). Late in succession, the increase in SLA would be partially generated by invasion by two species – *Microstegium vimineum* (SLA 416 cm^2g^{-1}) and *Alliaria petiolata*

(SLA 313 cm^{2}g^{-1}). Though the invasion of these two species would appear widespread enough to increase the SLA of most fields, it should not generate the huge range in community SLA seen across the system.

The susceptibility of SLA to the influence of individual species suggests that by itself it is not a good predictor of successional dynamics. Though there were clear differences among life forms, the diversity within each of those forms was sufficient to allow dispersal limitation or selective recruitment to generate large variation in the community-aggregated values. Perhaps there are just too many contingencies between SLA and other traits to allow a simple classification of its dynamics in succession. For example, perennial forest herbs and those of open habitats of mid-successional fields should have quite different SLA values. Similarly, Douma *et al.* (2012) found that the successional response of SLA varied with local environmental conditions in a large-scale study of trait–succession linkages.

Though some traits have been individually useful in understanding successional dynamics, this approach clearly offers a limited view of succession. The presence of complex interactions between traits and the continuous range of plant strategies suggests that no one trait is going to adequately capture successional dynamics. There are simply too many ways to be successful in a plant community. From the simplistic, single-trait approach above, we will now build towards a slightly more complex view of traits in succession – one that uses three traits that have been suggested to represent primary functions within plants. From there, we will broaden to an approach that encompasses all of our available trait data. While one can always argue that a more complete view is always going to be more accurate, there can also be diminishing returns and the potential to so finely categorize the system as to approach the complexity of following individual species. We will return to this idea and assess whether the application of additional data has improved our understanding of the system.

Plant strategies: leaf–height–seed

While Grimes c-s-r scheme starts with the range of environments that plants may live in and works towards strategies and traits, Westoby's (1998) leaf–height–seed scheme of plant strategies starts with traits. This makes the scheme much more tractable as it is clear which traits need to be measured to functionally characterize species. The three traits were selected to represent issues of regeneration (seed mass), resource acquisition (plant height) and physiology (specific leaf area) following the basic arguments discussed in the previous sections. The traits required are very easy to measure, making this an incredibly simple scheme to employ. This method for describing plant strategies is also a continuous one, not relying on dividing species up into categories. For this reason it is suitable for characterizing entire communities, as well as individual species. This also allows us to evaluate the role of species pools and differential performance in generating the functional transitions in succession.

We will use the LHS scheme to examine the trait structure of the BSS through two ordinations. First, we will examine the distribution of individual species within trait space to describe the foundation of successional dynamics and to assess the ability of the LHS scheme to differentiate species. As seed mass was strongly related to height in the BSS data, we ran a principal components analysis (PCA) to condense the three traits into a two-dimensional representation. In the second ordination, we will look at successional trajectories in LHS space. To do this, we calculated abundance-weighted trait values, again using relative cover to scale each trait by species composition and generate a single value for the community. This was done separately for each field in each year sampled. Community-aggregated trait values were then run through a PCA, as with the individual species.

In the ordination of individual species, the first axis represented increasing seed mass and plant height (Figure 12.5). The second PCA axis was positively associated with SLA alone. SLA was also negatively related to the first axis, but not nearly as strongly as plant height or seed mass. Not surprisingly for a three-variable data set with strong associations, the two axes explained 92% of the variation in LHS. Within the ordination, herbaceous species are primarily on the left side of the graph, as would be expected based on their shorter heights and smaller seeds relative to most woody plants. There is a much greater

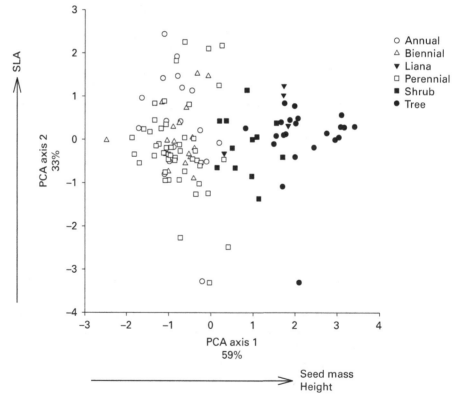

Figure 12.5 Species ordination examining distribution of species in the leaf–seed–height plant strategy scheme. Seed mass ($R = 0.91$; $P < 0.0001$) and height ($R = 0.93$; $P < 0.001$) were positively associated with PCA axis 1. Specific leaf area was positively associated with PCA axis 2 ($R = 0.96$; $P < 0.0001$) and to a lesser extent negatively associated with axis 1 ($R = -0.29$; $P = 0.0016$). Filled and open symbols represent woody and herbaceous species, respectively.

spread within life forms on the second axis, though little separation. This reflects the small differences among life forms described previously (Figure 12.3). Overall this analysis shows a continuum of plant strategies that differential performance can select upon.

The ordination of community-weighted values explained 95% of the variation in LHS over succession and was remarkably similar structurally to the ordination of individual species traits (Figure 12.6). Seed mass and plant height were strongly associated with axis 1, which explained almost 2/3 of the variation in LHS over

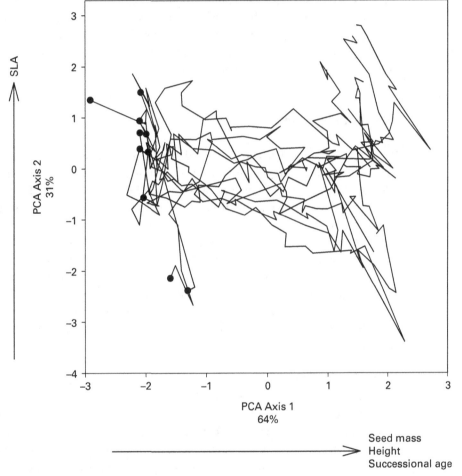

Figure 12.6 Ordination of successional communities based on the leaf–seed–height plant strategy scheme using principal components analysis. Values are based on the abundance-weighted trait values for each field during each year. Each line represents a different BSS field with the starting points indicated by filled circles. Seed mass ($R = 0.91$; $P < 0.0001$) and height ($R = 0.96$; $P < 0.0001$) were positively associated with PCA axis 1. Specific leaf area was positively associated with PCA axis 2 ($R = 0.90$; $P < 0.0001$) and to a lesser extent negatively associated with axis 1 ($R = -0.42$; $P < 0.0001$). Successional age was strongly correlated with PCA axis 1 ($R = 0.91$; $P < 0.0001$) but was not associated with PCA axis 2 ($R = 0.09$; $P = 0.11$).

succession. This axis was also strongly correlated with successional age, reflecting the overall importance of plant height and seed mass to successional dynamics. Specific leaf area was the primary determinant of axis 2, explaining nearly 1/3 of the variation. Again, SLA was also negatively correlated to axis 1. The pattern of trait associations with PCA axes was identical to that of the ordination of individual species. This similarity makes it easy to relate the two ordinations. Early successional communities varied dramatically in LHS space along the second axis and to a lesser extent on the first. The variation among fields decreased, as was seen in the community trajectory of SLA. Variation increased among fields late in succession as well, probably reflecting the influence of spatially variable invasion by non-native species and divergence in canopy composition. Nicely, the functional perspective gives a single successional axis. It also shows rapid changes in the functional characteristics of the community, particularly along axis 2. This axis may be particularly sensitive to functional changes brought on by species invasion or droughts.

Plant strategies: overall

With the limited trait data currently available for BSS species, we can attempt to form a broader and more integrative view of plant strategies by incorporating all the traits into a single analysis. Starting with Westoby's LHS scheme is useful, but does it truly capture the breadth of functional variation in the community? To assess functional variation, we again run a PCA of species traits, this time incorporating height, reproductive characteristics (seed mass, clonality, biotic pollination and biotic seed dispersal) and leaf characteristics (SLA, LA and LDMC). The ordination resulted in only two axes that explained more variation than would be expected by chance alone. Interestingly, these two axes clearly follow the same broad patterns found in the ordinations of LHS above (Figure 12.7). The BSS trait patterns largely follow the tradeoffs found in Mediterranean successional communities for the traits that are shared between studies (Navas *et al.*, 2010).

As seen in the previous two ordinations, seed mass and plant height are strongly and positively associated with the first axis, as are leaf area, dry matter content and biotic dispersal. The other end of the first axis is weakly associated with characteristics that are more closely associated with the second axis – SLA, clonality and biotic pollination. The second ordination axis represents a gradient from clonal species with high LDMC to species with high SLA and biotic pollination, and to a lesser extent, biotic seed dispersal. If we examine the positioning of the LHS traits in the ordination, we find, even down to a secondary negative association of SLA with the first axis, that these patterns are identical. The addition of traits allowed better resolution between woody and herbaceous species and to a lesser extent among life forms. However, the additional plant traits have not altered the overall structure of the analysis or revealed any new axes of variation. Instead, these additional traits are clearly aligned with the LHS traits. The incorporation of other traits along those fundamental axes strongly argues for the utility of the LHS scheme of plant strategies

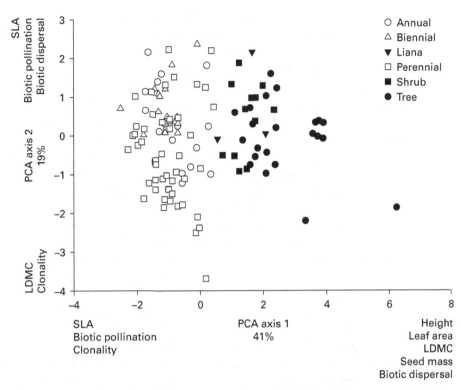

Figure 12.7 Principal components ordination of species traits using all of the life history and leaf physical traits (height, seed mass, clonality, biotic pollination, biotic dispersal, leaf area, specific leaf area and leaf dry matter content). Individual traits are positioned along axes based on correlation analyses with the PCA axes. Life form symbols as in Figure 12.5.

(Golodets *et al.*, 2009; Laughlin *et al.*, 2010). The successional trajectory of this suite of traits argues that these traits are capturing ecological characteristics important in community dynamics.

Life forms as functional groups: is there justification?

Throughout this text we have relied on plant life form as a primary way to functionally characterize and group species. As these basic forms have been used to describe species since people first started categorizing species, they may (and have) been criticized as being far too simplistic to adequately capture the complexity of ecological systems. We have found these groupings to be useful and the preceding chapters are full of examples of how these life forms vary in their dynamics and responses. As seen in other community-scale analyses of plant traits (e.g. Navas *et al.*, 2010), the BSS trait data clearly show that life forms do separate out cleanly. While the dividing line separating perennial herbs from biennial/short-lived perennial species is certainly subjective, as a whole we see consistent separation. Within the herbaceous groups we see that annuals and biennials differ in having less clonal reproduction and biotic seed dispersal than their perennial counterparts. Of course their population dynamics are completely different. To separate the short-lived species, annuals have a much greater SLA than the biennials and have a much lower incidence of biotic pollination.

Within the woody species, we also see support for the separation of species into life forms. Trees in the system are clearly more likely to be wind pollinated, have less biotic seed dispersal, are largely non-clonal and tend to have the largest seeds of the woody species. Leaf characteristics are less clear, but trees have the largest leaves and appear to be at the extremes of SLA and LDMC for the woody species. One could argue that shrubs and lianas could be merged into one opportunistic woody-plant category. Both life forms consistently have biotic pollination and seed dispersal, and have very similarly sized seeds. Their separation appears along the spectrum of leaf structure, with lianas having larger leaves, greater SLA and lower LDMC. These differences are not large, but are consistent with the strategies of lianas to expand rapidly when conditions favor their growth and establishment (Ladwig and Meiners, 2010b). Our data also show clear separation between woody and herbaceous taxa. Woody species produce much larger seeds, are much more likely to have vertebrate dispersal vectors for those seeds and have the greatest percentage of leaf dry matter content. In combination with the population dynamics discussed in Chapter 6, there is clear separation from the herbaceous species.

The arguments for functional separation of life forms are based on *average* differences. Individual species can and do vary widely, as should be expected, and this leads to broad overlap among life forms. This in no way detracts from the usefulness of such groupings. The utility of functional groups is in generating aggregations of species that share basic characteristics and some aspect of their ecology. To be useful these groups must be easily definable for any system. We could easily take any of the ordinations presented in this chapter, break the axes into quadrants and generate functional groups. These groupings would be based on quantitative and qualitative trait data, but would still be plagued by the arbitrariness of group delineation. Most problematically, these groups would mean nothing to researchers working in other systems and would take an appreciable amount of data just to form the functional groups. Based on our experience, plant life form is very useful in quickly understanding the dynamics of the system and often suggests more detailed analyses that can yield mechanistic insight. Within the BSS, legumes and grasses are less important than in other systems, such as prairies (e.g. Tilman *et al.* 1997b), so these species have been pooled into other groups – largely perennial herbs. This approach may not be appropriate for all communities. Overall, we argue that plant life form is a simple and quick way to gain real ecological insight into a system even when much more complex functional trait data is available (Raunkiaer, 1934).

The dichotomy between herbaceous and woody species

We have clearly shown that the separation of woody and herbaceous species is justified, based on population dynamics and functional characteristics. This is not surprising, given the fundamental difference in allocation patterns involved in the production of woody tissues. However, is the functional separation too great to combine woody and herbaceous taxa in our analyses of functional traits? Perhaps based on convenience, some of the early trait work focused on either woody or herbaceous taxa. Much of Grimes' comparative ecology involved herbaceous species, though clearly the underlying principles should apply to both (Grime, 1977). The global leaf economics spectrum (Wright *et al.*, 2004) also started out looking at woody species based on the large data sets that were available. Many studies focus on woody (e.g. Comas and Eissenstat, 2004; Norden *et al.*, 2009) or herbaceous (e.g Fukami *et al.*, 2005; Vile *et al.*, 2006) communities and are able to examine underlying tradeoffs in traits and mechanistically understand community dynamics. Since then, some functional ecology studies have combined woody and herbaceous taxa (e.g. Navas *et al.*, 2010; Tecco *et al.*, 2010; Douma *et al.*, 2012).

A potential problem with the fundamental shift in plant strategy associated with the production of woody tissues is whether the underlying ecological tradeoffs are the same for woody and herbaceous species. One may view plant life forms as a gradient in longevity and associated species traits. In this view, the progression from short-lived to long-lived herbs transitions continuously to the weedier woody species. The underlying tradeoffs in allocation and life history are the same, so that a continuum of plant strategies is formed. We see this sort of perspective in the resource ratio hypothesis of succession, where communities and species transition from limitation by soil resources to limitation by light (Tilman, 1985). The continuity of life forms is also an underlying assumption to many of the analyses presented in this chapter. However, if the shift towards woodiness represents a fundamental change in not only architecture, but also the underlying selective pressures that generate plant strategies, then the rules that apply to one suite of species may not apply to the other, and combining woody and herbaceous taxa in a study may mask tradeoffs and other patterns. For example, high specific leaf area is linked with rapid growth rates in species. However, a rapidly growing woody species and a rapidly growing herbaceous species are likely to have very different SLA values, though they may both have other characteristics that maximize resource acquisition. Studies that pool woody and herbaceous taxa run the risk of causally linking traits associated with early successional woody species to growth. If the tradeoffs between traits are different, then the issues become even more complex. As herbaceous species are much more numerous in the BSS data, as in many plant communities, their patterns will drive patterns at the community scale.

In an attempt to assess whether woody and herbaceous species can be analyzed together, we took all of the traits presented here and ran a separate set of correlations for each species pool. The purpose of this was to determine whether the tradeoffs and associations among traits were similar for each suite of species. To test for overall association in the patterns, we then correlated the pair-wise correlation coefficients. The pattern of association among traits was statistically significant ($R = 0.49$; $P = 0.0078$) suggesting that overall, the same tradeoffs operate across woody and herbaceous species. This supports our usage of pooled trait data. That being said, the analyses also found some very large and perhaps important discrepancies in associations that differ from the patterns seen in ordination analyses. Tests in woody species had less power

Table 12.1 Correlations between plant functional traits for herbaceous (top, unshaded) and woody (bottom, shaded) species. Values are Spearman rank–sum correlations. Significant correlations (at P <0.05) are indicated in bold. Values are not adjusted for multiple comparisons as we are interested in broad patterns only.

	Herbaceous species							
	Height	LA	LDMC	SLA	Seed mass	Biotic pollination	Biotic dispersal	Clonality
Height	—	**0.40**	0.14	−0.09	0.11	0.01	0.06	0.04
LA	**0.58**	—	−0.12	0.12	**0.31**	0.06	**0.24**	−0.10
LDMC	0.27	0.20	—	**−0.49**	**−0.30**	**−0.26**	−0.05	0.19
SLA	−0.12	−0.09	**−0.62**	—	0.20	0.08	0.16	−0.14
Seed mass	**0.50**	**0.39**	0.09	0.06	—	0.01	**0.31**	−0.23
Biotic pollination	**−0.56**	**−0.41**	**−0.44**	0.15	**−0.39**	—	0.23	−0.15
Vertebrate dispersal	−0.21	−0.25	0.03	−0.06	−0.02	**0.44**	—	−0.07
Clonality	−0.22	−0.11	−0.06	0.10	**−0.48**	0.21	0.18	—
	Woody species							

because of the smaller sample size, but this does not appear to be sufficient to explain the level of divergence seen.

Similarities: general plant strategies – Examining the common tradeoffs in traits across both woody and herbaceous species allows us to examine not only patterns that are invariant, but also allows us to form linkages among disparate traits (Table 12.1). This approach is critical to generating a comprehensive view of functional plant strategies. Leaf area was positively correlated with both height and seed mass. Taller, and therefore larger, plants would be able to overtop smaller plants and perhaps larger leaf surfaces would allow competitive suppression. Such large leaves would not be energetically feasible in species consistently in a subordinate position within the community. Having greater seed reserves would allow these species to consistently reach higher into the plant canopy. We also saw a consistent negative association between SLA and percentage leaf dry matter. This tradeoff follows the general leaf energetic spectrum with high SLA plants growing rapidly and allocating relatively little to the structural components of leaves.

Interestingly, biotic pollination and seed dispersal were linked in both herbaceous and woody species. Plants growing under conditions that favor abiotic pollination may also be expected to benefit from abiotic seed dispersal as well. While this pattern is consistent in woody and herbaceous plants, the situation under which the association occurred was quite different. Herbaceous species with abiotic pollination and seed dispersal are annual forbs, regardless of when they occur in succession, and grasses. Woody species with abiotic pollination and seed dispersal, in contrast, are composed of canopy trees that flower before the canopy closes and tend to disperse their seeds late in the season (with the exception of *Acer rubrum* and *A. negundo*, which disperse in the spring). In both cases, unique subsets of species generate the pattern, but these represent two very different ecological conditions. Also associated with biotic pollination was leaf dry matter. Biotically pollinated species tended to have lower leaf dry matter contents. Again, this may be linked with the identity of the species. For the herbaceous species, wind-pollinated grasses have very high dry matter content because of the silica and other structural compounds in their leaves. Similarly, the leaves of trees such as *Quercus* spp. and *Carya* spp. have large amounts of tannins and are physically tough.

Perhaps most interesting is the negative association between clonality and seed mass. We mentioned this difference briefly earlier in this chapter, but it is worth re-visiting. This pattern represents a tradeoff between two fundamentally different plant strategies. Plants that rely on vegetative expansion for the majority of their population growth within sites do not need to allocate resources to producing large, high-quality offspring (Handel, 1985). Instead, these plants produce small seeds that may lie dormant for long periods of time until conditions are favorable for germination, or disperse into new habitats. As an example, *Solidago canadensis* is one of the dominant clonal perennials in the BSS. Seedling establishment initially leads to a genetically diverse population, but clonal interactions quickly lead to the loss of genotypes and dominance of a few (Hartnett and Bazzaz, 1985). For this reason, the number of individual seedlings is not limiting to the population. We see similar patterns within woody species such as *Rhus* spp., *Rubus* spp. and *Toxicodendron radicans*, where small, bird-dispersed seeds produce expanding clones that form large thickets in successional communities.

Differences: form-specific strategies – Though similarities in associations and tradeoffs suggest commonalities in the underlying constraints that form plant strategies, differences between woody and herbaceous taxa suggest potential shifts in selective forces. If these shifts are sufficiently large, the development of separate woody and non-woody functional approaches may be necessary. Seed mass was strongly associated with potential height in the trait ordinations (Figures 12.5 and 12.7). When separated, seed mass had the same strong positive association with height in woody species, but there was no association in herbs (Table 12.1). The much greater range of heights in woody species may have driven that pattern in the ordination, though it makes sense from a successional perspective with the shift to tree dominance. Biotic pollination was also negatively associated with both height and leaf area in woody taxa, while there was no pattern in herbs. The decrease in biotic pollination with height represents the overall inappropriateness of wind pollination in understory environments and so makes sense from an ecological perspective. Wind-pollinated herbs largely occurred early in succession, when being overtopped by other plants was less of an issue. The ecological explanation for why biotic pollination should decrease with leaf area in woody species is not clear. This may be an artifact of the dominance of large-leaved canopy trees with wind pollination (e.g. *Carya* spp. and *Quercus* spp.) in the species pool, whereas biotically pollinated species have a much broader range of leaf areas.

Within herbaceous species, seed mass was negatively correlated with LDMC, in contrast to the pattern seen across all species. This pattern may be driven by the relatively large size of some annual seeds (e.g. *Abutilon theophrasti*, *Polygonum pensylvanicum* and *Raphanus raphanistrum*), which also have much less allocation to leaf structure. Seed mass and LDMC were unrelated in woody species. We also saw a positive association between biotic seed dispersal and both leaf area and seed mass in herbaceous species. Vertebrate-dispersed herbaceous species tended to have larger leaves, while vertebrate-dispersed woody species, if anything, had smaller leaves. Vertebrate dispersal was overall rare in herbaceous taxa so the pattern was largely driven by a few species such as fleshy fruited *Fragaria virginiana* and *Solanum carolinense* and adhesively dispersed species such as *Polygonum virginiana* and *Desmodium* spp. Abiotically dispersed woody plants were also relatively infrequent and would be driven by canopy trees such as *Acer* spp., *Fraxinus* spp. and *Ailanthus altissima*. Large-seeded herbaceous species would require dispersal by a mutualist as stem heights would greatly limit the spread of wind-dispersed seeds of a similar size. In contrast, canopy trees can effectively rely on either wind or vertebrates for dispersal with much less limitation from seed mass.

Do the differences really matter? – What does all of this mean for whether studies of functional traits should incorporate both woody and non-woody species simultaneously? As the overarching goal of this book is to understand the ecology of successional communities, there is a strong argument for including both fundamental types of species as the successional dynamics of the site include both. The differences between the trait relationships of woody and herbaceous species appear largely to be driven by the idiosyncrasies of the dominant species of the site, particularly the relatively small pool of canopy trees. This highlights the role of differential species availability in regulating the patterns seen within communities. Each of the differences was readily understandable and does not appear to represent a major shift in trait relationships. In contrast, the similarities appear to represent much more fundamental tradeoffs that should be consistent among sites. Variation in these associations would be much more troubling. Whether or not woody and herbaceous species form a continuum of plant strategies, they clearly transition in dominance within the BSS.

This does not mean that merging woody and herbaceous species in all functional studies is appropriate. When the goal is to understand broad-scale patterns in traits, and the physiological and evolutionary basis of these patterns, it may be necessary to focus on each separately (Tecco *et al.*, 2010). This approach will allow the inclusion of traits specific to each group. For example, wood density is linked with structural allocation in trees, while C3/C4 photosynthetic processes and timing of peak biomass captures the phenological aspects of herbaceous plant development. Including these traits may allow researchers to fine tune their studies to encompass the range of characteristics important to their system. In systems where a common pool of traits is desirable, researchers should, at the least, test whether the combination of woody and herbaceous species is justifiable. Ultimately, we need studies designed to address the functional shifts between woody and herbaceous species. Studies not dependent on the stochastic nature of differential species availability will give a broader view of trait relationships and provide the full range of species variation.

Conclusions

After all of the data and analyses, is a functional approach to succession a useful one? Though not all traits provided clear views of successional dynamics, there is a strong utility in examining succession functionally. The traits that have been explored here are some of the more common ones studied in plant communities and are the ones thought to be critical across broad spatial scales. The data show strong successional patterns that are now linked mechanistically with the underlying drivers that generate dynamics. On top of these functional patterns we see superimposed mechanisms of stochasticity, such as dispersal limitation and historical contingencies. Though these factors may at first glance appear to be annoying sources of noise, they are in fact commonplace drivers of local communities and belong in the discussion of functional ecology as constraints on more deterministic processes. Another useful feature of functional ecology was in examining species patterns to understand the linkages among traits. These patterns represent the strategies shaped and constrained by evolutionary pressures and provide opportunities to link community and evolutionary processes. If we return to Grimes's words that began this chapter, the value of these approaches lies in their ability to translate to other systems. The patterns generated here, though perhaps unique in temporal scale, should be testable in other communities. We need to fully explore the potential of functional approaches, while determining the boundaries of their utility in ecology.

Part 4

Synthesis

In the final section of this text we will take a step back and take a brief look at all we have done during the process of generating this book. We will highlight the lessons learned, the perspectives gained and the surprises uncovered. This section will not repeat the previous section's detail unless necessary as an example. Instead, we will use the perspective that we have gained to think more broadly about ecological systems. This book started out with a call for integration in ecology, which we have tried to adhere to in the preceding chapters. While we originally anticipated a single conclusion chapter to this text, the richness of the BSS data generated two different, and worthwhile, perspectives that we wanted to share.

The first of the synthetic chapters will address habitat restoration and management. We have discussed the applied utility of the BSS data in several publications over the years, but these have all been quite specific in their scope. Developing the preceding chapters led to a wealth of material that can be applied to the management of plant and maybe even animal communities. In this chapter, we will think about succession and restoration as related processes and highlight how management interventions interact within the conceptual framework. After all, restoration is perhaps the most integrative of ecological disciplines.

Finally, we will revisit our call for integration in ecology and the utility of our conceptual framework as a mechanism of integration. We will highlight a few of the key findings across this text, but will mostly focus on how the conceptual framework has allowed us great insight into how contingencies are critical to all aspects of succession. The framework explicitly portrays those contingencies in its structure. While successional systems inherently defy predictability in a strict sense, understanding the contingencies that relate processes allows understanding. We think that is even better.

13 Succession, habitat management and restoration

"From the very nature of climax and succession, development is immediately resumed when the disturbing cause ceases, and in this fact lies the basic principle of all restoration or rehabilitation."

Frederic Edward Clements, Experimental ecology in the public service (1935)

The linkage between succession and habitat remediation is an old one. If we go back to our view of succession as being a special case of community dynamics in a broader sense, this linkage becomes even more clear. By attempting to alter community composition, we are actually augmenting the forces that structure communities, successional in a strict sense or otherwise. Based on this supposition, the BSS and the conceptual framework that we have used to structure this book should provide useful insights into ecological restoration and management. The quote that starts this chapter illustrates the strong linkage that Clements viewed to exist between succession and restoration. With the relatively limited contemporary focus on Clements' work in succession, it is easy to forget the broad range of work that he was involved in and its societal context (Eliot, 2007). Clements worked extensively in and around the dust bowl – one of the largest ecological and social disasters to plague the United States. The combination of a naturally occurring multiyear drought and a history of irresponsible farming practices led to massive erosion, dust storms and human suffering. Clements' ecological philosophy, in particular his successional perspective, shaped his responses to these challenges. He advocated land-use practices more in alignment with local environmental limitations, including episodic droughts, and habitat modifications that would remediate and prevent further environmental degradation.

Following this rich tradition, we will explore the relationship between restoration and succession in this chapter. We will start by looking at how restoration and succession have been linked to provide a broader context for the importance of this approach. We will then re-visit the BSS data presented throughout this book to glean how this information may be translated from the scientific context that it was originally presented in towards a more applied milieu. Our goal here is not to provide any hard and fast rules, but rather to point out opportunities and challenges that may help to inform remediation efforts. Finally, we will re-visit our conceptual framework and see how that can directly inform restoration and management efforts. We will specifically explore how this framework will help to form linkages between succession and restoration practices, with the goal of highlighting contingencies in the restoration process (Pickett *et al.*,

2001). By developing the conceptual context for restoration, we hope to inform both fields of study to their mutual benefit.

Succession and vegetation management

Following the early work of Clements that integrated succession into applied ecology, there was a relatively long gap before succession was again identified as a major player in habitat management. Jim Luken's book, *Directing Ecological Succession* (1990) came out at a time when big ideas were developing in conservation biology, as major ecological concepts were being placed into an applied setting. Perhaps the newer, snazzier conservation approaches overshadowed his important work, but the baseline for working with successional processes was clearly set. While applied ecologists latched onto the conceptual groundwork within conservation biology, succession again appears to have been relegated to a minor role. However, as restoration ecology has emerged as a separate discipline of conservation biology, succession has re-surfaced as an important concept, though one that is still in need of further development. Part of this delay in integration may have been driven by the population perspective adopted by many conservation biologists – a contrast to the community focus of many restoration projects. Restorations also tend to focus on plants, further moving people's focus towards vegetation dynamics. Building from this, there have been many recent calls for the importance of succession to ecology (Prach *et al.*, 2001; Davis *et al.*, 2005; Cramer *et al.*, 2008; Prach and Walker, 2011), as well as several books that focus on the applied aspects of succession (Walker and Del Moral, 2003; Cramer and Hobbs, 2007; Walker *et al.*, 2007). All of these works continue to call for further integration of succession into restoration and management, a clear indication that there is much work yet to be done.

There are some broad themes and generalizations that are emerging. Remediation and vegetation dynamics on severely disturbed or environmentally harsh sites will likely be constrained by processes more similar to primary succession and will draw from that research baseline (Walker and del Moral, 2009). On the opposite end of the stress gradient, restoration of more fertile sites with moderate climates is more likely to be controlled by processes more strongly affiliated with secondary succession. This relationship is partly driven by the dominance of the ecological literature by secondary succession on mesic, fertile sites (Cramer *et al.*, 2008). Local contingencies may be stronger under more environmentally severe conditions, making prediction and perhaps control more difficult. Clearly this level of generalization is a beginning rather than a confirmed foundation.

In practice, there are two major approaches in how successional processes can be linked to restoration activities – active and passive (Prach and Hobbs, 2008). In the active approach, understanding succession can suggest direct management interventions; whereas in the passive successional approach, the natural processes of succession lead towards the target vegetation with no manipulation – typically referred to as spontaneous succession. The key to evaluating the active vs. passive dichotomy is in understanding the contingencies that constrain successional processes. The relative value of each approach depends on the likelihood of the passive approach resulting in the target community within an acceptable time frame (Prach et al., 2001; Prach and Hobbs, 2008).

Spontaneous succession has many specific qualities that make it an effective default mechanism for habitat modification – primarily that it is free and that succession intends to proceed whether we want it to or not. Much of the support for the hands-off approach to restoration comes from surveys of disturbed habitats and their recovery (Ruprecht, 2006; Reinecke et al., 2008; Řehounková and Prach, 2010; Prach et al., 2014). While such surveys document that vegetation dynamics does lead to some sort of successional trajectory with no remediation, it does not determine whether the development or composition of that target community would have been enhanced with remediation efforts. However, there are some clear tests of the benefits of an active approach. For example, restoration of grasslands can be achieved through direct seeding or through allowing species to recolonize naturally. In one study of 35 European sites (Lencová and Prach, 2011), seeding did produce more plant cover more quickly. However, the composition of targeted grassland species did not differ between active and passive remediation. This would seem to argue universally for a passive approach to restoration, but there are a few major caveats. The success of passive restoration is strongly dependent on the local availability of target species to colonize the site. In isolated restoration sites, this is not likely to be true. Furthermore, the site must not retain a legacy of fertilizer residues that increase soil fertility.

Key to all of the lessons that succession can provide to restoration is the perspective that contingencies are critical. For example, many studies have found legacy effects of previous land use on vegetation structure and composition that last for decades to a century (Dupouey et al., 2002) or more (Foster, 1993; Standish et al., 2006; Flinn and Marks, 2007; Kuhman et al., 2011). Historical contingencies are what a site inherits, but there are also contingencies introduced by adjacent vegetation, weather patterns and heterogeneity within a site. These contingencies make passive vegetation restoration difficult, and perhaps too risky in many situations (Pickett et al., 2001). Ultimately passive restoration is only useful where the restoration target is a self-replacing community that succession will lead to deciduous forests in the case of the BSS, semi-natural grasslands in the European example above. Wherever a successionally intermediate stage is the target community, passive restoration will be a tool early in the process and a hindrance once the target community has been achieved. We will return to the idea of contingencies at the end of this chapter to explicitly link successional processes and restoration activities so that contingencies may be highlighted.

Application of the BSS to restoration and habitat management

In addition to the simple assembly rules suggested in Chapter 9, there are other insights that the BSS can provide for restoration and habitat management. Most of these revolve around identifying factors that may limit the successional development of vegetation or challenges when retarding succession is the management goal. These issues are by no means an exhaustive list, but are the strongest signals that the BSS provides.

Assessment of habitat quality – Often agencies want rapid ways to evaluate a site so that the conservation priority or management needs can be assessed and compared to other sites. These typically involve the rankings of structures or species on a scale of perceived desirability. For example, stream fishes are ranked based on their tolerance to disturbances, and poor water and substrate quality by comparing impaired watersheds with more pristine reference streams. The species located within a site are then combined in some manner to generate a composite score for that site. Within the US, metrics have been developed for stream invertebrates (MBI- Macroinvertebrate Assessment Index), stream fish assemblages (IBI – Index of Biotic Integrity), habitat structure (SHAP – Stream Habitat Assessment Protocol) and plant species composition (FQI – Floristic Quality Index) among others. While these metrics were designed for, and work well in, rapid assessment, they are becoming applied in more and more situations, necessitating a better understanding of their responses.

Within the FQI, species are ranked based on their conservatism – how tightly they are associated with high-quality, undisturbed plant communities (Swink and Wilhelm, 1994; Taft *et al.*, 1997). Non-native species are assigned a score of 0, habitat generalists low scores and habitat specialists, particularly less common species, are assigned the highest scores. Of course, to be useful these rankings must be specific both geographically and to particular community types. Vegetation dynamics pose issues for the use of such metrics as the compositional changes will likely cause changes in a site's scores as early successional species tend to be widespread, and therefore rank low on the quality scale.

In a detailed analysis of BSS data using FQI measures, Spyreas *et al.* (2012) found that the exact method of calculating a site's score did not change how FQI changed with succession. However, there were quite large increases in FQI during succession, particularly over the first 40 years. The temporal dynamics of this metric poses a problem for its implementation. To adequately rank sites based on their floristic quality, it will also be necessary to know something of their successional status. Two sites with similar FQI values that differ in successional age may truly represent different conservation opportunities. A young site with a disproportionately high FQI for its age may represent a community on its way to being a conservation priority. In contrast, an older site with a lower than expected FQI would likely represent either the failure of restoration practices or the existence of significant management needs. Unfortunately, the necessity to place FQI into a successional context severely limits its applicability in disturbed habitats, as the history of a site is not frequently known. Similarly, using these metrics to assess the success of restoration activities is difficult – one would need to separate the successional increase in FQI that would naturally occur in a site from the changes imposed by management activities. These challenges make the use of such rapid assessment techniques more difficult to implement, and – ironically – less rapid.

Invasions of non-native species – One of the major issues facing most land managers is that of invasive non-native species. Crucial to understanding the role of non-native species invasions in succession is determining whether they inhibit key successional transitions (Meiners *et al.*, 2007). In the BSS, there were several species that attained massive dominance and are considered regionally problematic species, particularly *Lonicera japonica* and *Rosa multiflora*. Neither of these species prevented the successional transition from more open shrubland to closed canopy forests in the BSS. That by no means suggests that these or similar species should be ignored in other systems. The abandonment of the BSS fields may have happened earlier in the spread of these species in the local landscape and perhaps these species attained lower cover than would have occurred if the fields were abandoned more recently, or perhaps colonization of the invasives was slow enough to allow trees to become sufficiently abundant to dominate the community. Assessments of impacts would need to be done locally to determine whether a particular invader was both abundant enough and able to reduce tree regeneration sufficiently to warrant intervention. The site-specific nature of such decisions is highlighted by the relatively low abundance of two other regionally problematic non-native species, *Lonicera maackii* and *Berberis thunbergii* at HMFC. Neither of these species has become sufficiently abundant in the site to inhibit forest regeneration, though they have been shown to have massive effects in other systems (Ehrenfeld, 1997; Hutchinson and Vankat, 1997; Gorchov and Trisel, 2003). Whether it is the species assemblage that is resistant to these two invaders or site conditions that have limited their performance, their potential to dominate has not manifested itself over the first 50 years of succession.

While early- and mid-successional non-native invasives did not prevent canopy closure, a second suite of species does appear to pose serious threats to forest understory development and potentially tree recruitment. Two species, *Alliaria petiolata* and *Microstegium vimineum*, have spread throughout the understory of all of the BSS fields and dominate the understory flora. Because these species have the ability to persist in the community as the canopy continues to develop, they represent a continuing pressure on the plant community. Similarly, two non-native tree species at the site have been able to continue to regenerate following canopy closure – *Ailanthus altissima* and *Acer platanoides*. *Ailanthus* is a shade-intolerant species that has established in forests throughout the US (Gómez-Aparicio and Canham, 2008). This species appears able to regenerate in the gaps that are beginning to form in the young forests of the BSS. *Acer platanoides*, in contrast, is a very shade-tolerant species that is regenerating in undisturbed forest understories, very similar to the native *Acer saccharum*, which also occurs at the site (Kloeppel and Abrams, 1995; Martin, 1999; Meiners, 2005). Placing species appropriately in the context of succession and (ultimately) forest regeneration allows management priorities to be set.

In addition to all of the non-native plant species is the overarching impact of an overabundant herbivore – white-tailed deer. This herbivore has left no place within the BSS unimpacted. It is impossible to determine whether the dominance of the late successional non-natives is because of their inherent competitive superiority or because of the suppression of natives by selective foraging (Baiser *et al.*, 2008). As both of these processes reside within differential species performance, they would appear similar in the BSS data without direct manipulation. Mitigation of the non-native species is dependent on the successful recruitment of other species to replace them – an unknown. Removal of the current suite of non-native invaders may merely open up an opportunity for a second suite of species that did not originally colonize the site.

Dispersal limitation – Species availability to a site is one of the primary drivers of successional dynamics in our conceptual model, so it seems appropriate that it also is important in restoration. Within the data we have shown in this book, the lack of species availability, or dispersal limitation, has appeared a few times. The early, somewhat random, assemblage of species that appears immediately following abandonment is rapidly supplemented with additional species as the community assembles itself, posing no management concern. As early successional species are often not conservation targets, lack of diversity at this successional stage would likely not warrant management anyway. However, later in succession, the impacts of dispersal limitation are much more pronounced and long lived. Woody species, particularly larger-seeded ones, do not move well throughout the site and are therefore patchily distributed near reproductively mature trees in the adjacent forest. If these large-seeded species are a management target, they may require assisted dispersal to colonize, particularly if the site is large. Adding these species into the community early on may in fact speed successional transitions as the windows for establishment are often brief (Rankin and Pickett, 1989; Debussche and Lepart, 1992; Bartha *et al.*, 2003).

Because of their availability in the landscape, the colonization of early successional species is rarely a management concern. Late successional species may provide a greater management challenge as their source populations are likely less abundant. The BSS is somewhat fortunate in that it is located adjacent to an old-growth forest that can serve as a source of propagules for forest species, native and non-native alike. As the anthropogenic habitat alteration that generates the need for restoration often fragments and isolates source populations from restoration sites, remediation of late successional target species may depend on installation of plants or seeds into a restoration. The timing of this supplementation will depend on the target community. Grassland restoration can include late successional species from the beginning and may even have greater success when this is done. Forest understory species will require some level of canopy development and loss of shade-intolerant competitors before they can be successful. Forest understory shrubs and subcanopy trees will likely need to be introduced at an intermediate period. While these species persist in shaded forest understories, they often regenerate in openings. Introducing these components of mature forests will need to balance their growth requirements with the competitive reduction in growth they will provide to the canopy trees as they also mature.

Managing plant dominance – Management activities are often targeted towards maintaining a diversity of species in a habitat. However, most of the diversity in a community is often contained within species that occur in low frequency and low abundance. The main challenge to rare species is the tendency of a few species to become physical and structural dominants in plant communities. To maintain diversity, management activities may therefore need to focus on preventing dominance to the level that would unduly inhibit the diversity of the system (Bartha *et al.*, 2014). This is particularly true in smaller habitat remnants, where the development of monodominant patches would impact a large proportion of the area.

From a functional ecology perspective, the species most likely to dominate over relatively large spatial scales would be those with the ability to clonally spread (Grime, 1977; Bartha *et al.*, 2014). These species are not dependent on seeds for their expansion within a community following their initial colonization, but instead spread via stolons, rhizomes or other vegetative mechanisms. What makes control of these species difficult is that the presence of an intact plant canopy may provide little resistance to their spread. In contrast, a well-developed plant community can strongly inhibit the expansion of species via seeds by reducing light and soil resources for newly emerging seedlings. Clonal species, through their integration with fully established ramets, are able to grow up through an established community and competitively interact with established plants. For this reason they are both more likely to spread and more likely to impact adjacent plant species.

Management in this context will need to focus not on the entire community, but specifically on those species with the tendency to dominate. In a wonderful example of such a targeted approach Collins *et al.* (1998) show the interplay between a non-specific disturbance and a selective herbivore in maintaining the diversity of tallgrass prairies. Fires are necessary to the maintenance of grasslands globally as they limit the establishment of woody species and maintain the dominance of grasses. However, the vast majority of diversity in grasslands lies not within the grasses, but within the dicots and non-graminoid monocots that also grow there. As the growth and competitive ability of C4 grasses is enhanced by fire, it also acts to reduce the diversity of the system. Therefore, the application of the most common management strategy of grasslands to prevent woody encroachment also inhibits its diversity. However, fires did not naturally occur in isolation in native grasslands; they occured with native grazers. Bison (*Bison bison*) are major sources of heterogeneity through their feeding and other activities (Knapp *et al.*, 1999). Bison selectively forage on recently burned grasses, reducing their dominance and therefore their competitive influence in grasslands. While increasing fire frequency reduces diversity, fire frequency has no impact on diversity when these grazers are present.

Not all systems will have a natural targeted grazer. However, direct management of individual species is possible. Chemical or mechanical removal of clonal species can be used to maintain them at an acceptably low abundance. One time mowing of a clonal woody shrub such as *Rubus* may not be sufficient to reduce its dominance, but multiple years of mowing, particularly in the growing season, may inhibit growth sufficiently. This is particularly true when the species is still concentrated in distinct patches. Likewise, targeted herbicide application, where allowable, can maintain species at levels where they pose limited risk to the community. The key to these approaches is to maintain potentially dominant species at moderate densities continually. Once they dominate a system, the management intervention is likely to be detrimental to the entire community, rather than focused on individual patches.

Stalling succession – One of the most difficult challenges in restoration or habitat management is likely to be maintaining mid-successional environments. Managing for early successional communities is easy – the systems just need to be disturbed, likely in a mosaic to allow the target species to maintain themselves across the entire system. Likewise, restoring late successional habitats involves enhancing the growth of persistent species in the absence of disturbance. The rate of restoration in this case will be determined by the rate of growth and establishment of the dominant species. Managing for somewhere in between these two is likely to be much more difficult. For example, succession to deciduous forests often passes through an open shrub phase. Some animal species, particularly birds, specialize on such habitats, making them a management target. However, succession rapidly tends to move these systems towards continuous woody cover.

Opposing successional forces to maintain open shrub habitats can take one of two forms. In large habitat reserves, whole blocks of woody vegetation can be removed on a rotation to maintain a mosaic of different successional phases within the landscape. Each individual patch may be too early or too mature successionally, but on the whole the system will reach the target goals. Management in this context is purely by disturbance on a shorter than average return cycle. However, most land units under management are not such large parcels or there is not sufficient coordination among units to allow such a process. In these cases management will need to focus on maintaining each location as a mid-successional community. Management strategies would then involve the removal of individual or patches of woody plants to maintain an overall open community with shrubs of a target density. This situation is much more difficult to maintain. As woody species grow larger, so does their reproductive output. This would likely result in an acceleration of woody-plant establishment over time, regardless of the number of individuals that were removed from the system. Similar challenges are likely in any management scenario where an intermediate successional target is desired.

Conceptually linking succession and restoration

Vegetation management and restoration are really ends of a spectrum of human interventions used to guide natural processes to generate a desired community composition or structure. On one end of the spectrum remediation may begin by completely denuding the vegetation of an area to install a new assemblage of desired species. This *tabula rasa* approach would clearly fit under the restoration realm of interventions. On the other end of the spectrum, activities such as removing small populations of non-native invasive species, maintaining a fire regime or thinning tree stems in a young forest certainly fit within the job description of most land managers. The problem comes in when you try to separate the two activities cleanly. How much intervention is necessary before management becomes restoration? Similarly, if we focus on outcomes, how much compositional or structural change is needed to shift from management to restoration? Restoration activities often explicitly specify management protocols to ensure reaching the target community. The two activities are clearly related and will not be separated here. Informing restoration also informs management and vice versa. Protocols for restoration often acknowledge the long-term need for subsequent management to ensure the success of the restoration. For this reason, we will treat restoration and management as the same process, which tries to exploit the same underlying ecological principles (Luken, 1990).

We can conceptualize restoration activities as attempts to alter one of the three primary drivers of succession (Figure 13.1). All management techniques appear to fit nicely into one of these categories. This structure suggests a mode of action for the intervention and potential interactions with other drivers of community dynamics in the system. There are not interactions among management interventions as these are

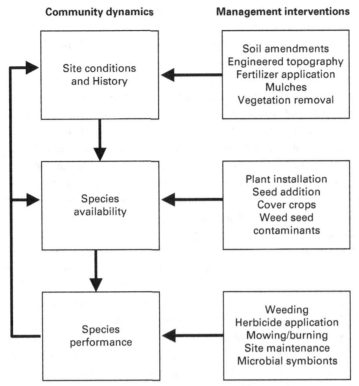

Figure 13.1 Relationship between management activities and the filter model of successional drivers. Management activities can be separated among each of the major successional drivers that they impact. Changes generated by these interventions can therefore be expressed across the entire system, but are also contingent upon other aspects of the system.

discreet activities employed in the site. However, these intervention activities can alter relationships among the drivers of community dynamics that may cascade throughout the system.

While site history is unalterable, the legacy of that history can be altered. In the most extreme conditions, this may represent physical alteration or removal of the soil. For example, to restore the diverse, low-productivity chalk grasslands of Europe, it is first necessary to remove the legacy of nutrient addition from agricultural inputs (Marrs, 1985; Walker et al., 2004). Similarly, to allow saltmarsh recovery following the removal of the invasive Brazilian Peppertree (*Schinus terebinthifolius*) in Florida, it is necessary to remove the accumulation of soil generated by the invasion, typically down to bedrock (Dalrymple et al., 2003). As agriculture often removes fine-scale topographic variation and increases drainage, re-contouring a site may be needed to generate fine-scale heterogeneity and encourage a diverse plant community to develop (Vivian-Smith, 1997). These are clearly more drastic interventions. Less intensive activities may include fertilization of installed plant material, application of mulch or other soil amendments, and watering. All of these activities set the stage for colonization and ultimate sorting of species within the restoration.

In unmanipulated systems, species availability sets the potential pool of species for the community. Management interventions often seek to control, or at least strongly influence, the pool of species in the community and their relative abundances. This is often achieved by seed additions, either into an existing community or to land that has had all vegetation removed. In situations where establishment from seed is deemed too slow or stochastic, the installation of established plants may be involved. There is an even greater potential to control species' relative abundances when installing plants instead of seeds. Remediation protocols will also often include the installation of cover crops intended to stabilize soils and facilitate the establishment of target species. In all of these mechanisms of intentional introduction, there is the potential for the introduction of non-target species. These may arrive as plants or seeds in containerized planting stock or as contaminants in seed mixes or soil amendments. These intentional and unintentional species introductions will augment the species already present in the site and those that will disperse into the site following treatment.

Activities considered to be restoration often focus on manipulating site conditions and species availability. Management, in contrast, often focuses on shifting the relative abundances of species away from a species composition that is undesirable, towards a composition that approaches the target community. This approach may utilize physical or chemical removal of unwanted species, weeding to reduce competition with installed or desirable plants, protection from herbivores and other types of site maintenance. For example, restoration of forests is often focused on the rapid establishment of trees. Installation of trees with little follow-up management yields poor plant growth and delayed canopy closure. For this reason, many forest restorations will mow or use weed fabric as a physical barrier to reduce competition with herbaceous plants and increase tree growth. Management efforts may also include removal of lianas that may reduce the growth of small trees (Dillenburgh et al., 1993; Fike and Niering, 1999; Schnitzer and Bongers, 2002; Ladwig and Meiners, 2009). Similarly, grassland restorations are typically burned to further reduce the abundance of unwanted species after an initial period of establishment. Interventions designed to influence species performance may need to be relatively short term, effecting results in one or a few treatments, such as mowing to reduce weeds early in the establishment of grasslands, or may transition into long-term management practices, such as the management of invasive species populations.

An active area of research in community dynamics is the role of soil microbial communities in regulating both the dynamics of a system and the relative abundance of species (Klironomos, 2002; Allison et al., 2005; Harris, 2009; Bever et al., 2010). As net plant–microbial interactions are often negative, they tend to reduce the dominance of species, leading to a more diverse plant community. The more soil microbial communities are investigated, the stronger the argument for their importance in understanding plant communities becomes. Unfortunately, most restorations do not take this potentially important interaction into account. In many cases the soil microbial community has been dramatically altered, or may be largely absent. In the restoration of former croplands, the physical disturbance of plowing and the long-term maintenance of a species-poor crop community may result in an altered soil microbial community. Succession and restoration in such sites will favor species that utilize mutualists that are associated with crop species and are not affected by microbes that are antagonistic to the crops (Reynolds et al., 2003; Anderson, 2008; Faber and Markham, 2012). Soil microbial communities may recover over time, or can be augmented with soil inoculations from undisturbed sites, though this is often not feasible (Middleton and Bever, 2012). While this potential driver is a part of site conditions, it manifests itself in differential species performance, highlighting again how these mechanisms interact to generate plant community dynamics.

Placing restoration and management activities into our conceptual framework highlights their role in shaping the community and how they may interact with the other successional drivers. Each management intervention is discreet in its direct impact, but has the potential to cascade throughout the system. Soil amendments at the initiation of a project will interact with species installed in the site to (hopefully) increase the performance of those species and decrease the performance of weeds or other non-target species. The establishment of those dominant species in the community may then go on to alter site conditions and species availability. For example, installing trees to restore a deciduous forest may involve the addition of mulch or tree tubes to reduce competitive effects on tree growth. As the trees mature, shade from the canopy will increase, altering resource availability and the understory community. The development of a tree canopy will then represent both an environmental opportunity for shade-tolerant forest species and perch sites for birds dispersing into the site, ushering in the next round of species interactions and compositional shifts.

One key feature of this conceptualization is the linkage of restoration with the inherent dynamics of a system. Restoration alters one or more individual pieces of the system, but the underlying dynamics of the system still operate. Such activities influence community dynamics, but do nothing to stop the other forces exerting influence and the resulting constant changes to the system. All restoration activities must be placed in the broader context of dynamic communities. The constantly changing nature of plant communities means that restorations will as well be constantly changing. Restoration planning must account for these dynamics so that variation in the resulting community will still meet restoration goals. Successional processes will continually shift systems, and in some cases, this change will be in opposition to restoration goals. For example, restoration of shrubby habitats or meadow/grassland communities in areas that can climatically support forests will constantly require intervention. Successional processes in these systems will lead towards the establishment of more and more woody species, and the loss of open environments. Similarly more xeric forest types naturally shift towards mesic species – compositional change without structural change (Auclair and Goff, 1971). Successional processes may lead towards or away from the target community. Planning for these changes and challenges in the context of community dynamics can inform management plans or generate more successionally viable target communities.

14 Where we stand: lessons and opportunities

> "Nothing in community ecology makes sense except in the light of succession."
> Steward T. A. Pickett. Symposium on Ecological Theory, presented at the
> Annual Meeting of the Ecological Society of America (2008)

We come full circle and finish this book with a quote included in the first chapter. This quote captures our view of succession as a process that can integrate much of ecology. This was originally presented with a multidimensional Venn diagram used to highlight the interactions and overlap with other areas of ecology (Figure 14.1). The utility of this diagram is that it illustrates the ability of different groups of scientists to look at the same body of knowledge and view it in a vastly different way. By rotating the figure, a population ecologist can place allocation front and center as the integrative theme that unites all of ecology. As community ecologists, we would of course protest, but the perspective is valid. Similarly, a landscape ecologist could also view their discipline as the primary focus of importance, and succession as an important component to understanding patch–patch interactions and temporal dynamics in the landscape. Rotating the diagram would also bring other processes towards the front; processes that are currently hidden by other components of the system. For example, selective pressures are important in regulating allocational differences among species, but would be secondary to the study of succession. Accepting such different perspectives is important to scientists and to science – as long as the connections are clearly made. This is how integration occurs.

As we near the end of this volume, we re-visit the initial ideas that spawned this book, as well as the goals that we set for ourselves. We start with another look at the conceptual framework that has formed the organizational structure for the treatment of succession we present. We then discuss the relevance of continued research on succession to modern ecology and societal concerns. Drawing from the BSS, we highlight a few of the key lessons that we have learned along the way. Finally, we look forward into succession to see what still needs to be done for our understanding of dynamic systems to continue to develop.

The utility of a conceptual framework

In the preceding chapters, we have developed a detailed picture of how this successional system has changed over time. However, it is not the detail that makes this study valuable, but rather the understanding and the conceptual framework that it has allowed us to develop. In using this framework, we can understand how such changes to the

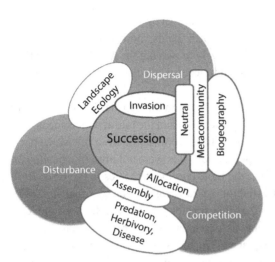

Figure 14.1 Multidimensional Venn diagram illustrating the interconnectedness of succession to other aspects of ecology. By rotating this diagram, different aspects of ecology become front-and-center, but are still connected to all other aspects. Appreciating these relationships is foundational to integration.

system may resonate through the plant community and lead to changes in composition and trajectory. Similarly, the conceptual framework can be applied to other successional systems with different conditions, species and physiognomies as a way to organize the processes that occur during succession. Theory is a way to organize thoughts to generate understanding (Scheiner and Willig, 2011) and does not necessarily always generate prediction. The conceptual structure of this book is our guiding theory to help us understand the intricate contingencies that thread throughout the system. Prediction would be nice, but understanding is better. So how has this framework allowed our understanding to grow?

Despite the unconstrained dynamics that followed the initial experimental abandonment of the BSS fields, we were still able to isolate the three broad drivers of community dynamics – history, differential species availability and differential species performance – and find patterns specific to each. Our ability to detect site history as a controlling factory was perhaps the weakest, but history still was important in understanding assembly rules. Differential species availability constrained some processes, but was unable to influence others. Species availability constrained patterning of woody species across the BSS fields, while dominance of non-native species towards short-lived species was unable to alter successional trajectories in life forms. Finally differential species performance, the area with the greatest contemporary and historical research concentration, resulted in the strong ecological sorting of species and life forms. It would be tempting to argue that differential performance was the primary determinant of community dynamics in this system. However, the framework highlights the importance of context. We documented one particular pattern of succession that was specific to this site history, the five decades of landscape change since 1958 and the particular species able to colonize. If, in a fit of optimism, we experimentally abandoned another 10 fields adjacent to the BSS fields, how would these new successional dynamics play out? Would it be simply re-playing the tape? Or would the hypothetical BSS II develop a new successional trajectory, one that is contingent on the new context of the system.

The presence of other successional communities (the BSS) close to the new fields would certainly alter the local species pool and the rate at which species may be able to colonize. This is particularly true for several non-native invasive species that have now formed persistent local populations that were not present when the original fields were abandoned. In contrast, the most dominant species in the BSS, *Rosa multiflora*, now has an established infection of rose rosette disease that has decimated local populations. It is unlikely that this species would be able to colonize and break up the dense canopy of herbaceous perennials that typically dominate after the short-lived species have been displaced. The landscape context of the site has also changed dramatically from a largely agricultural matrix to one dominated by housing and the different mix of non-native species that are favored in suburban developments (Chapter 2).

Site conditions have also certainly changed over the last 50 years. There have now been decades more acid and nutrient deposition to alter baseline soil conditions. If the new fields were in agricultural production in the intervening period, they would have also been subjected to decades more plowing and potential topsoil erosion. Perhaps more importantly, there would have been decades of inorganic fertilizer application, insecticides and fungicides, and likely crops genetically modified to resist herbicides. This certainly would change, not only abiotic soil conditions, but the microbial components of the soil that can be so important in regulating plant success. Finally, the decades since the BSS was started have witnessed a massive increase in white-tailed deer (*Odocoileus virginianus*) populations. The establishment of trees and other woody species may be much more constrained in the new site. Based on these changes to the system, even if patterns of differential performance remain unchanged, we would not expect the vegetation of the hypothetical BSS II to develop in similar ways to the BSS. This does not diminish the utility of long-term studies such as the BSS as a resource for understanding.

A further argument for the utility of our framework comes from its generality – similar to rotating the Venn diagram at the beginning of the chapter. The structure, comprising the three broad classes of successional drivers, is robust and can form the basis for examining other processes in plant communities. Within this volume, we have modified our framework to examine both non-native species invasions and habitat restoration. To understand invasions we mapped current theories that explain the success of non-native species onto the framework and found invasion theory to be largely analogous to drivers of succession. Similarly, we drew parallel frameworks for natural processes and management interventions to illustrate the interaction of restoration and succession. Again, this confirms the generality of succession to understanding plant communities. As all communities are dynamic at some temporal and spatial scale, succession provides key insights into the processes that regulate them.

The conceptual framework presented here is necessarily broad and so can form the basis to understanding many different systems. Starting with the most inclusive list, one can select the individual components within each of the three broad drivers that make sense for the individual system. For example, if understanding the dynamics of isolated patches of vegetation surrounding individual shrubs in an arid ecosystem is the target, the conceptual framework can be applied at a local scale. Site conditions would revolve around heterogeneity in (or differential) soil conditions cause by the shrub – likely enhanced deposition and capture of leaf litter, leading to a patch of higher organic matter in the soil, greater water retention and nutrient availability. Similarly the canopy of the shrub would generate variation in microclimate and perhaps even capture particulate nutrients. Differential species availability between shrub and non-shrub patches may be driven by the capture of wind-blown seeds or habitat selectivity by dispersal agents. For those species that colonize each type of patch, performance will be determined by species' response to local microenvironmental and soil conditions. Competition among the colonizing species as well as with the nucleating shrub may also be important in determining final composition. This multidriver approach is necessary to truly understanding the underlying dynamics of the system. Knowing whether colonization is equivalent across patches is necessary to deciding whether differential species availability, performance or both operate to generate patches of higher diversity associated with shrubs. If colonization is equivalent, then species performance in the patches must drive the system. If colonization is different, then performance differentials may further select species able to thrive under the local conditions or may be neutral in determining composition.

When such a research effort is completed, likely through several individual studies, a relatively complete picture of the system and its controllers will result. More importantly, from the perspective of deeper understanding, it is a model for how shrub–matrix interactions occur that can be applied to other systems. This leads to bigger questions, bigger ideas and (hold your breath) integration. One could address such questions as: Are the primary drivers of patch dynamics similar when grasses are the nucleating structural plant? Do the relative importance of drivers shift with latitude/rainfall/soil texture etc.? and How would the introduction/removal of grazers/dispersers alter patch dynamics and the structure of the system? Once this level is achieved, real predictions can be made about how systems are controlled. This is the ultimate value of a conceptual framework that is broad, inclusive and *not* system-specific.

Relevance of succession

Is succession still relevant today? In Chapter 1 we argued that it was relevant based on the large and increasing number of areas that are undergoing succession and their utility as model systems. Chapter 13 focused on the linkages of succession with restoration and management of plant communities and the animal communities that they support. The discussion of the utility and application of our conceptual framework just completed above also suggests that understanding succession leads to understanding of plant communities in general. These ideas are fairly well established. Looking forward, are there any other areas of ecological understanding that succession is likely to contribute to?

While Chapter 10 addressed how native and non-native species function in succession, it did not fully address the potential importance of successional systems to dealing with invasions. Because of the dynamic nature of successional systems, they can be used to rapidly assess the potential impacts of an invader on resident species. Based on these data, it may be possible to prioritize management of non-native species by their potential impacts. While plant–invader interactions in a successional community may be different from those found in older ecosystems, an aggressive invader in a successional system is likely to pose issues in conservation areas whenever disturbance is involved (Meiners and Pickett, 2013). Such prioritizations would not likely be perfect, but they would at least start from documented impacts.

Perhaps more importantly, successional communities can serve as bellwether systems for the detection and identification of the next suite of non-native plant invaders. Successional systems, with their disturbance history and frequent transitions in composition and structure, provide many opportunities for the establishment and spread of non-native species. Successional systems are also typically embedded in fragmented landscapes with anthropogenic disturbances and sources of new species introductions (planned or unplanned). For this reason, successional systems are often more likely to be initially colonized than remnant native habitats and may provide opportunities for early spread of species. Monitoring of successional habitats may allow early detection of non-native species before they begin to threaten areas of conservation concern.

More broadly, successional systems can act as the community equivalent of a phytometer (a phytocoenosis-ometer?) for a diversity of human-mediated changes to the environment. The dynamic nature of successional systems means that they should respond quickly to changes in conditions such as rainfall, nutrient deposition and growing season length that are expected to continue to change as a result of human activities (Likens, 1991). For example, while global climate change is expected to slowly lead to shifts in canopy tree distributions (e.g. Iverson and Prasad, 2002), successional opportunities will likely accelerate such changes by generating opportunities for regeneration of whole stands under the new climactic regime. Similarly, we know through manipulative experiments that fertilization, particularly with nitrogen, can alter successional dynamics (Vitousek, 2004). We may therefore expect successional transitions to be altered as atmospheric deposition continues. Changes in early- and mid-successional communities may be precursors to broader human-induced changes in vegetation. Therefore, succession can provide experimental opportunities for mechanistic studies now, and early warning signs later.

Lessons

Perhaps the major lesson from this exploration of the BSS long-term data is that contingencies rule processes (Box 14.1). While the conceptual framework specifically

> **BOX 14.1** The major lessons from exploring the BSS data using our conceptual framework
>
> (1) Contingencies rule – therefore succession continues to surprise
> (2) Population processes are non-additive
> (3) Non-native species exhibit complex, scale-dependent patterns through succession
> (4) Life forms work surprisingly well as an integrative functional group.

captures those contingencies in its structure, it was still surprising how often processes were dependent on each other. Thus, contingencies not only rule, but they are intertwined. More importantly, they are responsible for what appears on the surface to be the inherently unpredictable nature of community dynamics and succession. They also generate the fertile bed of opportunity that succession still represents for ecology both conceptually and empirically. Below are a few of the prime examples of the other lessons that a detailed foray into the BSS data has yielded.

Dynamics in the BSS revealed that the behavior of constituent populations is only a part of what generates the dynamics exhibited by the community as a whole. The non-additivity of population processes is perhaps why some of the broader generalizations made about succession were not supported in the BSS data. Odum and Egler, with their detailed expectations of what succession would generate, were partly right and partly wrong. This should perhaps not be surprising as the broad generalities expressed in their ideas on succession really could not capture the mechanistic contingencies inherent in such systems. Other expectations, such as the rate of succession and successional changes in stability were also not met by the data. Succession is clearly not as well-known an entity as has been suggested.

Non-native species exhibited complex and somewhat contradictory dynamics over succession. At the community scale, they became proportionately less abundant in the system, not because their total abundance decreased, but rather because the total abundance of native species increased. Their relatively constant absolute cover in the system is surprising, given the complete compositional turnover that occurred during succession. Diversity of native and non-native species also exhibited different successional trajectories, with native species following predictions of the intermediate disturbance hypothesis and non-native plants exhibiting the opposite patterns. These differences would suggest that native and non-native species are doing something quite different in succession. However, at the scale of individual species, we see overall equivalence of native and non-native species in the BSS. Of course, many North American native successional dominants are invasive elsewhere in the world, so this should perhaps not be surprising. Equivalence here does not argue that non-native species invasions are not a conservation concern. Dominance by a non-native always means that one or more native species have been displaced locally. That, in and of itself, is justification for management. Equivalence does suggest, however, that what it takes to be a successful species is not inherently different between the groups (Davis *et al.*, 2011). Luckily, remediation efforts are never based on whether a non-native species is functionally different from natives, but rather simply that the species is non-native.

Perhaps the most surprising contingencies are those that were imposed by the original design of the study. Experimental variation in abandonment conditions generated fairly large differences in composition. Woody species expanded at roughly the same rate in all fields, but were dominated by different life forms based on how the fields were abandoned. Similarly, abundances of some early successional species were strongly contingent on abandonment conditions. Abundances of large-seeded species were much patchier than small-seeded species, likely contingent on the position of reproductive individuals in the old-growth forest and dispersal limitation. While these effects diminished over time, there is the potential for subtle lingering effects of early compositional variation on the following successional stages. As succession proceeds, the life span of such legacy effects increases with the life span of the plants. This is why patchiness in tree dominance has persisted. There is a very real possibility that processes that are witnessed in one time are partly a combination of contemporary interactions and partly of historical effects. It is no wonder succession defies predictability.

In sharp contrast to all of the difficulties that the BSS posed to forming generalizations, there is one simplifying feature that was supported throughout this work. Important successional processes can be captured by examining plant life forms. Their successional dynamics are much more consistent across sites with differing histories than even the most abundant species. Life forms are consistently separable based on their traits, population dynamics and ecological responses. As Clements knew as early as 1916, life forms may be one of the simplest and most effective ways of classifying species so that sites with vastly different compositions can be directly compared. Long-term, it may be useful to explore variation within life forms to understand finer-scale ecological processes, but the utility of such a simple functional classification cannot be overstated.

Opportunities for integration: the path ahead

It is rare to have such a detailed and valuable data set as the BSS to explore. Writing this volume was the excuse to delve into the data to examine both its broad patterns and its subtleties. The previous decades of work had exploited the data to address contemporary ecological questions in the usual brief, hypothesis-driven mode of scientific manuscripts. These necessarily often focused on contemporary issues in ecology. Much of what we have learned here came from the necessity to document the basic patterns and dynamics of the system as illustrations of succession. Too often, ecological generalizations are based on what we think we know about a system and if people agree, it becomes conventional wisdom. Nothing compares with generalizations based on data for increasing our knowledge of systems.

This work, more than anything, argues for the importance for studies to address a multiplicity of mechanisms that may operate at different scales. This is the only way that contingencies can be identified. This does not mean we are calling for the end to detailed experimental studies of interactions or other ecological phenomena. Instead, researchers should place their experiments into a larger conceptual context. Over time, a series of studies may be developed that place individual interactions into a broader context, so that we know the conditions under which a particular interaction is expected to be important and what it may be contingent upon. Developing this context is necessary to building understanding and a necessary building block to developing an integrative view of the systems that we work in. Succession may still be one of the largest contexts that ecologists need to be familiar with.

Appendix 1

A summary of the people that have sampled the BSS plots over the span from 1958 to 2013. This list totals 198 samplers, but does not include people for whom only partial names were recorded or where the names were illegible. We apologize for any omissions, but do not take the blame for them.

Sampler	Years
Project Initiators	
Buell, Helen F.	1958–1986
Buell, Murray F.	1958–1971
Small, John A.	1960–1977
Current Project Leaders	
Pickett, Steward T. A.	1978–2013
Cadenasso, Mary L.	1992–2009
Meiners, Scott J.	1995–2013
Morin, Peter J.	1997–2013
Annual Samplers (Alphabetical order)	
Aagaard, Kevin	2011–2013
Adams-Krumins, Jen	2004
Adams-Manson, Robert H.	1991–1992
Anzelone, Marielle	1999
Armesto, Juan J.	1982–1984
Aronson, Myla F. J.	2001–2005
Avery, Julian	2009–2011
Bach, Edith	1974
Baiser, Ben	2008–2009
Banasiak, Steven	2006–2007
Banerji, Aabir	2004–2007
Barruli, John	1985
Bartha, Sandor	1993
Basnett, Khadga	1989–1990
Baumgarten, Joni	2013
Baxter, Jim	1993–1994
Decker, Karen	1982
Beimborn, William A.	1972–1973
Bell, Tim	1980–1981
Blake-Mahmud, Jennifer	2013
Bloechl, Jim	1990
Boarman, William	1987–1989
Boerner, Ralph	1979
Bond, William	1977
Bovitz, Paul	1986

Sampler	Years
Bresnahan, Pat	1981–1982
Brush, Ki	1983
Buchholz, Kenny	1979
Burkhalter, Curtis	2012–2013
Byers, Diane L.	1985–1992
Caiazza, Nicholas	1978
Calicotti, K	1970
Callanan, Brett	2002
Cantafio, Louis	1996–1998
Carlsward, Barbara	2011–2013
Carson, Walter P.	1985
Cartica, Bob	1977–1978
Challis, Melinda	1999
Chapman, Ellen	1979–1980
Cheplick, Gregory P.	1980
Chintala, Marty	1993
Cipollini, Martin	1986–1988
Cole, H	1971
Cole, Marlene	1997
Collins, Beverly S.	1978–1984
Collins, Scott L.	1984
Connott, Lisa	1984
Davis, Roger	1964
Davison, Sara E.	1979–1980
DeHart, Lisa	1994
DiGeley, Barb	1986
Dougherty, Kevin	1985
Duffin, Kirstin	2013
Dunne, Kenny	1984
Eident, Sam	2011
Elbin, Susan	1991
Elena, Tartaglia	2009–2012
Elfstrom, Bruce A.	1973–1974
Elgersma, Kenneth	2005–2008
Elle, Elizabeth	1993
Endres, Susan	1994–2002
Ensiya, Peter	1982
Facelli, José M.	1987–1990
Faillace, Cara	2011–2013
Fede, John	1994
Fenn, Tabby	2012
Ferrara, Lisa	1984–1987
Fiteni, Susan	1980
Foote, MaryAnne	1979–1981
Forgione, Helen	1991–1992
Foster, Melissa	1993
Fowler, Alexandra	2009
Frei, Karen R.	1960
Frye, Darlene	1973

Sampler	Years
Frye, Richard J.	1973–1977
Furtak, Doug	1978
Gallagher, Mark	1982
Gargiullo, Marge	1988–1989
Gates, Steve	1998–2004
Gibson, David	1985
Goodwillie, Carol	2001
Graham, John	1978–1979
Haines, Elizabeth M	1962–1965
Harrod, Jon	1991
Hart, Julie	1995
Hayslee, Simoza	1970
Hebbeger, JoAnna	1996
Hernandez, Alex	2004
Hofer, Charles	2007
Horlick, Kenny	1982
Jackson, James R.	1974–1975
Jennings, Rob	1996
Johns, Jennifer	1996–1999
Johnson, Brian	2012
Jonathon, Schramm	2003
Kaelon, John	1971
Kallay, Laura	1986
Kauzinger, Christina	1992
Klass, Jeremy	2005–2006
Kotliar, Tasha	1983
Krouer,	1972
Ladwig, Laura	2007–2009
Landesman, Bill	2006
Lang, Karlien	2010
Lang, Kimberly	2008–2010
LaPolla, John	2003–2004
LaPuma, David	2007–2008
LaPuma, Inga	2007
Leahy, Nick	1995
Learn, Niki	1998
Levin, Michael H.	1962–1963
Lewis, Alan J.	1970
LoGiudice, Kathleen	1997–1999
Long, Zac	2000–2004
Loveland, Linda	1993
Luo, Royce	2011
MacFarlane, David	1998
Martinkovic, Matt	2000
Matlack, Glenn	1986
McCarthy, Brian C.	1985–1988
McDonnell, Mark J.	1980–1983
McDonough, Walter	1967
Meadows, Marva	1985–1987

Sampler	Years
Meiners, Joseph	2006
Mekenian, Michael R.	1972–1974
Milne, Bruce	1983–1984
Monte, Judith A.	1969–1971
Morris, Natalie	2008
Muraoka, Joel	1983–1985
Myster, Randall W.	1985–1990
Nicola, Alexandra	1979–1980
Norrin, Carrie	2005
Nott, BreAnne	2006
Palmer, Matt	2000–2004
Panebianco, Robert J.	1975
Pehek, Ellen L.	1992
Perzley, Julia	2011–2013
Petchey, Owen	1999
Picard, Sarah	1999–2000
Pisula, Nikki	2007–2009
Price, Dana L.	2002–2004
Purandare, Uma	1992
Rankin, Duke	1981
Renda, Mike	1989–1990
Rietsma, Carol	1972
Robichaud, Beryl	1961
Robinson, George	1993
Robinson, Robbie	2013
Robison, Scott	1992
Rohleder, Linda	2005
Root, Peter G.	1960
Roth, Joan	1993
Royle, Denise	1992–1994
Rudis, Victor A.	1974
Ruhren, Scott	1996–1997
Rye, Timothy	2005–2006
Salvemini, Angie	1995
Scapati, Dominick	1981–1982
Schmalhoffer, Victoria	1991–1993
Schwarz, Kirsten	2002
Sigadel, Jeff	1990–1991
Shank, Leslie	1994–1998
Sizemore, Erin	2003
Skuta, Paul	1978
Slezak, William F.	1974–1975
Smith, Brendan	1995
Sorensen, Amanda	2013
Stanko, Maria	2006–2009
Stevens, Henry	1999–2000
Stouffer, Phil	1987
Sulser, Judy A. S.	1971
Timbrook, Troy	1990

Sampler	Years
Tuck, Jim	1976–1978
Tulloss, Elise	2005
Vivian-Smith, Gabrielle	1993–1996
Wachholder, Brent	2004
Walters, Lori	1969
Waynon, Rich	1988
Wein, Gary R.	1980–1982
Weissberger, Eric J	1993
Weldon, Orion	2011
White, Douglas W.	1979
Wijesinghe, Dushayantha K.	1990
Willemsen, Roger W.	1972–1976
Winfree, Rachael	1995
Worthen, Wade	1983
Wrobel, Betsy	1979
Yurkonis, Kathryn	2003–2004
Zurovchak, Joe	1990–1995

Appendix 2

Complete species list for the BSS data. For species that contribute 0.01% or more of the total cover from 1958–2011, their total abundance (all fields all years) and rank position are indicated. Nomenclature follows Gleason and Cronquist, 1991.

	Abundance	Rank
Abutilon theophrasti	0.01%	162
Acalypha gracilens	.	.
Acalypha rhomboidea	0.11%	81
Acalypha virginica	.	.
Acer negundo	1.13%	27
Acer platanoides	0.10%	85
Acer rubrum	5.25%	4
Acer saccharinum	0.01%	181
Acer saccharum	0.02%	141
Achillea millefolium	0.73%	38
Actaea pachypoda	.	.
Agrostis gigantea	0.03%	125
Agrostis capillaris	.	.
Agrostis hyemalis	0.10%	84
Agrostis perennans	.	.
Agrostis stolonifera	0.30%	60
Ailanthus altissima	0.81%	36
Alliaria petiolata	0.97%	30
Allium canadense	.	.
Allium vineale	0.04%	109
Amaranthus retroflexus	.	.
Ambrosia artemisiifolia	1.16%	26
Ambrosia trifida	.	.
Amelanchier canadensis	.	.
Anagallis arvensis	.	.
Anaphalis margaritacea	.	.
Andropogon virginicus	.	.
Antennaria neglecta	.	.
Anthemis arvensis	.	.
Anthemis cotula	.	.
Anthoxanthum odoratum	.	.
Apios americana	.	.
Apocynum androsaemifolium	.	.
Apocynum cannabinum	0.22%	68

	Abundance	Rank
Arctium minus	.	.
Arenaria serpyllifolia	.	.
Arisaema triphyllum	0.01%	178
Arrhenatherum elastius	.	.
Asclepias incarnata	.	.
Asclepias syriaca	0.01%	158
Asparagus officinalis	0.01%	160
Asplenium platyneuron	0.03%	127
Aster ericoides	0.58%	44
Aster lanceolatus	0.48%	49
Aster laterifolius	0.24%	65
Aster novae-angliae	0.01%	163
Aster pilosus	2.93%	9
Athyrium filix-femina	.	.
Barbarea vulgaris	0.64%	41
Berberis thunbergii	0.04%	106
Betula alleghaniensis	.	.
Bidens vulgata	0.01%	177
Botrychium virginianum	.	.
Bromus inermis	0.01%	170
Bromus racemosus	0.61%	42
Bromus tectorum	.	.
Calystegia sepium	1.33%	20
Capsella bursa-pastoris	.	.
Carex spp.	0.21%	69
Carex festucacea	.	.
Carex pensylvanica	.	.
Carex scoparia	.	.
Carya glabra	0.06%	101
Carya ovalis	0.09%	86
Carya ovata	0.07%	97
Carya tomentosa	0.04%	117
Celastrus orbiculatus	0.32%	59
Celastrus scandens	0.01%	149
Celtis occidentalis	.	.
Centaurea dubia	0.90%	33
Cerastium nutans	.	.
Cerastium vulgatum	0.15%	73
Chamaecrista nictitans	.	.
Chenopodium album	0.14%	74
Chrysanthemum leucanthemum	0.60%	43
Circaea lutetiana	0.19%	70
Cirsium arvense	0.08%	95
Cirsium discolor	0.09%	87
Cirsium vulgare	0.02%	144
Claytonia virginica	.	.
Clematis virginiana	0.01%	180
Collinsonia canadensis	.	.
Commelina communis	.	.

	Abundance	Rank
Conyza canadensis	0.04%	116
Cornus amomum	0.09%	90
Cornus florida	4.14%	6
Cornus racemosa	0.44%	51
Cyperus esculentus	.	.
Cystopteris fragilis	.	.
Dactylis glomerata	1.36%	18
Danthonia sericea	.	.
Danthonia spicata	0.08%	94
Daucus carota	1.12%	28
Desmodium glabellum	0.02%	145
Dianthus armeria	0.11%	82
Digitaria sanguinalis	0.36%	53
Draba verna	.	.
Dryopteris intermedia	.	.
Elaeagnus angustifolia	0.02%	138
Elytrigia repens	1.24%	23
Epilobium glandulosum	.	.
Erechtites hieracifolia	0.03%	129
Erigeron annuus	0.93%	32
Erigeron philadelphicus	.	.
Erigeron pulchellus	.	.
Erigeron strigosus	.	.
Euonymus alatus	0.01%	164
Euonymus americanus	.	.
Eupatorium nigra	.	.
Eupatorium perfoliatum	.	.
Eupatorium rugosum	1.34%	19
Euphorbia maculata	.	.
Euthamia graminifolia	2.12%	12
Fagus grandifolia	0.01%	182
Fragaria virginiana	3.53%	7
Fraxinus americana	1.25%	22
Fraxinus pennsylvanica	0.02%	137
Galinsoga parviflora	.	.
Galium aparine	0.03%	119
Galium circaezans	.	.
Galium triflorum	.	.
Geum canadense	0.04%	108
Glechoma hederacea	0.03%	123
Gleditsia triacanthos	.	.
Glycine max	.	.
Gnaphalium obtusifolium	.	.
Habenaria lacera	.	.
Hackelia virginiana	.	.
Hedeoma pulegioides	0.10%	83
Hedera helix	.	.
Helianthus annuus	.	.
Hibiscus trionum	.	.

	Abundance	Rank
Hieracium caespitosum	4.81%	5
Hieracium pilosella	.	.
Hieracium piloselloides	0.33%	55
Hypericum mutilum	.	.
Hypericum perforatum	0.01%	161
Hypericum punctatum	0.02%	143
Ilex glabra	.	.
Ilex opaca	.	.
Impatiens capensis	0.01%	148
Ipomoea hederacea	0.01%	156
Juglans nigra	1.45%	16
Juncus bufonius	.	.
Juncus secundus	0.01%	175
Juncus tenuis	0.03%	130
Juniperus virginiana	5.52%	3
Krigia biflora	.	.
Lactuca canadensis	0.04%	115
Lactuca serriola	0.03%	124
Leersia virginica	.	.
Lepidium campestre	0.12%	78
Lepidium sativum	.	.
Lepidium virginicum	.	.
Ligustrum vulgare	0.07%	98
Linaria vulgaris	0.46%	50
Lindera benzoin	0.05%	104
Lobelia inflata	0.04%	110
Lobelia puberula	.	.
Lolium perenne	0.03%	118
Lonicera japonica	6.86%	2
Lonicera maackii	0.28%	62
Lonicera tatarica	0.04%	114
Lotus corniculatus	0.01%	157
Lycopus americanus	.	.
Lysimachia quadrifolia	.	.
Medicago lupulina	.	.
Medicago sativa	.	.
Melilotus alba	0.01%	169
Melilotus officinalis	.	.
Menispermum canadense	.	.
Microstegium vimineum	1.94%	14
Mitchella repens	.	.
Mollugo verticillata	0.09%	88
Monotropa uniflora	.	.
Morus alba	.	.
Morus rubra	0.05%	105
Muhlenbergia frondosa	0.02%	139
Myrica pensylvanica	0.06%	102
Nyssa sylvatica	.	.
Oenothera biennis	0.29%	61

	Abundance	Rank
Oenothera perennis	.	.
Onoclea sensibilis	0.02%	146
Ophioglossum vulgatum	.	.
Osmunda regalis	.	.
Oxalis stricta	0.69%	39
Panicum capillare	.	.
Panicum clandestinum	.	.
Panicum depauperatum	.	.
Panicum dichotomiflorum	.	.
Panicum lanuginosum	.	.
Panicum leucothrix	0.04%	113
Panicum lindheimeri	.	.
Panicum linearifolium	.	.
Panicum virgatum	.	.
Parthenocissus quinquefolia	2.09%	13
Penstemon hirsutus	0.04%	111
Phalaris arundinacea	0.04%	112
Phleum pratense	0.25%	64
Phryma leptostachya	.	.
Physalis heterophylla	0.02%	140
Physalis longifolia	0.01%	150
Physalis virginiana	.	.
Phytolacca americana	0.03%	126
Pilea pumila	0.09%	91
Pinus strobus	.	.
Plantago lanceolata	0.57%	45
Plantago major	.	.
Plantago rugelii	1.18%	25
Plantago virginica	.	.
Platanus occidentalis	.	.
Poa annua	.	.
Poa compressa	1.42%	17
Poa pratensis	2.64%	10
Poa sylvestris	.	.
Podophyllum peltatum	.	.
Polygonatum pubescens	.	.
Polygonum aviculare	0.01%	153
Polygonum convolvulus	0.01%	173
Polygonum hydropiper	.	.
Polygonum pensylvanicum	0.01%	151
Polygonum persicaria	0.02%	135
Polygonum scandens	0.03%	134
Polygonum virginianum	0.13%	76
Polystichum acrostichoides	.	.
Portulaca oleracea	0.01%	152
Potentilla norvegica	0.01%	165
Potentilla recta	0.02%	147
Potentilla simplex	0.74%	37
Prunella vulgaris	0.15%	72

Appendix 2

	Abundance	Rank
Prunus avium	0.12%	79
Prunus hortulana	0.01%	168
Prunus serotina	0.49%	48
Prunus virginiana	.	.
Pyrus angustifolia	.	.
Pyrus coronaria	0.01%	176
Pyrus malus	0.35%	54
Quercus alba	0.12%	80
Quercus coccinea	0.13%	77
Quercus palustris	0.14%	75
Quercus rubra	0.54%	47
Quercus velutina	0.33%	57
Ranunculus abortivus	0.01%	159
Ranunculus bulbosus	.	.
Raphanus raphanistrum	0.23%	67
Rhamnus carthartica	0.03%	133
Rhus copallinum	0.01%	167
Rhus glabra	1.10%	29
Rhus typhina	0.01%	154
Robinia pseudoacacia	.	.
Rorippa palustris	.	.
Rosa multiflora	7.52%	1
Rosa virginiana	.	.
Rubus allegheniensis	0.83%	35
Rubus flagellaris	0.66%	40
Rubus hispidus	.	.
Rubus occidentalis	0.28%	63
Rubus phoenicolasius	0.03%	121
Rubus strigosus	.	.
Rudbeckia hirta	.	.
Rudbeckia triloba	.	.
Rumex acetosella	0.84%	34
Rumex crispus	0.03%	122
Sanicula gregaria	.	.
Sassafras albidum	0.08%	93
Schizachyrium scoparium	0.07%	96
Scleranthus annuus	.	.
Senecio vulgaris	.	.
Setaria faberii	0.03%	120
Setaria glauca	.	.
Setaria viridis	.	.
Silene antirrhina	.	.
Silene latifolia	0.18%	71
Silene vulgaris	.	.
Sinapis arvensis	.	.
Sisyrinchium angustifolium	.	.
Smilax rotundifolia	.	.
Solanum carolinense	0.57%	46
Solanum dulcamara	.	.

	Abundance	Rank
Solanum nigrum	.	.
Solidago canadensis	1.20%	24
Solidago gigantea	0.38%	52
Solidago juncea	2.48%	11
Solidago nemoralis	0.33%	58
Solidago rugosa	1.72%	15
Sorghum halepense	0.01%	172
Stellaria media	0.02%	142
Taraxacum officinale	0.07%	99
Taxus canadensis	.	.
Toxicodendron radicans	3.38%	8
Tragopogon dubius	.	.
Tridens flavus	0.03%	132
Trifolium arvense	.	.
Trifolium aureum	0.03%	128
Trifolium campestre	.	.
Trifolium hybridum	0.23%	66
Trifolium pratense	0.95%	31
Trifolium repens	0.02%	136
Triodanis perfoliata	.	.
Ulmus americana	0.08%	92
Ulmus pumila	.	.
Ulmus rubra	0.04%	107
Urtica dioica	.	.
Verbascum blattaria	0.05%	103
Verbascum thapsus	.	.
Verbena hastata	.	.
Verbena urticifolia	0.01%	155
Veronica arvensis	.	.
Veronica officinalis	0.03%	131
Veronica peregrina	.	.
Veronica serpyllifolia	.	.
Viburnum acerifolium	0.01%	166
Viburnum dentatum	0.09%	89
Viburnum prunifolium	0.33%	56
Vicia villosa	0.06%	100
Viola sororia	0.01%	171
Vitis sp.	1.29%	21
Vitis aestivalis	.	.
Vitis labrusca	0.01%	179
Vitis palmata	.	.
Vitis riparia	.	.
Vitis vulpina	0.01%	174

References

Agrawal, A. A. and Kotanen, P. M. (2003) Herbivores and the success of exotic plants: a phylogenetically controlled experiment. *Ecology Letters*, **6**, 712–715.

Allen, C. R. and Holling, C. S. (2002) Cross-scale structure and scale breaks in ecosystems and other complex systems. *Ecosystems*, **5**, 315–318.

Allen, E. B. and Allen, M. F. (1984) Competition between plants of different successional stages: mycorrhizae as regulators. *Canadian Journal of Botany*, **62**, 2625–2629.

Allison, V. J., Miller, R. M., Jastrow, J. D., Matamala, R. and Zak, D. R. (2005) Changes in soil microbial community structure in a tallgrass prairie chronosequence. *Soil Science Society of America Journal*, **69**, 1412–1421.

Ambler, M. A. (1965) Seven alien plant species. *William L. Hutcheson Memorial Forest Bulletin*, **2**, 1–8.

Anderson, K. J. (2007) Temporal patterns in rates of community change during succession. *American Naturalist*, **169**, 780–793.

Anderson, R. C. (2008) Growth and arbuscular mycorrhizal fungal colonization of two prairie grasses grown in soil from restorations of three ages. *Restoration Ecology*, **16**, 650–656.

Armas, C., Ordiales, R. and Pugnaire, F. I. (2004) Measuring plant interactions: a new comparative index. *Ecology*, **85**, 2682–2686.

Armesto, J. J. and Pickett, S. T. A. (1985) Experiments on disturbance in old-field plant communities: impact on species richness and abundance. *Ecology*, **66**, 230–240.

Armesto, J. J. and Pickett, S. T. A. (1986) Removal experiments to test mechanisms of plant succession in oldfields. *Vegetatio*, **66**, 85–93.

Armesto, J. J., Pickett, S. T. A. and McDonnell, M. J. (1991) Spatial heterogeneity during succession: a cyclic model of invasion and exclusion. In Kolasa, J. and Pickett, S. T. A. (Eds.) *Ecological Heterogeneity*. New York, Springer-Verlag.

Aronson, M. F. J., Handel, S. N. and Clemants, S. E. (2007) Fruit type, life form and origin determine the success of woody plant invaders in an urban landscape. *Biological Invasions*, **9**, 465–475.

Aubréville, A. (1938) Regeneration patterns in closed forest of Ivory Coast. In Eyre, S. R. (Ed.) *World Vegetation Types*. New York, Columbia University Press.

Auclair, A. N. D. and Goff, F. G. (1971) Diversity relations of upland forests in the western Great Lakes area. *American Naturalist*, **105**, 499–528.

Austin, M. P. (1981) Permanent quadrats: an interface for theory and practice. *Plant Ecology*, **46–47**, 1–10.

Baeten, L., Hermy, S., Van Daele, S. and Verheyen, K. (2010) Unexpected understory community development after 30 years in ancient and post-agricultural forests. *Journal of Ecology*, **98**, 1447–1453.

Baiser, B., Lockwood, J. L., La Puma, D. and Aronson, M. F. J. (2008) A perfect storm: two ecosystem engineers interact to degrade deciduous forests of New Jersey. *Biological Invasions*, **10**, 785–795.

Baker, H. G. (1965) Characteristics and modes of origin of weeds. *The Genetics of Colonizing Species*. New York, Academic Press.

Baker, H. G. (1972) Seed weight in relation to environmental conditions in California. *Ecology*, **53**, 997–1010.

Bakker, J. P., Olff, H., Willems, J. H. and Zobel, M. (1996) Why do we need permanent plots in the study of long-term vegetation dynamics? *Journal of Vegetation Science*, **7**, 147–156.

Banasiak, S. E. and Meiners, S. J. (2009) Long term dynamics of *Rosa multiflora* in a successional system. *Biological Invasions*, **11**, 215–224.

Banta, J., Stark, S., Stevens, M. *et al*. (2008) Light reduction predicts widespread patterns of dominance between asters and goldenrods. *Plant Ecology*, **199**, 65–76.

Bard, G. E. (1952) Secondary succession on the Piedmont of New Jersey. *Ecological Monographs*, **22**, 195–215.

Barrett, G. W., Peles, J. D. and Harper, S. J. (1995) Reflections on the use of experimental landscapes in mammalian ecology. In Lidicker, W. Z., Jr. (Ed.) *Landscape Approaches in Mammalian Ecology and Conservation*. Minneapolis, MN, University of Minnesota Press.

Bartha, S. (2001) Spatial relationships between plant litter, gopher disturbance and vegetation at different stages of old-field succession. *Applied Vegetation Science*, **4**, 53–62.

Bartha, S., Meiners, S. J., Pickett, S. T. A. and Cadenasso, M. L. (2003) Plant colonization windows in a mesic old field succession. *Applied Vegetation Science*, **6**, 205–212.

Bartha, S., Szentes, S., Horváth, A. *et al*. (2014) Impact of mid-successional dominant species on the diversity and progress of succession in regenerating temperate grasslands. *Applied Vegetation Science*, **17**, 201–2013.

Bartomeus, I., Sol, D., Pino, J., Vicente, P. and Font, X. (2012) Deconstructing the native–exotic richness relationship in plants. *Global Ecology and Biogeography*, **21**, 524–533.

Baskin, C. C. and Baskin, J. M. (1988) Germination ecophysiology of herbaceous plant species in a temperate region. *American Journal of Botany*, **75**, 286–305.

Baskin, C. C. and Baskin, J. M. (1998) *Seeds: Ecology, Biogeography, and Evolution of Dormancy and Germination*. New York, Academic Press.

Bastl, M., Kocÿr, P., Prach, K., Pyšek, P. and Brock, J. H. (1997) The effect of successional age and disturbance on the establishment of alien plants in man-made sites: an experimental approach. In *Plant Invasions: Studies from North America and Europe*. Leiden, The Netherlands, Backhuys Publishers.

Bazzaz, F. A. (1975) Plant species diversity in old-field successional ecosystems in southern Illinois. *Ecology*, **56**, 485–488.

Bazzaz, F. A. (1979) The physiological ecology of plant succession. *Annual Review of Ecology and Systematics*, **10**, 351–371.

Bazzaz, F. A. (1986) Life history of colonizing plants: some demographic, genetic, and physiological features. In Mooney, M. A. and Drake, J. A. (Eds.) *Ecology of Biological Invasions of North America and Hawaii*. New York, Springer-Verlag.

Bazzaz, F. A. (1996) *Plants in Changing Environments: Linking Physiological, Population, and Community Ecology*. New York, Cambridge University Press.

Bazzaz, F. A. and Carlson, R. W. (1982) Photosynthetic acclimation to variability in the light environment of early and late successional plants. *Oecologia*, **54**, 313–316.

Bazzaz, F. A. and Mooney, H. A. (1986) Life history characteristics of colonizing plants: some demographic, genetic, and physiological features. In Mooney, M. A. and Drake, J. A. (Eds.) *Ecology of Biological Invasions of North America and Hawaii*. New York, Springer-Verlag.

Ben Shahar, R. (1991) Successional patterns of woody plants in catchment areas in a semi-arid region. *Vegetatio*, **93**, 19–27.

Bengtsson, J., Angelstam, P., Elmqvist, T. *et al*. (2003) Reserves, resilience and dynamic landscapes. *Ambio*, **32**, 389–396.

Bennett, J. R., Dunwiddie, P. W., Giblin, D. E. and Arcese, P. (2012) Native versus exotic community patterns across three scales: roles of competition, environment and incomplete invasion. *Perspectives in Plant Ecology, Evolution and Systematics*, **14**, 381–392.

Benninger-Truax, M., Vankat, J. L. and Schaefer, R. L. (1992) Trail corridors as habitat and conduits for movement of plant species in Rocky Mountain National Park, Colorado, USA. *Landscape Ecology*, **6**, 269–278.

Bever, J. D., Dickie, I. A., Facelli, E. *et al*. (2010) Rooting theories of plant community ecology in microbial interactions. *Trends in Ecology and Evolution*, **25**, 468–478.

Biggs, R., Westley, F. R. and Carpenter, S. R. (2010) Navigating the back loop: fostering social innovation and transformation in ecosystem management. *Ecology and Society*, **15**, Article 9.

Billings, W. D. (1938) The structure and development of old field shortleaf pine stands and certain associated physical properties of the soil. *Ecological Monographs*, **8**, 437–500.

Biondini, M. E., Steuter, A. A. and Grygiel, C. E. (1989) Seasonal fire effects on the diversity patterns, spatial distribution and community structure of forbs in the northern mixed prairie, USA. *Plant Ecology*, **85**, 21–31.

Blatt, S. E., Crowder, A. and Harmsen, R. (2005) Secondary succession in two south-eastern Ontario old-fields. *Plant Ecology*, **177**, 25–41.

Bliss, L. C. and Linn, R. M. (1955) Bryophyte communities associated with old field succession in the North Carolina Piedmont. *The Bryologist*, **58**, 120–131.

Blossey, B. and Notzold, R. (1995) Evolution of increased competitive ability in invasive nonindigenous plants: a hypothesis. *Journal of Ecology*, **83**, 887–889.

Blumenthal, D., Jordan, M., Nicholas, R. and Russelle, M. P. (2003) Soil carbon addition controls weeds and facilitates prairie restoration. *Ecological Applications*, **13**, 605–615.

Boeken, B., Shachak, M., Gutterman, Y. and Brand, S. (1995) Patchiness and disturbance: plant community responses to porcupine diggings in the Central Negev. *Ecography*, **18**, 410–422.

Bonfil, C. (1998) The effects of seed size, cotyledon reserves, and herbivory on seedling survival and growth in *Quercus rugosa* and *Q. laurina* (Fagaceae). *American Journal of Botany*, **85**, 79–87.

Bonser, S. and Ladd, B. (2011) The evolution of competitive strategies in annual plants. *Plant Ecology*, **212**, 1441–1449.

Bonta, M. M. (1991) *Women in the Field: America's Pioneering Women Naturalists*. College Station, TX, Texas A & M Press.

Bormann, F. H. and Likens, G. E. (1979) *Pattern and Process in a Forested Ecosystem*. New York, Springer-Verlag.

Bosy, J. L. and Reader, R. J. (1995) Mechanisms underlying the suppression of forb seedling emergence by grass (*Poa pratensis*) litter. *Functional Ecology*, **9**, 635–639.

Braun, E. L. (1956) The development of association and climax concepts: their use in interpretation of the deciduous forest. *American Journal of Botany*, **43**, 906–911.

Briggs, J. M., Hoch, G. A. and Johnson, L. C. (2002) Assessing the rate, mechanisms, and consequences of the conversion of tallgrass prairie to *Juniperus virginiana* forest. *Ecosystems*, **5**, 578–586.

Brock, J. P. (2000) *The Evolution of Adaptive Systems*. San Diego, CA, Academic Press.
Brock, W. A. and Carpenter, S. R. (2006) Variance as a leading indicator of regime shift in ecosystem services. *Ecology and Society*, **11**, 9.
Brookshire, B. and Shifley, S. R. (Eds.) (1997) *Proceedings of the Missouri Ozark Forest Ecosystem Project Symposium: An Experimental Approach to Landscape Research*. USDA Forest Service, North Central Forest Experiment Station.
Brown, D. G. (1994) Beetle folivory increases resource availability and alters plant invasion in monocultures of goldenrod. *Ecology*, **75**, 1673–1683.
Brown, R. L. and Peet, R. K. (2003) Diversity and invasibility of southern Appalachian plant communities. *Ecology*, **84**, 32–39.
Brown, V. K. (1985) Insect herbivores and plant succession. *Oikos*, **44**, 17–22.
Brown, V. K. and Gange, A. C. (1992) Secondary plant succession: how is it modified by insect herbivory? *Vegetatio*, **101**, 3–13.
Buell, M. F. (1957) The mature oak forest of Mettler's woods. *William L. Hutcheson Memorial Forest Bulletin*, **1**, 16–19.
Buell, M. F., Buell, H. F. and Small, J. A. (1954). Fire in the history of Mettler's Woods. *Bulletin of the Torrey Botanical Club*, **81**, 253–255.
Buell, M. F., Buell, H. F., Small, J. A. and Siccama, T. G. (1971) Invasion of trees in secondary succession on the New Jersey Piedmont. *Bulletin of the Torrey Botanical Club*, **98**, 67–74.
Bugmann, H. (2001) A review of forest gap models. *Climatic Change*, **51**, 259–305.
Burke, M. J. W. and Grime, J. P. (1996) An experimental study of plant community invasibility. *Ecology*, **77**, 776–790.
Burroughs, J. (1896) *A Year in the Fields. Selections from the Writings of John Burroughs*. Boston, MA, Houghton, Mifflin and Company.
Burrows, C. J. (1990) *Processes of Vegetation Change*. Boston, MA, Unwin Hyman
Burton, P. J. and Bazzaz, F. A. (1995) Ecophysiological responses of tree seedlings invading different patches of old-field vegetation. *Journal of Ecology*, **83**, 99–112.
Cadenasso, M. L. and Pickett, S. T. A. (2000) Linking forest edge structure to edge function: mediation of herbivore damage. *Journal of Ecology*, **88**, 31–44.
Cadenasso, M. L. and Pickett, S. T. A. (2001) Effect of edge structure on the flux of species into forest interiors. *Conservation Biology*, **15**, 91–97.
Cadenasso, M. L., Traynor, M. M. and Pickett, S. T. A. (1997) Functional location of forest edges: gradients of multiple physical factors. *Canadian Journal of Forest Research*, **27**, 774–782.
Cadenasso, M. L., Pickett, S. T. A. and Morin, P. J. (2002) Experimental test of the role of mammalian herbivores on old field succession: community structure and seedling survival. *Journal of the Torrey Botanical Society*, **129**, 228–237.
Cadenasso, M. L., Pickett, S. T. A., Weathers, K. C. and Jones, C. G. (2003a) A framework for a theory of ecological boundaries. *BioScience*, **53**, 750–758.
Cadenasso, M. L., Pickett, S. T. A., Weathers, K. C., Bell, S. S., Benning, T. L., Carreiro, M. M. and Dawson, T. E. (2003b) An interdisciplinary and synthetic approach to ecological boundaries. *BioScience*, **53**, 717–722.
Cadenasso, M. L., Meiners, S. J. and Pickett, S. T. A. (2009) The success of succession: a symposium commemorating the 50th anniversary of the Buell-Small Succession Study. *Applied Vegetation Science*, **12**, 3–8.
Cadotte, M. W., Hamilton, M. A. and Murray, B. R. (2009) Phylogenetic relatedness and plant invader success across two spatial scales. *Diversity and Distributions*, **15**, 481–488.

Callaway, R. M., Mahall, B. E., Wicks, C., Pankey, J. and Zabinski, C. (2003) Soil fungi and the effects of an invasive forb on grasses: neighbor identity matters. *Ecology*, **84**, 129–135.

Canham, C. D. and Pacala, S. (1995) Linking tree population dynamics and forest ecosystem processes. In Jones, C. G. and Lawton, J. H. (Eds.) *Linking Species and Ecosystems*. New York, Chapman and Hall.

Carpenter, S. R. and Gunderson, L. H. (2001) Coping with collapse: ecological and social dynamics in ecosystem management. *BioScience*, **51**, 451–457.

Carson, W. P. and Barrett, G. W. (1988) Succession in old-field plant communities: effects of contrasting types of nutrient enrichment. *Ecology*, **69**, 984–994.

Carson, W. P. and Pickett, S. T. A. (1990) Role of resources and disturbance in the organization of an old-field plant community. *Ecology*, **71**, 226–238.

Carson, W. P. and Root, R. B. (2000) Herbivory and plant species coexistence: community regulation by an outbreaking phytophagous insect. *Ecological Monographs*, **70**, 73–99.

Castellano, S. M. and Gorchov, D. L. (2012) Reduced ectomycorrhizae on oak near invasive garlic mustard. *Northeastern Naturalist*, **19**, 1–24.

Catford, J. A., Daehler, C. C., Murphy, H. T. *et al*. (2012) The intermediate disturbance hypothesis and plant invasions: implications for species richness and management. *Perspectives in Plant Ecology, Evolution and Systematics*, **14**, 231–241.

Chabrerie, O., Loinard, J., Perrin, S., Saguez, R. and Decocq, G. (2010) Impact of *Prunus serotina* invasion on understory functional diversity in a European temperate forest. *Biological Invasions*, **12**, 1891–1907.

Chapin, F. S., III, Matson, P. A. and Mooney, H. A. (2002) *Principles of Terrestrial Ecosystem Ecology*. New York, Springer-Verlag.

Chazdon, R. L. (2008) Chance and determinism in tropical forest succession. In Carson, W. P. and Schnitzer, S. A. (Eds.) *Tropical Forest Community Ecology*. Chichester, Wiley-Blackwell.

Chen, J., Franklin, J. F. and Spies, T. A. (1995) Growing-season microclimate gradients from clearcut edges into old-growth Douglas-fir forests. *Ecological Applications*, **5**, 74–86.

Chesson, P. L. and Huntly, N. (1989) Short-term instabilities and long-term community dynamics. *Trends in Ecology and Evolution*, **4**, 293–300.

Christensen, N. L. and Peet, R. K. (1984) Convergence during secondary forest succession. *Journal of Ecology*, **72**, 25–36.

Christian, J. M. and Wilson, S. D. (1999) Long-term ecosystem impacts of an introduced grass in the northern great plains. *Ecology*, **80**, 2397–2407.

Cingolani, A. M., Cabido, M., Gurvich, D. E., Renison, D. and Díaz, S. (2007) Filtering processes in the assembly of plant communities: are species presence and abundance driven by the same traits. *Journal of Vegetation Science*, **18**, 911–920.

Clark, J. S., Carpenter, S. R., Barber, M. *et al*. (2001) Ecological forecasts: an emerging imperative. *Science*, **293**, 657–660.

Clay, K. and Holah, J. (1999) Fungal endophyte symbiosis and plant diversity in successional fields. *Science*, **285**, 1742–1744.

Claypole, E. W. (1877) On the migration of plants from Europe to America, with an attempt to explain certain phenomena connected therewith. *Report of the Montreal Horticultural Society*, **3**, 70–91.

Clements, F. E. (1905) *Research Methods in Ecology*, Lincoln, NE, University Publishing Company.

Clements, F. E. (1916) *Plant Succession: An Analysis of the Development of Vegetation*. Washington, D.C., Carnegie Institution.

Clements, F. E. (1935) Experimental ecology in the public service. *Ecology*, **16**, 342–363.
Colautti, R. I., Ricciardi, A., Grigorovich, I. A. and Macisaac, H. J. (2004) Is invasion success explained by the enemy release hypothesis? *Ecology Letters*, **7**, 721–733.
Coleman, D. C. (2010) *Big Ecology: The Emergence of Ecosystem Science*. Berkeley, University of California Press.
Coley, P. D., Bryant, J. P. and Chapin, F. S. (1985) Resource availability and plant antiherbivore defense. *Science*, **230**, 895–899.
Collins, B. and Wein, G. (1998) Soil resource heterogeneity effects on early succession. *Oikos*, **82**, 238–245.
Collins, B., Wein, G. and Philippi, T. (2001) Effects of disturbance intensity and frequency on early old-field succession. *Journal of Vegetation Science*, **12**, 721–728.
Collins, S. L. and Adams, D. E. (1983) Succession in grasslands: thirty-two years of change in a central Oklahoma tallgrass prairie. *Vegetatio*, **51**, 181–190.
Collins, S. L., Glenn, S. M. and Roberts, D. W. (1993) The hierarchical continuum concept. *Journal of Vegetation Science*, **4**, 149–156.
Collins, S. L., Glenn, S. M. and Gibson, D. J. (1995) Experimental analysis of intermediate disturbance and initial floristic composition: decoupling cause and effect. *Ecology*, **76**, 486–492.
Collins, S. L., Knapp, A. K., Briggs, J. M., Blair, J. M. and Steinauer, E. M. (1998) Modulation of diversity by grazing and mowing in native tallgrass prairie. *Science*, **280**, 745–747.
Comas, L. H. and Eissenstat, D. M. (2004) Linking fine root traits to maximum potential growth rate among 11 mature temperate tree species. *Functional Ecology*, **18**, 388–397.
Connell, J. H. (1978) Diversity in tropical rain forests and coral reefs. *Science*, **199**, 1302–1310.
Connell, J. H. and Slatyer, R. O. (1977) Mechanisms of succession in natural communities and their role in community stability and organization. *American Naturalist*, **111**, 1119–1144.
Cook, W. M., Yao, J., Foster, B. L., Holt, R. D. and Patrick, L. B. (2005) Secondary succession in an experimentally fragmented landscape: community patterns across space and time. *Ecology*, **86**, 1267–1279.
Cooper, W. S. (1913) The climax forest of Isle Royale, Lake Superior, and its development. *Botanical Gazette*, **55**, 1–44,115.
Cooper, W. S. (1926) The fundamentals of vegetational change. *Ecology*, **4**, 391–413.
Copeland, T. E., Sluis, W. and Howe, H. F. (2002) Fire season and dominance in an Illinois tallgrass prairie restoration. *Restoration Ecology*, **10**, 315–323.
Cornwell, W. K. and Ackerly, D. D. (2010) A link between plant traits and abundance: evidence from coastal California woody plants. *Journal of Ecology*, **98**, 814–821.
Côté, S. D., Rooney, T. P., Tremblay, J.-P., Dussault, C. and Waller, D. M. (2004) Ecological impacts of deer overabundance. *Annual Review of Ecology, Evolution, and Systematics*, **35**, 113–147.
Cowles, H. C. (1899) The ecological relations of the vegetation on the sand dune of Lake Michigan. *Botanical Gazette*, **27**, 95–117,167.
Craine, J. M. (2005) Reconciling plant strategy theories of Grime and Tilman. *Journal of Ecology*, **93**, 1041–1052.
Craine, J. M. (2009) *Resource Strategies of Wild Plants*. Princeton, NJ, Princeton University Press.
Cramer, V. A. and Hobbs, R. J. (Eds.) (2007) *Old Fields: Dynamics and Restoration of Abandoned Farmland*. Washington, D.C., Island Press.
Cramer, V. A., Hobbs, R. J. and Standish, R. J. (2008) What's new about old fields? Land abandonment and ecosystem assembly. *Trends in Ecology and Evolution*, **23**, 104–112.

Crawley, M. J., Brown, S. L., Heard, M. S. and Edwards, G. R. (1999) Invasion-resistance in experimental grassland communities: species richness or species identity? *Ecology Letters*, **2**, 140–148.

Culver, D. C. (1981) On using Horn's Markov succession model. *American Naturalist*, **117**, 572–574.

Daehler, C. C. (2003) Performance comparisons of co-occurring native and alien invasive plants: implications for conservation and restoration. *Annual Review of Ecology, Evolution, and Systematics*, **34**, 183–211.

Dalrymple, G., Doren, R., O'Hare, N., Norland, M. and Armentano, T. V. (2003) Plant colonization after complete and partial removal of disturbed soils for wetland restoration of former agricultural fields in Everglades National Park. *Wetlands*, **23**, 1015–1029.

D'Antonio, C. M. and Meyerson, L. A. (2002) Exotic plant species as problems and solutions in ecological restoration: a synthesis. *Restoration Ecology*, **10**, 703–713.

Darwin, C. R. (1859) *On the Origin of Species by Means of Natural Selection, or the Preservation of Favoured Races in the Struggle for Life*. London: John Murray.

Davidson, D. W. (1993) The effects of herbivory and granivory on terrestrial plant succession. *Oikos*, **68**, 23–35.

Davis, M. A., Wrage, K. J. and Reich, P. B. (1998) Competition between tree seedlings and herbaceous vegetation: support for a theory of resource supply and demand. *Journal of Ecology*, **86**, 652–661.

Davis, M. A., Grime, P. and Thompson, K. (2000) Fluctuating resources in plant communities: a general theory of invasibility. *Journal of Ecology*, **88**, 528–534.

Davis, M. A., Thompson, K. and Grime, J. P. (2001) Charles S. Elton and the disassociation of invasion ecology from the rest of ecology. *Diversity and Distributions*, **7**, 97–102.

Davis, M. A., Pergl, J., Truscott, A. *et al.* (2005) Vegetation change: a reunifying concept in plant ecology. *Perspectives in Plant Ecology, Evolution and Systematics*, **7**, 69–76.

Davis, M. A., Chew, M. K., Hobbs, R. J. *et al.* (2011) Don't judge species on their origins. *Nature*, **474**, 153–154.

Dawson, W., Fischer, M. and van Kleunen, M. (2012) Common and rare plant species respond differently to fertilisation and competition, whether they are alien or native. *Ecology Letters*, **15**, 873–880.

de Jong, T. J. (1995) Why fast-growing plants do not bother about defense. *Oikos*, **74**, 545–548.

Debussche, M. and Lepart, J. (1992) Establishment of woody plants in Mediterranean old fields: opportunity in space and time. *Landscape Ecology*, **6**, 133–145.

Debussche, M., Escarr, J., Lepart, J., Houssard, C. and Lavorel, S. (1996) Changes in Mediterranean plant succession: old-fields revisited. *Journal of Vegetation Science*, **7**, 519–526.

del Moral, R. (1993) Mechanisms of primary succession on volcanoes: a view from Mount St. Helens. In Miles, J. and Walton, D. W. H. (Eds.) *Primary Succession on Land*. Boston, MA, Blackwell Scientific Publications.

del Moral, R. (1998) Early succession on lahars spawned by Mount St. Helens. *American Journal of Botany*, **85**, 820–828.

del Moral, R. (2009) Increasing deterministic control of primary succession on Mount St. Helens, Washington. *Journal of Vegetation Science*, **20**, 1145–1154.

del Moral, R., Titus, J. H. and Cook, A. M. (1995) Early primary succession on Mount St. Helens, Washington, USA. *Journal of Vegetation Science*, **6**, 107–120.

del Moral, R., Thomason, L. A., Wenke, A. C., Lozanoff, N. and Abata, M. D. (2012) Primary succession trajectories on pumice at Mount St. Helens, Washington. *Journal of Vegetation Science*, **23**, 73–85.

Denslow, J. S. (1980) Patterns of plant species diversity during succession under different disturbance regimes. *Oecologia*, **46**, 18–21.

De Steven, D. (1991a) Experiments on mechanisms of tree establishment in old-field succession: seedling emergence. *Ecology*, **72**, 1066–1075.

De Steven, D. (1991b) Experiments on mechanisms of tree establishment in old-field succession: seedling survival and growth. *Ecology*, **72**, 1076–1088.

Diamond, J. M. (1975) Assembly of species communities. In Cody, M. L. and Diamond, J. M. (Eds.) *Ecology and Evolution of Communities*. Cambridge, Belknap Press.

Diaz, S., Hodgson, J. G., Thompson, K. et al. (2004) The plant traits that drive ecosystems: evidence from three continents. *Journal of Vegetation Science*, **15**, 295–304.

Dillenburgh, L. R., Whigham, D. F., Teramura, A. H. and Forseth, I. N. (1993) Effects of vine competition on availability of light, water, and nitrogen to a tree host (*Liquidambar styraciflua*). *American Journal of Botany*, **80**, 244–252.

Doak, D. F., Bigger, D., Harding, E. K. et al. (1998) The statistical inevitability of stability-diversity relationships in community ecology. *American Naturalist*, **151**, 264–276.

Douma, J. C., De Haan, M. W. A., Aerts, R., Witte, J. -P. M. and Van Bodegom, P. M. (2012) Succession-induced trait shifts across a wide range of NW European ecosystems are driven by light and modulated by initial abiotic conditions. *Journal of Ecology*, **100**, 366–380.

Dube, P., Fortin, M. J., Canham, C. D. and Marceau, D. J. (2001) Quantifying gap dynamics at the patch mosaic level using a spatially-explicit model of a northern hardwood forest ecosystem. *Ecological Modelling*, **142**, 39–60.

Dupouey, J. L., Dambrine, E., Laffite, J. D. and Moares, C. (2002) Irreversible impact of past land use on forest soils and biodiversity. *Ecology*, **83**, 2978–2984.

Dzwonko, Z. and Loster, S. (1990) Vegetation differentiation and secondary succession on a limestone hill in southern Poland. *Journal of Vegetation Science*, **1**, 615–622.

Edwards-Jones, G. and Brown, V. K. (1993) Successional trends in insect herbivore population densities: a field test of a hypothesis. *Oikos*, **66**, 463–471.

Egler, F. E. (1951) A commentary on American plant ecology, based on the textbooks of 1947–1949. *Ecology*, **32**, 673–695.

Egler, F. E. (1954) Vegetation science concepts. I. Initial floristic composition, a factor in old-field vegetation development. *Vegetatio*, **4**, 412–417.

Ehrenfeld, J. G. (1997) Invasion of deciduous forest preserves in the New York metropolitan region by Japanese barberry (*Berberis thunbergii* DC.). *Journal of the Torrey Botanical Society*, **124**, 210–215.

Eilts, J. A., Mittelbach, G. G., Reynolds, H. L. and Gross, K. L. (2011) Resource heterogeneity, soil fertility, and species diversity: effects of clonal species on plant communities. *American Naturalist*, **177**, 574–588.

Elger, A. and Barrat-Segretain, M. H. (2004) Plant palatability can be inferred from a single-date feeding trial. *Functional Ecology*, **18**, 483–488.

Eliot, C. (2007) Method and metaphysics in Clements's and Gleason's ecological explanations. *Studies in History and Philosophy of Science Part C: Studies in History and Philosophy of Biological and Biomedical Sciences*, **38**, 85–109.

Elkin, C., Reineking, B., Bigler, C. and Bugmann, H. (2012) Do small-grain processes matter for landscape scale questions? Sensitivity of a forest landscape model to the formulation of tree growth rate. *Landscape Ecology*, **27**, 697–711.

Elser, J. J. (2003) Biological stoichiometry: a theoretical framework connecting ecosystem ecology, evolution, and biochemistry for application in astrobiology. *International Journal of Astrobiology*, **2**, 185–193.

Elton, C. S. (1927) *Animal Ecology*. London, MacMillian.

Elton, C. S. (1958) *The Ecology of Invasions by Animals and Plants*. London, Methuen.

Emery, S. M. (2007) Limiting similarity between invaders and dominant species in herbaceous plant communities? *Journal of Ecology*, **95**, 1027–1035.

Faber, S. and Markham, J. (2012) Biotic and abiotic effects of remnant and restoration soils on the performance of tallgrass prairie species. *Ecological Restoration*, **30**, 106–115.

Facelli, J. M. and D'Angela, E. (1990) Directionality, convergence, and rate of change during early succession in the inland pampa, Argentina. *Journal of Vegetation Science*, **1**, 255–260.

Facelli, J. M. and Facelli, E. (1993) Interactions after death: plant litter controls priority effects in a successional plant community. *Oecologia*, **95**, 277–282.

Facelli, J. M. and Pickett, S. T. A. (1990) Markovian chains and the role of history in succession. *Trends in Ecology and Evolution*, **5**, 27–29.

Facelli, J. M. and Pickett, S. T. A. (1991a) Indirect effects of litter on woody seedlings subject to herb competition. *Oikos*, **62**, 129–138.

Facelli, J. M. and Pickett, S. T. A. (1991b) Plant litter: its dynamics and effects on plant community structure. *Botanical Review*, **57**, 1–32.

Facelli, J. M. and Pickett, S. T. A. (1991c) Plant litter: light interception and effects on an old-field plant community. *Ecology*, **72**, 1024–1031.

Falk-Petersen, J., Bøhn, T. and Sandlund, O. T. (2006) On the numerous concepts in invasion biology. *Biological Invasions*, **8**, 1409–1424.

Fay, P. A., Knapp, A. K., Blair, J. M. *et al.* (2002) Rainfall timing, soil moisture dynamics, and plant responses in a mesic tallgrass prairie ecosystem. In Weltzin, J. F. and Mcpherson, G. R. (Eds.) *Changing Precipitation Regimes and Terrestrial Ecosystems: A North American Perspective*. Tucson, AZ, University of Arizona Press.

Fenner, M. (1987) Seedlings. *New Phytologist*, **106**, 35–47.

Fenner, M., Hanley, M. E. and Lawrence, R. (1999) Comparison of seedling and adult palatability in annual and perennial plants. *Functional Ecology*, **13**, 546–551.

Fike, J. and Niering, W. A. (1999) Four decades of old field vegetation development and the role of *Celastrus orbiculatus* in the Northeastern United States. *Journal of Vegetation Science*, **10**, 483–492.

Finegan, B. and Delgado, D. (2000) Structural and floristic heterogeneity in a 30-year-old Costa Rican rain forest restored on pasture through natural secondary succession. *Restoration Ecology*, **8**, 380–393.

Fitter, A. H. (1982) Influence of soil heterogeneity on the coexistence of grassland species. *Journal of Ecology*, **70**, 139–148.

Flinn, K. M. and Marks, P. L. (2007) Agricultural legacies in forest environments: tree communities, soil properties, and light availability. *Ecological Applications*, **17**, 452–463.

Flinn, K. M., Vellend, M. and Marks, P. L. (2005) Environmental causes and consequences of forest clearance and agricultural abandonment in central New York, USA. *Journal of Biogeography*, **32**, 439–452.

Flory, S. L. and Clay, K. (2006) Invasive shrub distribution varies with distance to roads and stand age in eastern deciduous forests in Indiana, USA. *Plant Ecology*, **184**, 131–141.

Folke, C., Carpenter, S. R., Walker, B. *et al.* (2012) Resilience thinking: integrating resilience, adaptability and transformability. *Ecology and Society*, **15**, Article 20.

Forkner, R. E., Marquis, R. J. and Lill, J. T. (2004) Feeny revisited: condensed tannins as anti-herbivore defences in leaf-chewing herbivore communities of *Quercus*. *Ecological Entomology*, **29**, 174–187.

Foster, B. L., Questad, E. J., Collins, C. D. *et al.* (2011) Seed availability constrains plant species sorting along a soil fertility gradient. *Journal of Ecology*, **99**, 473–481.

Foster, D. R. (1993) Land-use history (1730–1990) and vegetation dynamics in central New England, USA. *Journal of Ecology*, **80**, 753–771.

Foster, D., Motzkin, G., O'Keefe, J. *et al.* (2004) The environmental and human history of New England. In Foster, D. R. and Aber, J. D. (Eds.) *Forests in Time*. New Haven, CT, Yale University Press.

Fox, J. W. (2013) The intermediate disturbance hypothesis should be abandoned. *Trends in Ecology and Evolution*, **28**, 86–92.

Fraser, E. C., Lieffers, V. J. and Landhäusser, S. M. (2006) Carbohydrate transfer through root grafts to support shaded trees. *Tree Physiology*, **26**, 1019–1023.

Frelich, L. E. and Reich, P. B. (1995) Spatial patterns and succession in a Minnesota southern-boreal forest. *Ecological Monographs*, **65**, 325–346.

Fridley, J. and Wright, J. (2012) Drivers of secondary succession rates across temperate latitudes of the Eastern USA: climate, soils, and species pools. *Oecologia*, **168**, 1069–1077.

Fukami, T., Martijn, B. T., Mortimer, S. R. and Putten, W. H. (2005) Species divergence and trait convergence in experimental plant community assembly. *Ecology Letters*, **8**, 1283–1290.

Fynn, R. W. S., Morris, C. D. and Kirkman, K. P. (2005) Plant strategies and trait trade-offs influence trends in competitive ability along gradients of soil fertility and disturbance. *Journal of Ecology*, **93**, 384–394.

Garnier, E., Cortez, J., Billès, G. *et al.* (2004) Plant functional markers capture ecosystem properties during secondary succession. *Ecology*, **85**, 2630–2637.

Gavier-Pizarro, G. I., Radeloff, V. C., Stewart, S. I., Huebner, C. D. and Keuler, N. S. (2010) Housing is positively associated with invasive exotic plant species richness in New England, USA. *Ecological Applications*, **20**, 1913–1925.

Gering, J. C., Crist, T. O. and Veech, J. A. (2003) Additive partitioning of species diversity across multiple spatial scales: implications for regional conservation of biodiversity. *Conservation Biology*, **17**, 488–499.

Getzin, S., Wiegand, T., Wiegand, K. and He, F. L. (2008) Heterogeneity influences spatial patterns and demographics in forest stands. *Journal of Ecology*, **96**, 807–820.

Gibson, D. J., Middleton, B. A., Foster, K., Honu, Y. A. K., Hoyer, E. W. and Mathis, M. (2005) Species frequency dynamics in an old-field succession: effects of disturbance, fertilization and scale. *Journal of Vegetation Science*, **16**, 415–422.

Gilbert, B. and Lechowicz, M. J. (2005) Invasibility and abiotic gradients: the positive correlation between native and exotic plant diversity. *Ecology*, **86**, 1848–1855.

Gill, D. S. and Marks, P. L. (1991) Tree and shrub seedling colonization of old fields in central New York. *Ecological Monographs*, **61**, 183–205.

Gleason, H. A. (1917) The structure and development of the plant association. *Bulletin of the Torrey Botanical Club*, **44**, 463–481.

Gleason, H. A. (1926) The individualistic concept of the plant association. *Bulletin of the Torrey Botanical Club*, **53**, 7–26.

Gleason, H. A. (1927) Further views on the succession-concept. *Ecology*, **8**, 299–326.

Gleason, H. A. and Cronquist, A. (1991) *Manual of Vascular Plants of Northeastern United States and Adjacent Canada*. New York, New York Botanical Garden, Bronx.

Glenn-Lewin, D. C. and Van Der Maarel, E. (1992) Patterns and processes of vegetation dynamics. In Glenn-Lewin, D. C., Peet, R. K. and Veblen, T. T. (Eds.) *Plant Succession: Theory and Prediction*. New York, Chapman and Hall.

Glitzenstein, J. S., Harcombe, P. A. and Streng, D. R. (1986) Disturbance, succession, and maintenance of species diversity in an East Texas forest. *Ecological Monographs*, **56**, 243–258.

Goldberg, D. E. (1987) Seedling colonization of experimental gaps in two old-field communities. *Bulletin of the Torrey Botanical Club*, **114**, 139–148.

Goldberg, D. E. and Gross, K. L. (1988) Disturbance regimes of midsuccessional old fields. *Ecology*, **69**, 1677–1688.

Goldberg, D. E. and Werner, P. A. (1983) The effects of size of opening in vegetation and litter cover on seedling establishment of goldenrods (*Solidago* spp.). *Oecologia*, **60**, 149–155.

Golley, F. B. (1965) Structure and function of an old-field broomsedge community. *Ecological Monographs*, **35**, 113–137.

Golley, F. B., Pinder, J. E., III, Smallidge, P. J. and Lambert, N. J. (1994) Limited invasion and reproduction of loblolly pines in a large South Carolina old field. *Oikos*, **69**, 21–27.

Golodets, C., Sternberg, M. and Kigel, J. (2009) A community-level test of the leaf-height-seed ecology strategy scheme in relation to grazing conditions. *Journal of Vegetation Science*, **20**, 392–402.

Gómez-Aparicio, L. and Canham, C. D. (2008) Neighbourhood analyses of the allelopathic effects of the invasive tree *Ailanthus altissima* in temperate forests. *Journal of Ecology*, **96**, 447–458.

Gorchov, D. L. and Trisel, D. E. (2003) Competitive effects of the invasive shrub, Lonicera maackii (Rupr.) Herder (Caprifoliaceae), on the growth and survival of native tree seedlings. *Plant Ecology*, **166** 13–24.

Gorchov, D. L., Cornejo, F., Ascorra, C. and Jaramillo, M. (1993) The role of seed dispersal in the natural regeneration of rain forest after strip-cutting in the Peruvian Amazon. *Vegetatio*, **107/108**, 339–349.

Gorchov, D. L., Rondon, X. J., Cornejo, F., Schaefer, R. L., Janosko, J. M. and Slutz, G. (2013) Edge effects in recruitment of trees, and relationship to seed dispersal patterns, in cleared strips in the Peruvian Amazon. *Ecological Research*, **28**, 53–65.

Gratzer, G., Canham, C. D., Dieckmann, U. *et al.* (2004) Spatio-temporal development of forests – current trends in field methods and models. *Oikos*, **107**, 3–15.

Gravel, D., Canham, C. D., Beaudet, M. and Messier, C. (2010) Shade tolerance, canopy gaps and mechanisims of coexistence of forest trees. *Oikos*, **119**, 475–484.

Gray, A. J. (1879) The pertinacity and predominance of weeds. *American Journal of Science and Arts*, **18**, 161–167.

Greene, D. F. and Johnson, E. A. (1996) Wind dispersal of seeds from a forest into a clearing. *Ecology*, **77**, 595–609.

Grime, J. P. (1973) Competitive exclusion in herbaceous vegetation. *Nature*, **242**, 344–347.

Grime, J. P. (1977) Evidence for the existence of three primary strategies in plans and its relevence to ecological and evolutionary theory. *American Naturalist*, **111**, 1169–1194.

Grime, J. P. (1979) *Plant Strategies and Vegetation Processes*. New York, John Wiley & Sons

Grime, J. P. (1998) Benefits of plant diversity to ecosystems: immediate, filter and founder effects. *Journal of Ecology*, **86**, 902–910.

Grime, J. P. (2001) *Plant Strategies, Vegetation Processes, and Ecosystem Properties*. Chichester, John Wiley & Sons.

Grime, J. P., Macpherson-Stewart, S. F. and Dearman, R. S. (1993) Palatability. In Hendry, G. A. F. and Grime, J. P. (Eds.) *Methods in Comparative Plant Ecology*. London, Chapman and Hall.

Grime, J. P., Cornelissen, J., Thompson, K. and Hodgson, J. G. (1996) Evidence of a causal connection between anti-herbivore defence and the decomposition rate of leaves. *Oikos*, **77**, 489–494.

Gross, K. L., Willig, M. R., Gough, L., Inouye, R. and Cox, S. B. (2000) Patterns of species density and productivity at different spatial scales in herbaceous plant communities. *Oikos*, **89**, 417–427.

Grubb, P. J. (1977) The maintenance of species-richness in plant communities: the importance of the regeneration niche. *Biological Reviews*, **52**, 107–145.

Gunckel, H. A. 1978. John Alvin Small obituary. *Bulletin of the Torrey Botanical Club*, **105**, 70–72.

Gunderson, L. H. (2000) Ecological resilience: in theory and application. *Annual Review of Ecology and Systematics*, **31**, 425–439.

Gunderson, L. H. and Holling, C. S. (Eds.) (2002) *Panarchy: Understanding Transformations in Human and Natural Systems*. Washington DC, Island Press.

Hager, H. A. and McCoy, K. D. (1998) The implications of accepting untested hypotheses: a review of the effects of purple loosestrife (*Lythrum salicaria*) in North America. *Biodiversity and Conservation*, **7**, 1069–1079.

Hale, A. and Kalisz, S. (2012) Perspectives on allelopathic disruption of plant mutualisms: a framework for individual- and population-level fitness consequences. *Plant Ecology*, **213**, 1991–2006.

Hale, A. N., Tonsor, S. J. and Kalisz, S. (2011) Testing the mutualism disruption hypothesis: physiological mechanisms for invasion of intact perennial plant communities. *Ecosphere*, **2**, 12.

Halpern, C. B., Frenzen, P. M., Means, J. E. and Franklin, J. F. (1990) Plant succession in areas of scorched and blown-down forest after the 1980 eruption of Mount Saint Helens, Washington. *Journal of Vegetation Science*, **1**, 181–194.

Handel, S. N. (1985) The intrusion of clonal growth patterns on plant breeding systems. *American Naturalist*, **125**, 367–384.

Hanley, M. E., Fenner, M. and Edwards, P. J. (1995) The effect of seedling age on the likelihood of herbivory by the slug *Deroceras reticulatum*. *Functional Ecology*, **9**, 754–759.

Harper, J. L. (1977) The contributions of terrestrial plant studies to the development of the theory of ecology. In Goulden, C. E. (Ed.) *Changing Scenes in the Natural Sciences*. Philadelphia, PA, Academy of Natural Sciences of Philadelphia.

Harrelson, S. M. and Matlack, G. R. (2006) Influence of stand age and physical environment on the herb composition of second-growth forest, Strouds Run, Ohio, USA. *Journal of Biogeography*, **33**, 1139–1149.

Harris, J. (2009) Soil microbial communities and restoration ecology: facilitators or followers? *Science*, **325**, 573–574.

Harrison, J. S. and Werner, P. A. (1982) Colonization by oak seedlings into a heterogeneous successional habitat. *Canadian Journal of Botany*, **62**, 559–563.

Harrison, S. (1999) Native and alien species diversity at the local and regional scales in a grazed California grassland. *Oecologia*, **121**, 99–106.

Hartnett, D. C. and Bazzaz, F. A. (1983) Physiological integration among intraclonal ramets in *Solidago canadensis*. *Ecology*, **64**, 779–788.

Hartnett, D. C. and Bazzaz, F. A. (1985) The genet and ramet population dynamics of *Solidago canadensis* in an abandoned field. *Journal of Ecology*, **74**, 407–413.

Hatna, E. and Bakker, M. (2011) Abandonment and expansion of arable land in Europe. *Ecosystems*, **14**, 720–731.

Heinselman, M. L. (1973) Fire in the virgin forests of the Boundary Waters Canoe Area, Minnesota. *Journal of Quaternary Research*, **3**, 329–382.

Hille Ris Lambers, J., Adler, P. B., Harpole, W. S., Levine, J. M. and Mayfield, M. M. (2012) Rethinking community assembly through the lens of coexistence theory. *Annual Review of Ecology, Evolution, and Systematics*, **43**, 227–248.

Hils, M. H. and Vankat, J. L. (1982) Species removals from a first-year old-field plant community. *Ecology*, **63**, 705–711.

Hobbs, R. J. and Drake, J. A. (1989) The nature and effects of disturbance relative to invasions. *Biological Invasions: A Global Perspective*. Chichester, John Wiley and Sons Ltd.

Holling, C. S. (1973) Resilience and stability of ecological systems. *Annual Review of Ecology and Systematics*, **4**, 1–23.

Holling, C. S. (1994) Simplifying the complex – the paradigms of ecological function and structure. *Futures*, **26**, 598–609.

Holling, C. S. and Gunderson, L. H. (2002) Resilience and adaptive cycles. In Gunderson, L. H. and Holling, C. S. (Eds.) *Panarchy: Understanding Transformations in Human and Natural Systems*. Washington, DC, Island Press.

Holm, J. A., Shugart, H. H., Van Bloem, S. J. and Larocque, G. R. (2012) Gap model development, validation, and application to succession of secondary subtropical dry forests of Puerto Rico. *Ecological Modelling*, **233**, 70–82.

Holthuijzen, A. M. A. and Sharik, T. L. (1984) The avian seed dispersal system of eastern red cedar (*Juniperus virginiana*). *Canadian Journal of Botany*, **63**, 1508–1515.

Honu, Y. A. K., Gibson, D. J. and Middleton, B. A. (2006) Response of *Tridens flavus* (L.) A. S. Hitchc. to soil nutrients and disturbance in an early successional old field. *Journal of the Torrey Botanical Society*, **133**, 421–428.

Horn, H. S. (1975) Markovian properties of forest succession. In Cody, M. L. (Ed.) *Ecology and Evolution of Communities*. Cambridge, MA, Harvard University Press.

Horn, H. S., Shugart, H. H. and Urban, D. L. (1989) Simulators as models of forest dynamics. In Roughgarden, J., May, R. M. and Levin, S. A. (Eds.) *Perspectives in Ecological Theory*. Princeton, NJ, Princeton University Press.

Horton, J. L. and Hart, S. C. (1998) Hydraulic lift: a potentially important ecosystem process. *Trends in Ecology and Evolution*, **13**, 232–235.

Houle, G. (1992) Spatial relationship between seed and seedling abundance and mortality in a deciduous forest of north-eastern North American. *Journal of Ecology*, **80**, 99–108.

Houseman, G. R. and Gross, K. L. (2011) Linking grassland plant diversity to species pools, sorting and plant traits. *Journal of Ecology*, **99**, 464–472.

Hughes, J. W. and Bechtel, D. A. (1997) Effect of distance from forest edge on regeneration of red spruce and balsam fir in clearcuts. *Canadian Journal of Forest Research*, **27**, 2088–2096.

Hughes, J. W. and Fahey, T. J. (1988) Seed dispersal and colonization in a disturbed northern hardwood forest. *Bulletin of the Torrey Botanical Club*, **115**, 89–99.

Hulme, P. E. (1994) Post-dispersal seed predation in grassland: its magnitude and sources of variation. *Journal of Ecology*, **82**, 645–652.

Huston, M. (1979) A general hypothesis of species diversity. *American Naturalist*, **113**, 81–101.

Huston, M. A. (1994) *Biological Diversity: The Coexistence of Species on Changing Landscapes*. Cambridge, Cambridge University Press.

Huston, M. A. (1997) Hidden treatments in ecological experiments: re-evaluating the ecosystem function of biodiversity. *Oecologia*, **110**, 449–460.

Huston, M. A. (2004) Management strategies for plant invasions: manipulating productivity, disturbance, and competition. *Diversity and Distributions*, **10**, 167–178.

Hutcheson, M. A. 1957. Presentation of the deed and trust. *The William L. Hutcheson Memorial Forest Bulletin*, **1**, 11–12.

Hutchings, M. J. and Wijesinghe, D. K. (1997) Patchy habitats, division of labour and growth dividends in clonal plants. *Trends in Ecology and Evolution*, **12**, 390–394.

Hutchinson, G. E. (1959) Homage to Santa Rosalia or why are there so many kinds of animals? *American Naturalist*, **93**, 145–159.

Hutchinson, T. F. and Vankat, J. L. (1997) Invasibility and effects of Amur honeysuckle in southwestern Ohio forests. *Conservation Biology*, **11**, 1117–1124.

Hutchinson, T. F. and Vankat, J. L. (1998) Landscape structure and spread of the exotic shrub *Lonicera maackii* (Amur honeysuckle) in southwestern Ohio forests. *American Midland Naturalist*, **139**, 383–390.

Hytteborn, H. and Verwijst, T. (2011) The importance of gaps and dwarf trees in the regeneration of Swedish spruce forests: the origin and content of Sernander's (1936) gap dynamics theory. *Scandinavian Journal of Forest Research*, **26**, 3–16.

Inouye, R. S., Huntly, N. J., Tilman, D. et al. (1987) Old-field succession on a Minnesota sand plain. *Ecology*, **68**, 12–26.

Inouye, R. S., Allison, T. D. and Johnson, N. C. (1994) Old field succession on a Minnesota sand plain: effects of deer and other factors on invasion by trees. *Bulletin of the Torrey Botanical Club*, **121**, 266–276.

Iverson, L. R. and Prasad, A. M. (2002) Potential redistribution of tree species habitat under five climate change scenarios in the eastern US. *Forest Ecology and Management*, **155**, 205–222.

Jackson, J. R. and Willemsen, R. W. (1976) Allelopathy in the first stages of secondary succession on the Piedmont of New Jersey. *American Journal of Botany*, **63**, 1015–1023.

Jackson, S. T., Futyma, R. P. and Wilcox, D. A. (1988) A paleoecological test of a classical hydrosere in the Lake Michigan dunes. *Ecology*, **69**, 928–936.

Jensen, K. and Gutekunst, K. (2003) Effects of litter on establishment of grassland plant species: the role of seed size and successional status. *Basic and Applied Ecology*, **4**, 579–587.

Johnson, E. A. and Miyanishi, K. (Eds.) (2007) *Plant Disturbance Ecology: The Process and the Response*. Burlington, MA, Academic Press.

Johnson, E. A. and Miyanishi, K. (2008) Testing the assumptions of chronosequences in succession. *Ecology Letters*, **11** 1–13.

Johnson, M. P. (2000) The influence of patch demographics on metapopulations, with particular reference to successional landscapes. *Oikos*, **88**, 67–74.

Johnson, N. C., Zak, D. R., Tilman, D. and Pfleger, F. L. (1991) Dynamics of vesicular-arbuscular mycorrhizae during old field succession. *Oecologia*, **86**, 349–358.

Johnson, W. C. (1988) Estimating dispersibility of *Acer, Fraxinus* and *Tilia* in fragmented landscapes from patterns of seedling establishment. *Landscape Ecology*, **1**, 175–187.

Johnson, W. C. and Adkisson, C. S. (1985) Dispersal of beech nuts by blue jays in fragmented landscapes. *American Midland Naturalist*, **113**, 319–324.

Johnstone, I. M. (1986) Plant invasion windows: a time-based classification of invasion potential. *Biological Reviews*, **61**, 369–394.

Jurado, E. and Westoby, M. (1992) Seedling growth in relation to seed size among species of arid Australia. *Journal of Ecology*, **80**, 407–416.

Kahmen, S., Poschlod, P. and Schreiber, K.-F. (2002) Conservation management of calcareous grasslands. Changes in plant species composition and response of functional traits during 25 years. *Biological Conservation*, **104**, 319–328.

Kaligarič, M., Meister, M. H., Škornik, S. *et al.* (2011) Grassland succession is mediated by umbelliferous colonizers showing allelopathic potential. *Plant Biosystems*, **145**, 688–698.

Keane, R. M. and Crawley, M. J. (2002) Exotic plant invasions and the enemy release hypothesis. *Trends in Ecology and Evolution*, **17**, 164–170.

Keddy, P. A. (1992) A pragmatic approach to functional ecology. *Functional Ecology*, **6**, 621–626.

Keever, C. (1950) Causes of succession on oldfields of the Piedmont, North Carolina. *Ecological Monographs*, **20**, 229–250.

Keever, C. (1983) A retrospective view of old-field succession after 35 years. *American Midland Naturalist*, **110**, 397–404.

Kidson, R. and Westoby, M. (2000) Seed mass and seedling dimensions in relation to seedling establishment. *Oecologia*, **125**, 11–17.

Kim, Y. and Lee, E. (2011) Comparison of phenolic compounds and the effects of invasive and native species in East Asia: support for the novel weapons hypothesis. *Ecological Research*, **26**, 87–94.

Kimmerer, R. W. (2005) Patterns of dispersal and establishment of bryophytes colonizing natural and experimental treefall mounds in northern hardwood forests. *The Bryologist*, **108**, 391–401.

Klironomos, J. N. (2002) Feedback with soil biota contributes to plant rarity and invasiveness in communities. *Nature*, **417**, 67–70.

Kloeppel, B. D. and Abrams, M. D. (1995) Ecophysiological attributes of the native *Acer saccharum* and the exotic *Acer platanoides* in urban oak forests in Pennsylvania, USA. *Tree Physiology*, **15**, 739–746.

Knapp, A. K., Blair, J. M., Briggs, J. M. *et al.* (1999) The keystone role of bison in North American tallgrass prairie. *BioScience*, **49**, 39–50.

Knops, J. M. H., Tilman, D., Haddad, N. M. *et al.* (1999) Effects of plant species richness on invasion dynamics, disease outbreaks, insect abundances and diversity. *Ecology Letters*, **2**, 286–293.

Kolb, T. E. and Steiner, K. C. (1990) Growth and biomass partitioning response of northern red oak genotypes to shading and grass root competition. *Forest Science*, **36**, 293–303.

Kreyling, J., Jentsch, A. and Beierkuhnlein, C. (2011) Stochastic trajectories of succession initiated by extreme climatic events. *Ecology Letters*, **14**, 758–764.

Kuhman, T. R., Pearson, S. M. and Turner, M. G. (2011) Agricultural land-use history increases non-native plant invasion in a southern Appalachian forest a century after abandonment. *Canadian Journal of Forest Research*, **41**, 920–929.

Kuiters, A. T., Kramer, K., Van Der Hagen, G. J. M. and Schaminée, J. H. J. (2009) Plant diversity, species turnover and shifts in functional traits in coastal dune vegetation: results from permanent plots over a 52-year period. *Journal of Vegetation Science*, **20**, 1053–1063.

Ladwig, L. M. and Meiners, S. J. (2009) Impacts of temperate lianas on tree growth in young deciduous forests. *Forest Ecology and Management*, **259**, 195–200.

Ladwig, L. M. and Meiners, S. J. (2010a) Liana host preference and implications for deciduous forest regeneration. *The Journal of the Torrey Botanical Society*, **137**, 103–112.

Ladwig, L. M. and Meiners, S. J. (2010b) Spatio-temporal dynamics of lianas during 50 years of succession to temperate forest. *Ecology*, **91**, 671–680.

Ladwig, L. M., Meiners, S. J., Pisula, N. L. and Lang, K. A. (2012) Conditional allelopathic potential of temperate lianas. *Plant Ecology*, **213**, 1927–1935.

Lambdon, P. W., Pyšek, P., Basnou, C. *et al.* (2008) Alien flora of Europe: species diversity, temporal trends, geographical patterns and research needs. *Preslia*, **80**, 101–149.

Lambin, E. F., Geist, H. J. and Lepers, E. (2003) Dynamics of land-use and land-cover change in tropical regions. *Annual Review of Environment and Resources*, **28**, 205–241.

Landres, P. B., Knight, R. L., Pickett, S. T. A. and Cadenasso, M. L. (1998) Ecological effects of administrative boundaries. In Knight, R. L. and Landres, P. B. (Eds.) *Stewardship Across Boundaries*. Washington, D.C., Island Press.

Lankau, R. (2010) Soil microbial communities alter allelopathic competition between *Alliaria petiolata* and a native species. *Biological Invasions*, **12**, 2059–2068.

Lankau, R. A. (2011) Resistance and recovery of soil microbial communities in the face of *Alliaria petiolata* invasions. *New Phytologist*, **189**, 536–548.

Laughlin, D. C., Leppert, J. J., Moore, M. M. and Sieg, C. H. (2010) A multi-trait test of the leaf-height-seed plant strategy scheme with 133 species from a pine forest flora. *Functional Ecology*, **24**, 493–501.

Laurance, W. F., Ferreira, L. V., Rankin-De Merona, J. M. and Laurance, S. G. (1998) Rain forest fragmentation and the dynamics of Amazonian tree communities. *Ecology*, **79**, 2032–2040.

Lavoie, M. and Mack, M. C. (2012) Spatial heterogeneity of understory vegetation and soil in an Alaskan upland boreal forest fire chronosequence. *Biogeochemistry*, **107**, 227–239.

Lavorel, S. and Garnier, E. (2002) Predicting changes in community composition and ecosystem functioning from plant traits: revisiting the Holy Grail. *Functional Ecology*, **16**, 545–556.

Lavorel, S. and Grigulis, K. (2011) How fundamental plant functional trait relationships scale-up to trade-offs and synergies in ecosystem services. *Journal of Ecology*, **100**, 128–140.

Lavorel, S., Lepart, J., Debussche, M., Lebreton, J.-D. and Beffy, J.-L. (1994) Small scale disturbances and the maintenance of species diversity in Mediterranean old fields. *Oikos*, **70**, 455–473.

Lavorel, S., Mcintyre, S., Landsberg, J. and Forbes, T. D. A. (1997) Plant functional classifications: from general groups to specific groups based on response to disturbance. *Trends in Ecology and Evolution*, **12**, 474–478.

Lawson, D., Inouye, R. S., Huntly, N. and Carson, W. P. (1999) Patterns of woody plant abundance, recruitment, mortality, and growth in a 65 year chronosequence of old-fields. *Plant Ecology*, **145**, 267–279.

Leck, M. A. and Leck, C. F. (1998) A ten-year seed bank study of old field succession in central New Jersey. *Journal of the Torrey Botanical Society*, **125**, 11–32.

Leck, M. A., Parker, V. T. and Simpson, R. L. (Eds.) (1989) *Ecology of Soil Seed Banks*. San Diego, CA, Academic Press.

Leicht, S. A., Silander, J. A. and Greenwood, K. (2005) Assessing the competitive ability of Japanese stilt grass, *Microstegium vimineum* (Trin.) A. Camus. *Journal of the Torrey Botanical Society*, **132**, 573–580.

Leishman, M. R., Wright, I. J., Moles, A. T. and Westoby, M. (2000) The evolutionary ecology of seed size. In Fenner, M. (Ed.) *Seeds: The Ecology of Regeneration in Plant Communities*. 2nd edn. Oxon, CABI Publishing.

Lencová, K. and Prach, K. (2011) Restoration of hay meadows on ex-arable land: commercial seed mixtures vs. spontaneous succession. *Grass and Forage Science*, **66**, 265–271.

Lenda, M., Skórka, P., Knops, J. M. H., Moroń, D., Tworek, S. and Woyciechowski, M. (2012) Plant establishment and invasions: an increase in a seed disperser combined with land abandonment causes an invasion of the non-native walnut in Europe. *Proceedings of the Royal Society B: Biological Sciences*, **279**, 1491–1497.

LePage, P. T., Canham, C. D., Coates, K. D. and Bartemucci, P. (2000) Seed abundance versus substrate limitation of seedling recruitment in northern temperate forests of British Columbia. *Canadian Journal of Forest Research*, **30**, 415–427.

Levin, S. A. (1976) Population dynamic models in heterogeneous environments. *Annual Review of Ecology and Systematics*, **7**, 287–310.

Levine, J. M. (2000) Species diversity and biological invasions: relating local processes to community pattern. *Science*, **288**, 852–854.

Levins, R. (1966) The strategy of model building in population biology. *American Scientist*, **54**, 421–431.

Li, X. and Wilson, S. D. (1998) Facilitation among woody plants establishing in an old field. *Ecology*, **79**, 2694–2705.

Lichter, J. (1998) Primary succession and forest development on coastal Lake Michigan sand dunes. *Ecological Monographs*, **68**, 487–510.

Likens, G. E. (1984) Beyond the shoreline: a watershed-ecosystem approach. *Verhandlungen der Internationalen Vereinigung für Theoretische und Angewandte Limnologie*, **22**, 1–22.

Likens, G. E. (1991) Human-accelerated environmental change. *BioScience*, **41**, 130.

Liste, H.-H. and White, J. (2008) Plant hydraulic lift of soil water – implications for crop production and land restoration. *Plant and Soil*, **313**, 1–17.

Lockwood, J. L., Cassey, P. and Blackburn, T. (2005) The role of propagule pressure in explaining species invasions. *Trends in Ecology and Evolution*, **20**, 223–228.

Lonsdale, W. M. (1999) Global patterns of plant invasions and the concept of invasibility. *Ecology*, **80**, 1522–1536.

Lord, J., Westoby, M. and Leishman, M. (1995) Seed size and phylogeny in six temperate floras: constraints, niche conservatism, and adaptation. *American Naturalist*, **146**, 349–364.

Loucks, O. L. (1970) Evolution of diversity, efficiency, and community stability. *American Zoologist*, **10**, 17–25.

Luken, J. O. (1990) *Directing Ecological Succession*. London, Chapman and Hall.

Luken, J. O. and Thieret, J. W. (1987) Sumac-directed patch succession on northern Kentucky roadside embankments. *Transactions of the Kentucky Academy of Science*, **48**, 51–54.

Lundholm, J. T. (2009) Plant species diversity and environmental heterogeneity: spatial scale and competing hypotheses. *Journal of Vegetation Science*, **20**, 377–391.

MacArthur, R. and Levins, R. (1967) The limiting similarity, convergence, and divergence of coexisting species. *American Naturalist*, **101**, 377–385.

MacArthur, R. H. and Wilson, E. O. (1967) *The Theory of Island Biogeography*. Princeton, NJ, Princeton University Press.

Mack, R. (2000) Cultivation fosters plant naturalization by reducing environmental stochasticity. *Biological Invasions*, **2**, 111–122.

Mack, R. N., Simberloff, D., Lonsdale, W. M., Evans, H., Clout, M. and Bazzaz, F. A. (2000) Biotic invasions: causes, epidemiology, global consequences, and control. *Ecological Applications*, **10**, 689–710.

Mackey, R. L. and Currie, D. J. (2001) The diversity-disturbance relationship: is it generally strong and peaked? *Ecology*, **82**, 3479–3492.

MacMahon, J. A. (1981) Successional processes: comparisons among biomes with special reference to probable role of and influences on animals. In West, D. C., Shugart, H. H. and Botkin, D. B. (Eds.) *Forest Succession: Concepts and Applications*. New York, Springer-Verlag.

MacMahon, S. M., Cadotte, M. W. and Fukami, T. (2006) Tracking the tractable: using invasions to guide the exploration of conceptual ecology. In Cadotte, M. W., Macmahon, S. M. and Fukami, T. (Eds.) *Conceptual Ecology and Invasion Biology*. Dordrecht, Springer.

Maestre, F. T., Callaway, R. M., Valladares, F. and Lortie, C. J. (2009) Refining the stress-gradient hypothesis for competition and facilitation in plant communities. *Journal of Ecology*, **97**, 199–205.

Manson, R. H. and Stiles, E. W. (1998) Links between microhabitat preferences and seed predation by small mammals in old fields. *Oikos*, **82** 37–50.

Manson, R. H., Ostfeld, R. S. and Canham, C. D. (1998) The effects of tree seed and seedling density on predation rates by rodents in old fields. *Ecoscience*, **5**, 183–190.

Manson, R. H., Ostfeld, R. S. and Canham, C. D. (2001) Long-term effects of rodent herbivores on tree invasion dynamics along forest-field edges. *Ecology*, **82**, 3320–3329.

Marks, P. L. (1974) The role of pin cherry (*Prunus pensylvanica* L.) in the maintenance of stability in northern hardwood ecosystems. *Ecological Monographs*, **44**, 73–88.

Marks, P. L. (1983) On the origin of the field plants of the northeastern United States. *American Naturalist*, **122**, 210–228.

Marks, P. L. and Gardescu, S. (1998) A case study of sugar maple (*Acer saccharum*) as a forest seedling bank species. *Journal of the Torrey Botanical Society*, **125**, 287–296.

Marrs, R. H. (1985) Techniques for reducing soil fertility for nature conservation purposes: a review in relation to research at Roper's Heath, Suffolk, England. *Biological Conservation*, **34**, 307–332.

Martin, P. H. (1999) Norway maple (*Acer platanoides*) invasion of a natural forest stand: understory consequence and regeneration pattern. *Biological Invasions*, **1**, 215–222.

Mason, N. W. H., Carswell, F. E., Richardson, S. J. and Burrows, L. E. (2010) Leaf palatability and decomposability increase during a 200-year-old post-cultural woody succession in New Zealand. *Journal of Vegetation Science*, **22**, 6–17.

Mason, P. A., Wilson, J., Last, F. T. and Walker, C. (1983) The concept of succession in relation to the spread of sheathing mycorrhizal fungi on inoculated tree seedlings growing in unsterile soils. *Plant and Soil*, **71**, 247–256.

Matlack, G. R. (1993) Microenvironment variation within and among forest edge sites in the eastern United States. *Biological Conservation*, **66**, 185–194.

Matlack, G. R. (1994) Plant species migration in a mixed-history forest landscape in Eastern North America. *Ecology*, **75**, 1491–1502.

Matthews, J. W. and Spyreas, G. (2010) Convergence and divergence in plant community trajectories as a framework for monitoring wetland restoration progress. *Journal of Applied Ecology*, **47**, 1128–1136.

Mazer, S. J. (1989) Ecological, taxonomic, and life history correlates of seed mass among Indiana Dune angiosperms. *Ecological Monographs*, **59**, 153–175.

McCarthy, B. C. and Facelli, J. M. (1990) Microdisturbances in oldfields and forests: implications for woody seedling establishment. *Oikos*, **58**, 55–60.

McClanahan, T. R. and Wolfe, R. W. (1993) Accelerating forest succession in a fragmented landscape: the role of birds and perches. *Conservation Biology*, **7**, 279–288.

McConnaughay, K. D. M. and Bazzaz, F. A. (1987) The relationship between gap size and performance of several colonizing annuals. *Ecology*, **68**, 411–416.

McCormick, J. T. and Meiners, S. J. (2000) Season and distance from forest – old field edge effect and seed predation by white footed mice. *Northeastern Naturalist*, **7**, 7–16.

McDonnell, M. J. (1986) Old field vegetation height and the dispersal pattern of bird-disseminated woody plants. *Bulletin of the Torrey Botanical Club*, **113**, 6–11.

McDonnell, M. J. and Stiles, E. W. (1983) The structural complexity of old field vegetation and the recruitment of bird-dispersed plant species. *Oecologia*, **56**, 109–116.

McGill, B. J., Enquist, B. J., Weiher, E. and Westoby, M. (2006) Rebuilding community ecology from functional traits. *Trends in Ecology and Evolution*, **21**, 178–185.

McIntosh, R. P. (1985) *The Background of Ecology*. Cambridge, Cambridge University Press.

McLane, C. R., Battaglia, L. L., Gibson, D. J. and Groninger, J. W. (2012) Succession of exotic and native species assemblages within restored floodplain forests: a test of the parallel dynamics hypothesis. *Restoration Ecology*, **20**, 202–210.

McLendon, T. and Redente, E. F. (1991) Nitrogen and phosphorus effects on secondary succession dynamics on a semi-arid sagebrush site. *Ecology*, **72**, 2016–2024.

Meiners, S. J. (2005) Seed and seedling ecology of *Acer saccharum* and *Acer platanoides*: a contrast between native and exotic congeners. *Northeastern Naturalist*, **12**, 23–32.

Meiners, S. J. (2007) Native and exotic plant species exhibit similar population dynamics during succession. *Ecology*, **88**, 1098–1104.

Meiners, S. J. and Cadenasso, M. L. (2005) The relationship between diversity and exotic plants: cause or consequence of invasion? In Inderjit (Ed.) *Invasive Plants: Ecological and Agricultural Aspects*. Basel, Birkhauser.

Meiners, S. J. and Handel, S. N. (2000) Additive and non-additive effects of herbivory and competition on tree seedling mortality, growth and allocation. *American Journal of Botany*, **87**, 1821–1826.

Meiners, S. J. and LoGiudice, K. (2003) Temporal consistencey in the spatial pattern of seed predation across a forest – old field edge. *Plant Ecology*, **168**, 45–55.

Meiners, S. J. and Martinkovic, M. J. (2002) Survival of and herbivore damage to a cohort of *Quercus rubra* planted across a forest – old field edge. *American Midland Naturalist*, **147**, 247–256.

Meiners, S. J. and Pickett, S. T. A. (1999) Changes in community and population responses across a forest-field gradient. *Ecography*, **22**, 261–267.

Meiners, S. J. and Pickett, S. T. A. (2011) Succession. In Simberloff, D. and Rejmánek, M. (Eds.) *Encyclopedia of Invasive Introduced Species*. Oakland, CA, University of California Press.

Meiners, S. J. and Pickett, S. T. A. (2013) Plant invasion in protected landscapes: exception or expectation? In Foxcroft, L. C., Richardson, D. M., Pyšek, P. and Genovesi, P. (Eds.), *Plant Invasions in Protected Areas*. New York, Springer.

Meiners, S. J., Handel, S. N. and Pickett, S. T. A. (2000) Tree seedling establishment under insect herbivory: edge effects and inter-annual variation. *Plant Ecology*, **151**, 161–170.

Meiners, S. J., Pickett, S. T. A. and Cadenasso, M. L. (2001) Effects of plant invasions on the species richness of abandoned agricultural land. *Ecography*, **24**, 633–644.

Meiners, S. J., Pickett, S. T. A. and Cadenasso, M. L. (2002a) Exotic plant invasions over 40 years of old field succession: community patterns and associations. *Ecography*, **25**, 215–223.

Meiners, S. J., Pickett, S. T. A. and Handel, S. N. (2002b) Probability of tree seedling establishment changes across a forest-old field edge gradient. *American Journal of Botany*, **89**, 466–471.

Meiners, S. J., Cadenasso, M. L. and Pickett, S. T. A. (2004) Beyond biodiversity: individualistic controls of invasion in a self-assembled community. *Ecology Letters*, **7**, 121–126.

Meiners, S. J., Cadenasso, M. L. and Pickett, S. T. A. (2007) Succession on the Piedmont of New Jersey and its implication for ecological restoration. In Cramer, V. A. and Hobbs, R. J. (Eds.) *Old Fields: Dynamics and Restoration of Abandoned Farmland*. Washington, D.C., Island Press.

Meiners, S. J., Rye, T. A. and Klass, J. R. (2009) On a level field: the utility of studying native and non-native species in successional systems. *Applied Vegetation Science*, **12**, 45–53.

Menard, A., Dube, P., Bouchard, A., Canham, C. D. and Marceau, D. J. (2002) Evaluating the potential of the SORTIE forest succession model for spatio-temporal analysis of small-scale disturbances. *Ecological Modelling*, **153**, 81–96.

Middleton, E. L. and Bever, J. D. (2012) Inoculation with a native soil community advances succession in a grassland restoration. *Restoration Ecology*, **20**, 218–226.

Miles, J. (1979) *Vegetation Dynamics*. New York, Wiley.

Miller, K. E. and Gorchov, D. L. (2004) The invasive shrub, *Lonicera maackii*, reduces growth and fecundity of perennial forest herbs. *Oecologia*, **139**, 359–375.

Moles, A. T., Flores-Moreno, H., Bonser, S. P. et al. (2011) Invasions: the trail behind, the path ahead, and a test of a disturbing idea. *Journal of Ecology*, **100**, 116–127.

Molinari, N. and Knight, C. (2010) Correlated evolution of defensive and nutritional traits in native and non-native plants. *Botanical Journal of the Linnean Society*, **163**, 1–13.

Moran, M. A. (1984) Influence of adjacent land use on understory vegetation of New York forests. *Urban Ecology*, **8**, 329–340.

Murray, B. R., Rice, B. L., Keith, D. A., Myerscough, P. J., Howell, J., Floyd, A. G., Mills, K. and Westoby, M. (1999) Species in the tail of rank-abundance curves. *Ecology*, **80**, 1806–1816.

Myers, J. A. and Harms, K. E. (2009) Seed arrival, ecological filters, and plant species richness: a meta-analysis. *Ecology Letters*, **12**, 1250–1260.

Myers, J. A., Vellend, M., Gardescu, S. and Marks, P. L. (2004) Seed dispersal by white-tailed deer: implications for long-distance dispersal, invasion, and migration of plants in eastern North America. *Oecologia*, **139**, 35–44.

Myster, R. W. (2003) Using biomass to model disturbance. *Community Ecology*, **4**, 101–105.

Myster, R. W. (Ed.) (2008) *Post-Agricultural Succession in the Neotropics*, New York, Springer.

Myster, R. W. and Fernández, D. S. (1995) Spatial gradients and patch structure on two Puerto Rican landslides. *Biotropica*, **27**, 149–159.

Myster, R. W. and McCarthy, B. C. (1989) Effects of herbivory and competition on survival of *Carya tomentosa* (Juglandaceae) seedlings. *Oikos*, **56**, 145–148.

Myster, R. W. and Pickett, S. T. A. (1988) Individualistic patterns of annuals and biennials in early successional oldfields. *Vegetatio*, **78**, 53–60.

Myster, R. W. and Pickett, S. T. A. (1990) Initial conditions, history and successional pathways in ten contrasting old fields. *American Midland Naturalist*, **124**, 231–238.

Myster, R. W. and Pickett, S. T. A. (1992a) Dynamics of associations between plants in ten old fields during 31 years of succession. *Journal of Ecology*, **80**, 291–302.

Myster, R. W. and Pickett, S. T. A. (1992b) Effect of palatability and dispersal mode on spatial patterns of trees in oldfields. *Bulletin of the Torrey Botanical Club*, **119**, 145–151.

Myster, R. W. and Pickett, S. T. A. (1993) Effects of litter, distance, density, and vegetation patch type on postdispersal tree seed predation. *Oikos*, **66**, 381–388.

Myster, R. W. and Pickett, S. T. A. (1988). Individualistic patterns of annuals and biennials in early successional oldfields. *Vegetatio*, **78**, 53–60.

Myster, R. W. and Walker, L. R. (1997) Plant successional pathways on Puerto Rican landslides. *Journal of Tropical Ecology*, **13**, 165–173.

Naeem, S., Knops, J. M. H., Tilman, D., Howe, K. M., Kennedy, T. and Gale, S. (2000) Plant diversity increases resistance to invasion in the absence of covarying extrinsic factors. *Oikos*, **91**, 97–108.

Navas, M. L., Roumet, C., Bellmann, A., Laurent, G. and Garnier, E. (2010) Suites of plant traits in species from different stages of a Mediterranean secondary succession. *Plant Biology*, **12**, 183–196.

New Jersey Department of Environmental Protection – New Jersey Geological Survey (2006). *Information Circular: Physiographic Provinces of New Jersey*. Prepared by Richard Dalton in 2003 and reprinted in 2006.

Niering, W. A. and Dreyer, G. D. (1989) Effects of prescribed burning on *Andropogon scoparius* in postagricultural grasslands in Connecticut. *American Midland Naturalist*, **122**, 88–102.

Niering, W. A. and Egler, F. E. (1955) A shrub community of *Viburnum lentago*, stable for twenty-five years. *Ecology*, **36**, 356–360.

Niering, W. A., Dreyer, G. D., Egler, F. E. and Anderson, J. P. (1986) Stability of *Viburnum lentago* shrub community after 30 years. *Bulletin of the Torrey Botanical Club*, **113**, 23–27.

Norden, N., Daws, M. I., Antoine, C., Gonzalez, M. A., Garwood, N. D. and Chave, J. (2009) The relationship between seed mass and mean time to germination for 1037 tree species across five tropical forests. *Functional Ecology*, **23**, 203–210.

Nurse, R. E., Darbyshire, S. J., Bertin, C. and DiTommaso, A. (2009) The biology of Canadian weeds. 141. *Setaria faberi* Herrm. *Canadian Journal of Plant Science*, **89**, 379–404.

Nuttle, T., Royo, A. A., Adams, M. B. and Carson, W. P. (2013) Historic disturbance regimes promote tree diversity only under low browsing regimes in eastern deciduous forest. *Ecological Monographs*, **83**, 3–17.

Odum, E. P. (1960) Organic production and turnover in old field succession. *Ecology*, **41**, 34–49.

Odum, E. P. (1969) The strategy of ecosystem development. *Science*, **164**, 262–270.

Ogden, L., Heynen, N., Oslender, U., West, P., Kassam, K.-A. and Robbins, P. (2013) Global assemblages, resilience, and Earth stewardship in the Anthropocene. *Frontiers in Ecology and the Environment*, **11**, 341–347.

Olsson, E. G. (1987) Effects of dispersal mechanisms on the initial pattern of old-field forest succession. *Acta Oecologia/Oecologia Generalis*, **8**, 379–390.

Omancini, M., Chaneton, E. J., Leon, R. S. C. and Batista, W. B. (1995) Old-field successional dynamics on the Inland Pampa, Argentina. *Journal of Vegetation Science*, **6**, 309–316.

Oosting, H. J. (1957) Plant ecology and natural areas. *The William L. Hutcheson Memorial Forest Bulletin*, **1**, 26–32.

Oosting, H. J. and Humphreys, M. E. (1940) Buried viable seeds in a successional series of old field and forest soils. *Bulletin of the Torrey Botanical Club*, **67**, 254–273.

Ortega, Y. K. and Pearson, D. E. (2005) Weak vs. strong invaders of natural plant communities: assessing invasibility and impact. *Ecological Applications*, **15**, 651–661.

Ostrom, E. (2005) *Understanding Institutional Diversity*. Princeton, NJ, Princeton University Press.

Pacala, S. W. and Tilman, D. (1994) Limiting similarity in mechanistic and spatial models of plant competition in heterogeneous environments. *American Naturalist*, **143**, 222–257.

Pacala, S. W., Canham, C. D. and Silander, J. A., Jr. (1993) Forest models defined by field measurements: the design of a northeastern forest simulator. *Canadian Journal of Forest Research*, **23**, 1980–1988.

Pacala, S. W., Canham, C. D., Saponara, J. *et al.* (1996) Forest models defined by field measurements: estimation, error analysis and dynamics. *Ecological Monographs*, **66** 1–43.

Pakeman, R. J., Garnier, E., Lavorel, S. *et al.* (2008) Impact of abundance weighting on the response of seed traits to climate and land use. *Journal of Ecology*, **96**, 355–366.

Pardini, E. A., Drake, J. M., Chase, J. M. and Knight, T. M. (2009) Complex population dynamics and control of the invasive biennial *Alliaria petiolata* (garlic mustard). *Ecological Applications*, **19**, 387–397.

Parker, I. M. and Gilbert, G. S. (2007) When there is no escape: the effects of natural enemies on native, invasive, and noninvasive plants. *Ecology*, **88**, 1210–1224.

Parker, I. M., Simberloff, D., Lonsdale, W. M. *et al.* (1999) Impact: toward a framework for understanding the ecological effects of invaders. *Biological Invasions*, **1**, 3–19.

Parker, J. D., Burkepile, D. E. and Hay, M. E. (2006) Opposing effects of native and exotic herbivores on plant invasions. *Science*, **311**, 1459–1461.

Parrish, J. A. D. and Bazzaz, F. A. (1982) Responses of plants from three successional communities to a nutrient gradient. *Journal of Ecology*, **70**, 233–248.

Patterson, D. T. (1976) The history and distribution of five exotic weeds in North Carolina. *Castanea*, **41**, 177–180.

Pauchard, A. and Alaback, P. B. (2004) Influence of elevation, land use, and landscape context on patterns of alien plant invasions along roadsides in protected areas of South-Central Chile. *Conservation Biology*, **18**, 238–248.

Peet, R. K. and Christensen, N. L. (1980) Succession: a population process. *Vegetatio*, **43**, 131–140.

Pellmyr, O. (2002) Pollination by animals. In Herrera, C. M. and Pellmyr, O. (Eds.) *Plant–Animal Interactions*. Malden, Blackwell Publishing.

Pemberton, R. W. and Liu, H. (2009) Marketing time predicts naturalization of horticultural plants. *Ecology*, **90**, 69–80.

Peroni, P. A. (1994) Invasion of red maple (*Acer rubrum* L.) during old field succession in the North Carolina Piedmont: age structure of red maple in young pine stands. *Bulletin of the Torrey Botanical Club*, **121**, 357–359.

Peters, D. P. C., Lugo, A. E., Chapin, F. S., III et al. (2011) Cross-system comparisons elucidate disturbance complexities and generalities. *Ecosphere*, **2**, art 81.

Peters, H. A., Baur, B., Bazzaz, F. A. and Körner, C. (2000) Consumption rates and food preferences of slugs in a calcareous grassland under current and future CO_2. *Oecologia*, **125**, 72–81.

Peterson, C. J. (2000) Catastrophic wind damage to North American forests and the potential impact of climate change. *Science of the Total Environment*, **262**, 287–311.

Peterson, G., Allen, C. R. and Holling, C. S. (1998) Ecological resilience, biodiversity, and scale. *Ecosystems*, **1**, 6–18.

Petraitis, P. S., Latham, R. E. and Niesenbaum, R. A. (1989) The maintenance of species diversity by disturbance. *Quarterly Review of Biology*, **64**, 393–418.

Petranka, J. W. and Mcpherson, J. K. (1979) The role of *Rhus copallina* in the dynamics of the forest prairie ecotone in north-central Oklahoma. *Ecology*, **60**, 956–965.

Pianka, E. R. (1970) On r- and K- selection. *American Naturalist*, **104**, 592–597.

Pickett, S. T. A. (1976) Succession: an evolutionary interpretation. *American Naturalist*, **110**, 107–119.

Pickett, S. T. A. (1980) Non-equilibrium coexistence of plants. *Bulletin of the Torrey Botanical Club*, **107**, 238–248.

Pickett, S. T. A. (1983) The absence of an *Andropogon* stage in old-field succession at the Hutcheson Memorial Forest. *Bulletin of the Torrey Botanical Club*, **110**, 533–535.

Pickett, S. T. A. (1989) Space-for-time substitution as an alternative to long-term studies. In Likens, G. E. (Ed.) *Long-Term Studies in Ecology: Approaches and Alternatives*. New York, Springer-Verlag.

Pickett, S. T. A. and Bazzaz, F. A. (1978) Germination of co-occurring annual species on a soil moisture gradient. *Bulletin of the Torrey Botanical Club*, **105**, 312–316.

Pickett, S. T. A. and Cadenasso, M. L. (2005) Vegetation dynamics. In Van Der Maarel, E. (Ed.) *Vegetation Ecology*. Malden, MA, Blackwell Publishing.

Pickett, S. T. A. and Thompson, J. N. (1978) Patch dynamics and the design of nature reserves. *Biological Conservation*, **13**, 27–37.

Pickett, S. T. A. and White, P. S. (1985) *The Ecology of Natural Disturbance and Patch Dynamics*. San Diego, CA, Academic Press.

Pickett, S. T. A., Collins, S. L. and Armesto, J. J. (1987) Models, mechanisms and pathways of succession. *Botanical Review*, **53**, 335–371.

Pickett, S. T. A., Cadenasso, M. L. and Bartha, S. (2001) Implications from the Buell-Small Succession Study for vegetation restoration. *Applied Vegetation Science*, **4**, 41–52.

Pickett, S. T. A., Cadenasso, M. L. and Meiners, S. J. (2009) Ever since Clements: from succession to vegetation dynamics and understanding to intervention. *Applied Vegetation Science*, **12**, 9–21.

Pickett, S. T. A., Meiners, S. J. and Cadenasso, M. L. (2011) Domain and propositions of succession theory. In Scheiner, S. M. and Willig, M. R. (Eds.) *The Theory of Ecology*. Chicago, IL, University of Chicago Press.

Pickett, S. T. A., Cadenasso, M. L. and McGrath, B. (Eds.) (2013a) *Resilience in Ecology and Urban Design: Linking Theory and Practice for Sustainable Cities*. New York, Springer.

Pickett, S. T. A., Cadenasso, M. L. and Meiners, S. J. (2013b) Vegetation dynamics. In Van Der Maarel, E. and Franklin, J. (Eds.) *Vegetation Ecology*. 2nd edn. Malden, MA, Blackwell Science Ltd.

Pillar, V. D., Duarte, L. D. S., Sosinski, E. E. and Joner, F. (2009) Discriminating trait-convergence and trait-divergence assembly patterns in ecological community gradients. *Journal of Vegetation Science*, **20**, 334–348.

Pimentel, D., Lach, L., Zuninga, R. and Morrison, D. (2000) Environmental and economic costs of nonindigenous species in the United States. *BioScience*, **50**, 53–65.

Pinder, J. E., III, Golley, F. B. and Lide, R. F. (1995) Factors affecting limited reproduction of loblolly pine in a large old field. *Bulletin of the Torrey Botanical Club*, **122**, 306–311.

Pisula, N. L. and Meiners, S. J. (2010a) Allelopathic effects of goldenrod species on turnover in successional communities. *American Midland Naturalist*, **163**, 161–172.

Pisula, N. L. and Meiners, S. J. (2010b) Relative allelopathic potential of invasive plant species in a young disturbed woodland. *Journal of the Torrey Botanical Society*, **137**, 81–87.

Planty-Tabacchi, A. M., Tabacchi, E., Naiman, R. J., Deferrari, C. and Decamps, H. (1996) Invasibility of species-rich communities in riparian zones. *Conservation Biology*, **10**, 598–607.

Plue, J., Verheyen, K., Van Calster, H. et al. (2010) Seed banks of temperate deciduous forests during secondary succession. *Journal of Vegetation Science*, **21**, 965–978.

Pound, C. E. and Egler, F. E. (1953) Brush control in southeastern New York: fifteen years of stable tree-less communities. *Ecology*, **34**, 63–73.

Prach, K. and Hobbs, R. J. (2008) Spontaneous succession versus technical reclamation in the restoration of disturbed sites. *Restoration Ecology*, **16**, 363–366.

Prach, K. and Pyšek, P. (1999) How do species dominating in succession differ from others? *Journal of Vegetation Science*, **10**, 383–392.

Prach, K. and Walker, L. R. (2011) Four opportunities for studies of ecological succession. *Trends in Ecology and Evolution*, **26**, 119–123.

Prach, K., Pyšek, P. and Smilauer, P. (1993) On the rate of succession. *Oikos*, **66**, 343–346.

Prach, K., Pyšek, P. and Smilauer, P. (1997) Changes in species traits during succession: a search for pattern. *Oikos*, **79**, 201–205.

Prach, K., Bartha, S., Joyce, C. B. et al. (2001) The role of spontaneous vegetation succession in ecosystem restoration: a perspective. *Applied Vegetation Science*, **4**, 111–114.

Prach, K., Řehounková, K., Lencová, K. et al. (2014) Vegetation succession in restoration of disturbed sites in Central Europe: the direction of succession and species richness across 19 seres. *Applied Vegetation Science*, **17**, 193–200.

Pyšek, P., Sádlo, J., Mandák, B. and Jarošík, V. (2003) Czech alien flora and the historical pattern of its formation: what came first to central Europe? *Oecologia*, **135**, 122–130.

Pyšek, P., Křivánek, M. and Jarošík, V. (2009) Planting intensity, residence time, and species traits determine invasion success of alien woody species. *Ecology*, **90**, 2734–2744.

Quinn, J. A. and Meiners, S. J. (2004) Growth rates, survivorship, and sex ratios of *Juniperus virginiana* on the New Jersey Piedmont from 1963 to 2000. *Journal of the Torrey Botanical Society*, **131**, 187–194.

Raevel, V., Violle, C. and Munoz, F. (2012) Mechanisms of ecological succession: insights from plant functional strategies. *Oikos*, **121**, 1761–1770.

Rankin, W. T. and Pickett, S. T. A. (1989) Time of establishment of red maple (*Acer rubrum*) in early oldfield succession. *Bulletin of the Torrey Botanical Club*, **116**, 182–186.

Raunkiaer, C. (1934) *The Life Forms of Plants and Statistical Plant Geography*. Oxford, The Clarendon Press.

Reader, R. J. (1991) Control of seedling emergence by ground cover: a potential mechanism involving seed predation. *Canadian Journal of Botany*, **69**, 2084–2087.

Reader, R. J. and Beisner, B. E. (1991) Species-dependent effects of seed predation and ground cover on seedling emergence of old-field forbs. *American Midland Naturalist*, **126**, 279–286.

Redman, C. L. and Kinzig, A. P. (2003) Resilience of past landscapes: resilience theory, society, and the longue duree. *Conservation Ecology*, **7**, Article 14.

Rees, M., Condit, R., Crawley, M., Pacala, S. and Tilman, D. (2001) Long-term studies of vegetation dynamics. *Science*, **293**, 650–655.

Řehounková, K. and Prach, K. (2010) Life-history traits and habitat preferences of colonizing plant species in long-term spontaneous succession in abandoned gravel-sand pits. *Basic and Applied Ecology*, **11**, 45–53.

Reich, P. B., Ellsworth, D. S. and Walters, M. B. (1998) Leaf structure (specific leaf area) modulates photosynthesis–nitrogen relations: evidence from within and across species and functional groups. *Functional Ecology*, **12**, 948–958.

Reinecke, M. K., Pigot, A. L. and King, J. M. (2008) Spontaneous succession of riparian fynbos: is unassisted recovery a viable restoration strategy? *South African Journal of Botany*, **74**, 412–420.

Rejmánek, M. (1996) A theory of seed plant invasiveness: the first sketch. *Biological Conservation*, **78**, 171–181.

Rejmánek, M. and Drake, J. A. (1989) Invasibility of plant communities. *Biological Invasions: A Global Perspective*. Chichester, John Wiley & Sons.

Rejmánek, M. and Rosen, E. (1992) Cycles of heterogeneity during succession: a premature generalization? *Ecology*, **73**, 2329–2331.

Rejmánek, M. and Sandlund, O. Y. (1999) Invasive plant species and invasible ecosystems. In *Invasive Species and Biodiversity Management*. Dordrecht, Kluwer Academic Publishers.

Reynolds, H. L., Packer, A., Bever, J. D. and Clay, K. (2003) Grassroots ecology: plant-microbe-soil interactions as drivers of plant community structure and dynamics. *Ecology*, **84**, 2281–2291.

Ribbens, E., Silander, J. A. and Pacala, S. W. (1994) Seedling recruitment in forests: calibrating models to predict patterns of tree seedling dispersion. *Ecology*, **75**, 1794–1806.

Ricklefs, R. E. (2008) Disintegration of the ecological community. *The American Naturalist*, **172**, 741–750.

Roberts, K. J. and Anderson, R. C. (2005) Effect of garlic mustard [*Alliaria petiolata* (Beib, Cavarra and Grande)] extracts on plants and arbuscular Mycorrhizal (AM) fungi. *American Midland Naturalist*, **146**, 146–152.

Roberts, T. L. and Vankat, J. L. (1991) Floristics of a chronosequence corresponding to old field-deciduous forest succession in southwestern Ohio. II. Seed banks. *Bulletin of the Torrey Botanical Club*, **118**, 377–384.

Robertson, D. J., Robertson, M. C. and Tague, T. (1994) Colonization dynamics of four exotic plants in a Northern Piedmont natural area. *Bulletin of the Torrey Botanical Club*, **121**, 107–118.

Robertson, G. P. and Vitousek, P. M. (1981). Nitrification potentials in primary and secondary succession. *Ecology*, **62**, 376–386.

Robinson, G. R. and Handel, S. N. (1993) Forest restoration on a closed landfill: rapid addition of new species by bird dispersal. *Conservation Biology*, **7**, 271–278.

Robinson, G. R. and Handel, S. N. (2000) Directing spatial patterns of recruitment during an experimental urban woodland reclamation. *Ecological Applications*, **10**, 174–188.

Robinson, G. R., Holt, R. D., Gaines, M. S. *et al.* (1992) Diverse and contrasting effects of habitat fragmentation. *Science*, **257**, 524–526.

Rogers, W. E. and Hartnett, D. C. (2001) Temporal vegetation dynamics and recolonization mechanisms on different-sized soil disturbances in tallgrass prairie. *American Journal of Botany*, **88**, 1634–1642.

Rouget, M. and Richardson, D. M. (2003) Inferring process from pattern in plant invasions: a semimechanistic model incorporating propagule pressure and environmental factors. *American Naturalist*, **162**, 713–724.

Roxburgh, S. H., Shea, K. and Wilson, J. B. (2004) The intermediate disturbance hypothesis: patch dynamics and mechanisms of species coexistence. *Ecology*, **85**, 359–371.

Royo, A. A. and Carson, W. P. (2006) On the formation of dense understory layers in forests worldwide: consequences and implications for forest dynamics, biodiversity, and succession. *Canadian Journal of Forest Research*, **36**, 1345–1362.

Runkle, J. R. (1982) Patterns of disturbance in some old-growth mesic forests of Eastern North America. *Ecology*, **63**, 1533–1546.

Ruprecht, E. (2006) Successfully recovered grassland: a promising example from Romanian old-fields. *Restoration Ecology*, **14**, 473–480.

Ruprecht, E., Bartha, S., Botta-Dukát, Z. and Szabó, A. (2007) Assembly rules during old-field succession in two contrasting environments. *Community Ecology*, **8**, 31–40.

Saverimuttu, T. and Westoby, M. (1996) Seedling longevity under deep shade in relation to seed size. *Journal of Ecology*, **84**, 681–689.

Schaffner, U., Ridenour, W. M., Wolf, V. C. *et al.* (2011) Plant invasions, generalist herbivores, and novel defense weapons. *Ecology*, **92**, 829–835.

Scheffer, M., Westley, F., Brock, W. A. and Holmgren, M. (2002) Dynamic interaction of societies and ecosystems – linking theories from ecology, economy, and sociology. In Gunderson, L. H. and Holling, C. S. (Eds.) *Panarchy: Understanding Transformations in Human and Natural Systems*. Washington DC, Island Press.

Scheiner, S. and Willig, M. (2005) Developing unified theories in ecology as exemplified with diversity gradients. *American Naturalist*, **166**, 458–469.

Scheiner, S. M. and Willig, M. R. (2011) A general theory of ecology. *The Theory of Ecology*. Chicago, IL, University of Chicago Press

Schlaepfer, D. R., Glättli, M., Fischer, M. and Van Kleunen, M. (2010) A multi-species experiment in their native range indicates pre-adaptation of invasive alien plant species. *New Phytologist*, **185**, 1087–1099.

Schmitz, O. J. (2010) *Resolving Ecosystem Complexity*. Princeton, NJ, Princeton University Press.

Schnitzer, S. A. and Bongers, F. (2002) The ecology of lianas and their role in forests. *Trends in Ecology and Evolution*, **17**, 223–230.

Schoonmaker, P. and McKee, A. (1988) Species composition and diversity during secondary succession of coniferous forests in the Western Cascade Mountains of Oregon. *Forest Science*, **34**, 960–979.

Seifan, M., Seifan, T., Schiffers, K., Jeltsch, F. and Tielbörger, K. (2013) Beyond the competition-colonization trade-off: linking multiple trait response to disturbance characteristics. *The American Naturalist*, **181**, 151–160.

Shea, K. and Chesson, P. (2002) Community ecology theory as a framework for biological invasions. *Trends in Ecology and Evolution*, **17**, 170–176.

Shea, K., Roxburgh, S. H. and Rauschert, E. S. J. (2004) Moving from pattern to process: coexistence mechanisms under intermediate disturbance regimes. *Ecology Letters*, **7**, 491–508.

Sheil, D. (1999) Developing tests of successional hypotheses with size-structured populations, and an assessment using long-term data from a Ugandan rain forest. *Plant Ecology*, **140**, 117–127.

Shipley, B. (2010) *From Plant Traits to Vegetation Structure*. Cambridge, UK, Cambridge University Press.

Shugart, H. H. and West, D. C. (1980) Forest succession models. *BioScience*, **30**, 308–313.

Shuman, J. K., Shugart, H. H. and O'Halloran, T. L. (2011) Sensitivity of Siberian larch forests to climate change. *Global Change Biology*, **17**, 2370–2384.

Shure, D. J. (1971) Insecticide effects on early succession in an old field ecosystem. *Ecology*, **52**, 271–279.

Silvertown, J., Dodd, M. and Gowing, D. (2001) Phylogeny and the niche structure of meadow plant communities. *Journal of Ecology*, **89**, 428–435.

Silvertown, J., Dodd, M., Gowing, D., Lawson, C. and McConway, K. (2006) Phylogeny and the hierarchical organization of plant diversity. *Ecology*, **87**, S39–S49.

Simard, S. W., Jones, M. D., Durall, D. M., Perry, D. A., Myrold, D. D. and Molina, R. (1997a) Reciprocal transfer of carbon isotopes between ectomycorrhizal *Betula papyrifera* and *Pseudotsuga menziesii*. *New Phytologist*, **137**, 529–542.

Simard, S. W., Perry, D. A., Jones, M. D., Myrold, D. D., Durall, D. M. and Molina, R. (1997b) Net transfer of carbon between ectomycorrhizal tree species in the field. *Nature*, **388**, 579–582.

Simberloff, D. (1980) A succession of paradigms in ecology: essentialism to materialism and probabilism. *Synthese*, **43**, 3–39.

Simberloff, D. (2011) Native invaders. In Simberloff, D. and Rejmánek, M. (Eds.) *Encyclopedia of Biological Invasions*. Berkeley, CA, University of California Press.

Simberloff, D., Souza, L., Nuñez, M. A., Barrios-Garcia, M. N. and Bunn, W. (2012) The natives are restless, but not often and mostly when disturbed. *Ecology*, **93**, 598–607.

Sipe, T. W. and Bazzaz, F. A. (1994) Gap partitioning among maples (*Acer*) in central New England: shoot architecture and photosynthesis. *Ecology*, **75**, 2318–2332.

Sipe, T. W. and Bazzaz, F. A. (1995) Gap partitioning among maples (*Acer*) in central New England: survival and growth. *Ecology*, **76**, 1587–1602.

Small, C. J. and McCarthy, B. C. (2003) Spatial and temporal variability of herbaceous vegetation in an eastern deciduous forest. *Plant Ecology*, **164**, 37–48.

Small, J. A. (1961) Drought response in William L. Hutcheson Memorial Forest, 1957. *Bulletin of the Torrey Botanical Club*, **88**, 180–183.

Small, J. A. (1975) Murray Fife Buell: October 5, 1905 – July 3, 1975. *Bulletin of the Torrey Botanical Club*. **102**, 201–206.

Smith, D. E. (1986) *The Practice of Silviculture*. New York, NY, John Wiley and Sons.

Spyreas, G., Meiners, S. J., Matthews, J. W. and Molano-Flores, B. (2012) Successional trends in floristic quality. *Journal of Applied Ecology*, **49**, 339–348.

Standish, R. J., Cramer, V. A., Hobbs, R. J. and Kobryn, H. T. (2006) Legacy of land-use evident in soils of Western Australia's wheatbelt. *Plant and Soil*, **280**, 189–207.

Stapanian, M. A., Sundberg, S. D., Baumgardner, G. A. and Liston, A. (1998) Alien plant species composition and associations with anthropogenic disturbance in North America. *Plant Ecology*, **139**, 49–62.

Stearns, S. C. (1992) *The Evolution of Life Histories*. New York, Oxford University Press.

Steffan-Dewenter, I. and Tscharntke, T. (2001) Succession of bee communities on fallows. *Ecography*, **24**, 83–93.

Steffan-Dewenter, I., Münzenberg, U., Bürger, C., Thies, C. and Tscharntke, T. (2002) Scale-dependent effects of landscape context on three pollinator guilds. *Ecology*, **83**, 1421–1432.

Stevens, M. H. H. and Carson, W. P. (1999) Plant density determines species richness along an experimental fertility gradient. *Ecology*, **80** 455–465.

Stevenson, J. R. (1957). The life and work of William L. Hutcheson. *The William L. Hutcheson Memorial Forest Bulletin*, **1**, 5–10.

Stohlgren, T. J., Bull, K. A., Otsuki, Y., Villa, C. A. and Lee, M. (1998) Riparian zones as havens for exotic plant species in the central grasslands. *Plant Ecology*, **138**, 113–125.

Stohlgren, T. J., Binkley, D., Chong, G. W. *et al.* (1999) Exotic plant species invade hot spots of native plant diversity. *Ecological Monographs*, **69**, 25–46.

Stohlgren, T. J., Otsuki, Y., Villa, C. A., Lee, M. and Belnap, J. (2001) Patterns of plant invasions: a case example in native species hotspots and rare habitats. *Biological Invasions*, **3**, 37–50.

Studlar, S. M. (1980) Trampling effects on bryophytes: trail surveys and experiments. *The Bryologist*, **83**, 301–313.

Studlar, S. M. (1983) Recovery of trampled bryophyte communities near Mountain Lake, Virginia. *Bulletin of the Torrey Botanical Club*, **110**, 1–11.

Sutherland, S. (2004) What makes a weed a weed: life history traits of native and exotic plants in the USA. *Oecologia*, **141**, 24–39.

Svensson, J. R., Lindegarth, M., Jonsson, P. R. and Pavia, H. (2012) Disturbance-diversity models: what do they really predict and how are they tested? *Proceedings of the Royal Society B: Biological Sciences*, **279**, 2163–2170.

Swink, F. and Wilhelm, G. (1994) *Plants of the Chicago Region*. Indianapolis, IN, Indiana Academy of Science.

Symstad, A. J. (2000) A test of the effects of functional group richness and composition on grassland invasibility. *Ecology*, **81**, 99–109.

Taft, J. B., Wilhelm, G. S., Ladd, D. M. and Masters, L. A. (1997) *Floristic Quality Assessment for Vegetation in Illinois: A Method for Assessing Vegetation Integrity*. Westville, IL, Illinois Native Plant Society.

Tainter, J. A. (1988) *The Collapse of Complex Societies*. New York, Cambridge University Press.

Tainter, J. A. (2006) Social complexity and sustainability. *Ecological Complexity*, **3**, 91–103.

Tansley, A. G. (1935) The use and abuse of vegetational concepts and terms. *Ecology*, **16**, 284–307.

Tecco, P. A., Díaz, S., Cabido, M. and Urcelay, C. (2010) Functional traits of alien plants across contrasting climatic and land-use regimes: do aliens join the locals or try harder than them? *Journal of Ecology*, **98**, 17–27.

Temperton, V. M. and Hobbs, R. J. (2004) The search for ecological assembly rules and its relevance to restoration ecology. In Temperton, V. M., Hobbs, R. J., Nuttle, T. and Halle, S. (Eds.) *Assembly Rules and Restoration Ecology*. Washington, D.C., Island Press.

Thompson, K., Hodgson, J. G. and Rich, T. C. G. (1995) Native and alien invasive plants: more of the same? *Ecography*, **18**, 390–402.

Thorpe, A. S., Thelen, G. C., Diaconu, A. and Callaway, R. M. (2009) Root exudate is allelopathic in invaded community but not in native community: field evidence for the novel weapons hypothesis. *Journal of Ecology*, **97**, 641–645.

Tilman, D. (1983) Plant succession and gopher disturbance along an experimental gradient. *Oecologia*, **60**, 285–292.

Tilman, D. (1985) The resource-ratio hypothesis of plant succession. *American Naturalist*, **125**, 827–852.

Tilman, D. (1987) Secondary succession and the pattern of plant dominance along experimental nitrogen gradients. *Ecological Monographs*, **57** 189–214.

Tilman, D. (1990) Constraints and tradeoffs: toward a predictive theory of competition and succession, *Oikos*, **58**, 3–15.

Tilman, D. (1994) Competition and biodiversity in spatially structured habitats. *Ecology*, **75**, 2–16.

Tilman, D. and Downing, J. A. (1994) Biodiversity and stability in grasslands. *Nature*, **367**, 363–365.

Tilman, D. and Wedin, D. (1991) Dynamics of nitrogen competition between successional grasses. *Ecology*, **72**, 1038–1049.

Tilman, D., Pacala, S. and Ricklefs, R. E. (1993) The maintenance of species richness in plant communities. In *Species Diversity in Ecological Communities*. Chicago, IL, University of Chicago Press.

Tilman, D., Knops, J., Wedin, D. *et al.* (1997a) The influence of functional diversity and composition on ecosystem processes. *Science*, **277**, 1300–1302.

Tilman, D., Lehman, C. L. and Thomson, K. T. (1997b) Plant diversity and ecosystem productivity: theoretical considerations. *Proceedings of the National Academy of Science*, **94**, 1857–1861.

Tognetti, P. M., Chaneton, E. J., Omacini, M., Trebino, H. J. and León, R. J. C. (2010) Exotic vs. native plant dominance over 20 years of old-field succession on set-aside farmland in Argentina. *Biological Conservation*, **143**, 2494–2503.

Trinder, C. J., Brooker, R. W. and Robinson, D. (2013) Plant ecology's guilty little secret: understanding the dynamics of plant competition. *Functional Ecology*, **27**, 918–929.

Troumbis, A. Y., Galanidis, A. and Kokkoris, G. D. (2002) Components of short-term invasibility in experimental Mediterranean grasslands. *Oikos*, **98**, 239–250.

Turnbull, L. A., Rees, M. and Crawley, M. J. (1999) Seed mass and the competition/colonization trade-off: a sowing experiment. *Journal of Ecology*, **87**, 899–912.

Turner, M. G., Gardner, R. H., Dale, V. H. and O'Neill, R. V. (1989) Predicting the spread of disturbance across heterogeneous landscapes. *Oikos*, **55**, 121–129.

Turner, M. G., Hargrove, W. W., Gardner, R. H. and Romme, W. H. (1994) Effects of fire on landscape heterogeneity in Yellowstone National Park, Wyoming. *Journal of Vegetation Science*, **5**, 731–742.

Uesugi, A. and Kessler, A. (2013) Herbivore exclusion drives the evolution of plant competitiveness via increased allelopathy. *New Phytologist*, **198**, 916–924.

Ugolini, F. C. (1964). Soil development on the red beds of New Jersey. *The William L. Hutcheson Memorial Forest Bulletin*, **2**, 1–34.

Valéry, L., Fritz, H. and Lefeuvre, J.-C. (2013) Another call for the end of invasion biology. *Oikos*, **122**, 1143–1146.

van der Maarel, E. (1996) Pattern and process in the plant community: fifty years after A. S. Watt. *Journal of Vegetation Science*, **7**, 19–28.

van der Maarel, E. and Franklin, J. (Eds.) (2013) *Vegetation Ecology*, 2nd edn. Chichester, Wiley-Blackwell.

van der Maarel, E. and Sykes, M. T. (1993) Small-scale plant species turnover in a limestone grassland: the carousel model and some comments on the niche concept. *Journal of Vegetation Science*, **4**, 179–188.

van Kleunen, M. and Fischer, M. (2009) Release from foliar and floral fungal pathogen species does not explain the geographic spread of naturalized North American plants in Europe. *Journal of Ecology*, **97**, 385–392.

van Kleunen, M., Dawson, W., Schlaepfer, D., Jeschke, J. M. and Fischer, M. (2010) Are invaders different? A conceptual framework of comparative approaches for assessing determinants of invasiveness. *Ecology Letters*, **13**, 947–958.

van Kleunen, M., Schlaepfer, D. R., Glaettli, M. and Fischer, M. (2011) Preadapted for invasiveness: do species traits or their plastic response to shading differ between invasive and non-invasive plant species in their native range? *Journal of Biogeography*, **38**, 1294–1304.

Vankat, J. L. (1991) Floristics of a chronosequence corresponding to old field-deciduous forest succession in southwestern Ohio. IV. intra- and inter-stand comparisons and their implications for succession mechanisms. *Bulletin of the Torrey Botanical Club*, **118**, 392–398.

Vankat, J. L. and Carson, W. P. (1991) Floristics of a chronosequence corresponding to old field-deciduous forest succession in southwestern Ohio. III. post-disturbance vegetation. *Bulletin of the Torrey Botanical Club*, **118**, 385–391.

Veblen, T. T. (1992) Regeneration dynamics. In Glenn-Lewin, D. C., Peet, R. K. and Veblen, T. T. (Eds.) *Plant Succession: Theory and Prediction*. New York, Chapman & Hall.

Veech, J. A. and Crist, T. O. (2009) PARTITION: software for hierarchical partitioning of species diversity, version 3.0.

Veech, J. A., Summerville, K. S., Crist, T. O. and Gering, J. C. (2002) The additive partitioning of species diversity: recent revival of an old idea. *Oikos*, **99**, 3–9.

Vilà, M. and Weiner, J. (2004) Are invasive plant species better competitors than native plant species? – evidence from pair-wise experiments. *Oikos*, **105**, 229–238.

Vilà, M., Espinar, J. L., Hejda, M. *et al.* (2011) Ecological impacts of invasive alien plants: a meta-analysis of their effects on species, communities and ecosystems. *Ecology Letters*, **14**, 702–708.

Vile, D., Shipley, B. and Garnier, E. (2006) A structural equation model to integrate changes in functional strategies during old-field succession. *Ecology*, **87**, 504–517.

Violle, C., Navas, M. L., Vile, D., Kazakou, E., Fortunel, C., Hummel, I. and Garnier, E. (2007) Let the concept of trait be functional! *Oikos*, **116**, 882–892.

Vitousek, P. M. (2004) *Nutrient Cycling and Limitation: Hawai'i as a Model System*. Princeton, NJ, Princeton University Press.

Vitousek, P. M. and Reiners, W. A. (1975) Ecosystem succession and nutrient retention: a hypothesis. *BioScience*, **25** 376–381.

Vitousek, P. M. and White, P. S. (1981) Process studies in succession. In West, D. C., Shugart, H. H. and Botkin, D. B. (Eds.) *Forest Succession: Concepts and Applications*. New York, Springer-Verlag.

Vivian-Smith, G. (1997) Microtopographic heterogeneity and floristic diversity in experimental wetland communities. *Journal of Ecology*, **85**, 71–82.

Wagner, H. H., Wildi, O. and Ewald, K. C. (2000) Additive partitioning of plant species diversity in an agricultural mosaic landscape. *Landscape Ecology*, **15**, 219–227.

Wales, B. A. (1972) Vegetation analysis of north and south edges in a mature oak-hickory forest. *Ecological Monographs*, **42**, 451–471.

Walker, K. J., Stevens, P. A., Stevens, D. P. *et al.* (2004) The restoration and re-creation of species-rich lowland grassland on land formerly managed for intensive agriculture in the UK. *Biological Conservation*, **119**, 1–18.

Walker, L. R. (Ed.) (1999) *Ecosystems of Disturbed Ground*. New York, Elsevier.

Walker, L. R. and Chapin, F. S., III (1987) Interactions among processes controlling successional change. *Oikos*, **50**, 131–135.

Walker, L. R. and del Moral, R. (2003) *Primary Succession and Ecosystem Rehabilitation*. Cambridge, Cambridge University Press.

Walker, L. R. and del Moral, R. (2009) Lessons from primary succession for restoration of severely damaged habitats. *Applied Vegetation Science*, **12**, 55–67.

Walker, L. R., Zarin, D. J., Fetcher, N., Myster, R. W. and Johnson, A. H. (1996) Ecosystem development and plant succession on landslides in the Caribbean. *Biotropica*, **28**, 566–576.

Walker, L. R., Smith, S. D. and Luken, J. O. (1997) Impacts of invasive plants on community and ecosystem properties. *Assessment and Management of Plant Invasions*. New York, Springer.

Walker, L. R., Walker, J. and Hobbs, R. J. (2007) *Linking Restoration and Ecological Succession*, New York, Springer.

Walker, L. R., Wardle, D. A., Bardgett, R. D. and Clarkson, B. D. (2010) The use of chronosequences in studies of ecological succession and soil development. *Journal of Ecology*, **98**, 725–736.

Wardle, D. A., Walker, L. R. and Bardgett, R. D. (2004) Ecosystem properties and forest decline in contrasting long-term chronosequences. *Science*, **305**, 509–523.

Watt, A. S. (1947) Pattern and process in the plant community. *Journal of Ecology*, **35**, 1–22.

Weatherhead, P. J. (1986) How unusual are unusual events? *American Naturalist*, **128**, 150–154.

Webb, C. O., Ackerly, D. D., McPeek, M. A. and Donoghue, M. J. (2002) Phylogenies and community ecology. *Annual Review of Ecology and Systematics*, **33**, 475–505.

Wedin, D. and Tilman, D. (1993) Competition among grasses along a nitrogen gradient: initial conditions and mechanisms of competition. *Ecological Monographs*, **63**, 199–229.

Weiher, E. and Keddy, P. A. (Eds.) (1999) *Ecological Assembly Rules*. Cambridge, Cambridge University Press.

Weiher, E., Van Der Werf, A., Thompson, K., Roderick, M., Garnier, E. and Eriksson, O. (1999) Challenging Theophrastus: a common core list of plant traits for functional ecology. *Journal of Vegetation Science*, **10**, 609–620.

Werner, P. A., Gross, R. S. and Bradbury, I. K. (1980) The biology of Canadian weeds: 45. *Solidago canadensis* L. *Canadian Journal of Plant Science*, **60**, 1393–1409.

West, D. C., Shugart, H. H. and Botkin, D. B. (Eds.) (1981) *Forest Succession: Concepts and Applications*. New York, Springer-Verlag.

Westoby, M. (1998) A leaf-height-seed (LHS) plant ecology strategy scheme. *Plant and Soil*, **199**, 213–227.

Westoby, M. and Wright, I. J. (2006) Land-plant ecology on the basis of functional traits. *Trends in Ecology and Evolution*, **21**, 261–268.

Westoby, M., Jurado, E. and Leishman, M. (1992) Comparative evolutionary ecology of seed size. *Trends in Ecology and Evolution*, **7**, 368–372.

Westoby, M., Falster, D. S., Moles, A. T., Vesk, P. A. and Wright, I. J. (2002) Plant ecological strategies: some leading dimensions of variation between species. *Annual Review of Ecology and Systematics*, **33**, 125–159.

White, P. S. and Jentsch, A. (2001) The search for generality in studies of disturbance and ecosystem dynamics. *Progress in Botany*, **62**, 399–449.

White, P. S. and Pickett, S. T. A. (1985) Natural disturbance and patch dynamics: an introduction. In Pickett, S. T. A. and White, P. S. (Eds.) *The Ecology of Natural Disturbance and Patch Dynamics*. Orlando, FL, Academic Press.

White, T. A., Campbell, B. D. and Kemp, P. D. (2001) Laboratory screening of the juvenile responses of grassland species to warm temperature pulses and water deficits to predict invasiveness. *Functional Ecology*, **15**, 103–112.

Whitford, P. C. and Whitford, P. B. (1978) Effects of trees on ground cover in old-field succession. *American Midland Naturalist*, **99**, 435–443.

Whittaker, R. H. (1965) Dominance and diversity in land plant communities. *Science*, **147**, 250–260.

Whittaker, R. H. (1975) *Communities and Ecosystems*. New York, Macmillan.

Wilby, A. and Brown, V. K. (2001) Herbivory, litter and soil disturbance as determinants of vegetation dynamics during early old-field succession under set-aside. *Oecologia*, **127**, 259–265.

Wilkinson, D. M. (1999) The disturbing history of intermediate disturbance. *Oikos*, **84**, 145–147.

Willemsen, R. W. (1975) Dormancy and germination of common ragweed seeds in the field. *American Journal of Botany*, **62**, 639–643.

Williams, J. and Crutzen, P. J. (2013) Perspectives on our planet in the Anthropocene. *Environmental Chemistry*, **10**, 269–280.

Willson, M. F. and Crome, F. H. J. (1989) Patterns of seed rain at the edge of a tropical Queensland rain forest. *Journal of Tropical Ecology*, **5**, 301–308.

Wilsey, B. J. and Polley, H. W. (2002) Reductions in grassland species evenness increase dicot seedling invasion and spittle bug infestation. *Ecology Letters*, **5**, 676–684.

Wilsey, B. J. and Potvin, C. (2000) Biodiversity and ecosystem functioning: importance of species evenness in an old field. *Ecology*, **81**, 887–892.

Wilson, J. B. (1999) Assembly rules in plant communities. In Weiher, E. and Keddy, P. A. (Eds.) *Ecological Assembly Rules*. Cambridge, Cambridge University Press.

Wilson, J. B., Gitay, H., Roxburgh, S. H., King, W. M. and Tangney, R. S. (1992) Egler's concept of "Initial floristic composition" in succession: ecologists citing it don't agree what it means. *Oikos*, **64**, 591–593.

Wilson, J. B., Wells, T. E. C., Trueman, I. C. et al. (1996) Are there assembly rules for plant species abundance? An investigation in relation to soil resources and successional trends. *Journal of Ecology*, **84**, 527–538.

Wilson, S. D. and Tilman, D. (1991) Interactive effects of fertilization and disturbance on community structure and resource availability in an old-field plant community. *Oecologia*, **88**, 61–71.

Wilson, S. D. and Tilman, D. (2002) Quadratic variation in old-field species richness along gradients of disturbance and nitrogen. *Ecology*, **83**, 492–504.

Wiser, S. K., Allen, R. B., Clinton, P. W. and Platt, K. H. (1998) Community structure and forest invasion by an exotic herb over 23 years. *Ecology*, **79**, 2071–2081.

Wolfe, R. W. (2002) Why alien invaders succeed: support for the escape-from-enemy hypothesis. *American Naturalist*, **160** 705–711

Wright, I. J., Reich, P. B., Westoby, M. et al. (2004) The worldwide leaf economics spectrum. *Nature*, **428**, 821–827.

Wright, J. P. and Fridley, J. D. (2010) Biogeographic synthesis of secondary succession rates in eastern North America. *Journal of Biogeography*, **37**, 1584–1596.

Wu, J. and Wu, T. (2013) Ecological resilience as a foundation for urban design and sustainability. In Pickett, S. T. A., Cadenasso, M. L. and McGrath, B. (Eds.) *Resilience in Ecology and Urban Design: Linking Theory and Practice for Sustainable Cities*. New York, Springer.

Yachi, S. and Loreau, M. (1999) Biodiversity and ecosystem productivity in a fluctuating environment: the insurance hypothesis. *Proceedings of the National Academy of Science*, **96**, 1463–1468.

Yu, D. W. and Wilson, H. B. (2001) The competition-colonization trade-off is dead; long live the competition-colonization trade-off. *The American Naturalist*, **158**, 49–63.

Yurkonis, K. A. and Meiners, S. J. (2004) Invasion impacts local species turnover in a successional system. *Ecology Letters*, **7**, 764–769.

Yurkonis, K. A. and Meiners, S. J. (2006) Drought impacts and recovery are driven by variation in local species turnover. *Plant Ecology*, **184**, 325–336.

Yurkonis, K. A., Meiners, S. J. and Wachholder, B. E. (2005) Invasion impacts diversity through altered community dynamics. *Journal of Ecology*, **93** 1053–1061.

Index

α and β diversity, 134
abandonment conditions, 77, 82, 155, 157, 158, 160, 167, 193
abandonment season, 161
Acer negundo, 174
Acer platanoides, 173
Acer rubrum, 52, 59, 91, 195, 196, 218
Acer saccharum, 195
Achillea millefolium, 108
adaptive cycle, 40
Ailanthus altissima, 173, 199
allelopathy, 58, 199
Alliaria petiolata, 74, 91, 122, 173, 222
allocation patterns, 97, 123
Ambrosia artemisiifolia, 52, 58, 91, 159, 174, 194, 198, 199, 215
annuals, 68, 91, 158, 201, 215
Asplenium platyneuron, 85
assembly rules, 151, 181
  and succession, 167
  dispersal- vs. interaction-based, 157
  trait-based, 154
assimilation, 55
Aster lanceolatus, 174
Aster pilosus, 91, 108

Barbarea vulgaris, 91, 116, 159
Bard, G. E., 80
bare soil, 70, 127, 129
  and rainfall variability, 117
Berberis thunbergii, 173
biennials, 69, 158, 202, 215
Brassicaceae, 158
Braun, E. L., 111
Buell, H. F., 22
Buell, M. F., 17, 22
Buell–Small Succession Study
  history of, 15
  lesson from, 253
  sampling methods, 24
  successional phases, 89
Burroughs, J., 169

canopy closure, 69, 75
Carya spp., 194, 218
Carya tomentosa, 197
causal framework
  function of, 60
Celastrus orbiculatus, 91, 173, 200
Chenopodium album, 215
chronosequence, 22, 80, 81
Chrysanthemum leucanthemum, 116, 173
Clements, F. E., 21, 31, 33, 45, 51, 67, 87, 153, 210, 237
climax community, 7, 21
clonal reproduction. *see* vegetative reproduction
colonization, 105, 120, 166, 168, 172
  affected by species richness, 185
  and drought, 122
community assembly, 147, 242
competition, 39, 57, 142, 197, 199, 246
competition-colonization tradeoff, 100, 128, 145
conceptual framework, v, 135, 149, 195, 212, 237
  and invasion, 186
  and restoration, 244
  as an assembly rule, 152
  utility of, 3, 248
Connell and Slatyer
  successional processes, 33, 46
convergence, 79, 152, 155
  and scale, 153
  functional, 153
Conyza canadensis, 174
Cooper, W. S., 32, 151, 190
core–satellite hypothesis, 137
Cornus florida, 59, 79, 91, 194, 196
Cornus racemosa, 91
Cowles, H. C., 21, 31
Craine, J. M., 40
c-s-r strategies, 39, 41, 57, 104, 217, 223

Dactylis glomerata, 157, 173, 193, 201
Danthonia spicata, 116
Darwin, C. R., 1, 31
Daucus carota, 91, 205
Davis, W. M., 31

decomposition rate, 181
deterministic processes, 152, 155, 167
Digitaria sanguinalis, 91, 159
dispersal
  bird, 194
dispersal limitation, 82, 85, 135, 153, 155, 189, 241
disturbance, 38, 39, 40, 41, 43, 49, 71, 142, 198
  adaptive response to, 44
  and heterogeneity, 191
  and management, 243
  by mammals, 127
  characteristics of, 112, 192
  definition of, 7
divergence, 155
diversity, 75, 131, 143, 195
  and disturbance, 128
  as a management goal, 242
  partitioning of, 134
  spatial turnover in, 140
dominant species, 137, 170
  and management, 242
drought, 113, 196, 237
  recovery from, 119
  return interval, 113
  seasonality of, 115
  species replacement, 121

ecological theory
  nature of, 249
edge effects, 195, 197
  on succession, 163
Egler, F. E., 21, 51, 85
Elaeagnus umbellata, 173
Elytrigia repens, 91, 107, 173
episodic reproduction, 116
Erigeron annuus, 91, 159, 174, 194, 199, 205
Eupatorium rugosum, 75, 91
Euthamia graminifolia, 108
evenness, 75, 84, 143
extinction, 105, 120
  and drought, 122

facilitation, 33
forest understory, 122, 125, 155, 218, 241, 247
Fragaria virginiana, 91, 198, 218
Fraxinus americana, 52, 91, 196, 218
functional traits, 211
  and population dynamics, 100
  correlations among, 229
  native vs. non-native, 177
  ordination of, 224, 226
  tradeoffs, 226
  utility of, 211
  woody vs. herbaceous species, 228

gap partitioning, 38
gap phase models, 37

gap regeneration, 112, 129
germination, 54, 128, 195, 196
Gleason, H. A., 21, 32, 51, 87, 189
grass abundance, 157
grasses, 159
Gray, A., 169
Grime, J. P., 39, 104, 211

habitat quality metrics, 240
  successional context, 240
Hackelia virginiana, 75, 85
Harper, J. L., 189
hay fields, 159, 193, 201, 215
height, 101, 223
herbivory, 59, 163, 164, 191, 196
  and competition, 197
  exclosure studies, 196
  insect, 196
  mammal vs. insect, 197
  palatability, 179, 196
heterogeneity, 38, 40, 41, 82, 119, 135, 163, 189, 245
  and herbivory, 243
  and restoration, 245
  as a cause and consequence of succession, 189
  biotically derived, 193
  compositional, 195
  in interactions, 197
  population patterns, 201
  soil conditions, 193, 201
  successional pattern, 190, 208
  temporal cycle of, 207
Hieracium caespitosum, 91, 173
Hieracium pratense, 198
historical contingencies, 45, 141, 153, 158, 192, 239, 249
Horn, H. S., 37
Hutcheson Memorial Forest Center, 16, 193
  landscape context of, 19, 26
  soils of, 19
Hutchinson, G. E., 131

individual-based models, 40
inhibition, 34, 198
initial floristic composition, 21, 51, 85
integration, v, 60, 238, 248, 251
  need for, 3, 45
  opportunities for, 254
  scale and, 131
  vs. reductionism, 2
intermediate disturbance hypothesis, 132, 142, 146
invasion, 252
  dynamics, 105
  escape from herbivores, 179
  impacts, 107, 197, 252
  in restoration, 240
  successional transitions, 240
island biogeography, 105, 190

Juglans nigra, 155
Juniperus virginiana, 59, 79, 91, 174, 196, 200

Keever, C., 57

lag time, 95, 175
latitude, 79
leaf area, 102, 220
   successional pattern, 222
leaf dry matter, 219
   successional pattern, 219
leaf water content, 102
leaf–height–seed scheme, 223
Lepidium campestre, 159
Levin, S., 190
lianas, 58, 69, 92, 93, 161, 199, 203
   impacts on trees, 200
life form, 79, 90, 173, 175, 217
   and heterogeneity, 201, 207
   justification for, 227
   population dynamics, 95
   successional transition in, 68
life history, 56, 171
light availability, 198, 199, 243
   heterogeneity in, 201
limiting similarity, 133, 146
litter, 198
Lonicera japonica, 77, 81, 91, 107, 108, 173, 197, 200
Lonicera maackii, 173
Loucks, O. L., 82

MacArthur, R. H., 38
Markov models, 37
Mettler's Woods, 16
Microstegium vimineum, 74, 91, 140, 173, 185, 222
mosses, 71
mycorrhizae, 56

nitrogen, 43, 198, 201
NMS ordination, 77, 154
non-native species, 105, 195
   and intermediate disturbance, 146
   bias in species pool, 173
   community-level approach, 170
   dominance in successional systems, 171
   equivalence to native species, 176
   equivalence to natives, 170, 185
   invasive species, 173
   palatability to herbivores, 180
   population dynamics, 174
   successional patterns, 72, 146
nutrient deposition, 82, 250

observational approach, 9
Odocoileus virginianus, 71, 81, 191, 241, 250
Odum, E. P., 35, 46, 82, 99, 122

Oenothera biennis, 174
Oosting, H. J., 15
Oxalis stricta, 108, 174

Panicum dichotomiflorum, 199
Parthenocissus quinquefolia, 91, 174, 200
patch dynamics, 190
peak abundance, 95, 176
perennials, 69, 92, 203
Peromyscus leucopus, 195
photosynthetic rate, 55
Pilea pumila, 85
plant strategies, 97, 211, 212, 226, 230
Plantago lanceolata, 173
plowing, 159, 161, 250
Poa compressa, 173
pollination, 214
   insect, 103, 215
   linkage with seed dispersal, 218
   wind, 102
population dynamics, 93
   and edges, 165
   measures of, 94
   ordination of, 97, 176
propagule pressure, 186
Prunus serotina, 174

Quercus alba, 199
Quercus palustris, 195, 196
Quercus rubra, 174, 196
Quercus spp., 91, 194, 218

R* theory, 39, 57
rainfall, 19
   variability in, 113, 117
rank–abundance curves, 75, 136
   native vs. non-native, 181
Raphanus raphanistrum, 58, 91, 159
rare species, 137
rate of increase, 95, 175
Reiners, W. A., 36
relay floristics, 51, 85
resilience, 40
resource availability, 7, 41, 43, 50, 56, 79, 128, 141, 192, 198, 238, 243, 247
resource ratio hypothesis, 39
restoration, 237
   active vs. passive, 238
   landscape context, 242
   soil microbial communities, 246
Rhus glabra, 91
r-K selection, 38, 41, 44, 84, 97, 99
   support for, 99
Rosa multiflora, 71, 81, 91, 105, 107, 108, 155, 161, 173, 196, 197
Rosa virginiana, 91
Rubus allegheniensis, 91, 94

Rubus flagellaris, 91
Rubus phoenicolasius, 74, 173
Rubus spp., 194
Rumex acetosella, 173, 198
Rutgers University, 17, 22, 193

Sanicula gregaria, 85
scale, 45, 111, 131, 146
  and diversity, 132
  plot vs. field, 73, 117, 132, 134, 183
Schizachyrium scoparium, 81
seed availability, 163
seed bank, 43, 52, 116, 129, 143, 160, 194, 202
seed dispersal, 163, 164, 177, 191, 217
  bird, 52, 191
  landscape context, 53
  mammal, 52
  vertebrate, 104, 177
  wind, 104
seed mass, 57, 102, 155, 177, 218, 223, 242
  and clonality, 217, 231
  successional pattern, 219
seed predation, 53, 59, 163, 191, 195, 198
  edge effects, 195
  temporal variation in, 195
Sernander, R., 33
Setaria faberii, 52, 194, 199
shade tolerance, 38, 40, 80, 93
shrubs, 69, 92, 161, 203
Shugart, H. H., 37
site availability, 51, 61
site conditions, 201
Small, J. A., 22
social-ecological systems, 4, 43
Solanum carolinense, 218
Solidago canadensis, 174, 185, 198, 199
Solidago gigantea, 174
Solidago juncea, 91, 206
Solidago rugosa, 108
Solidago spp., 77, 174
SORTIE, 40
space-for-time substitution, 22, 80
span of abundance, 96
species availability, 52, 61, 85, 128, 135, 148, 167, 172, 194, 201, 249
  and edges, 166
  and landscape context, 153
species complementarity, 159
species introduction
  patterns of, 171
species performance, 53, 63, 121, 128, 135, 141, 148, 164, 167, 172, 194, 204, 213, 249
  and edges, 166
species pool, 137, 146, 168, 171, 174, 215
  constraints imposed by, 213

species richness, 73, 82, 117, 128, 132, 143, 195, 198
  and rainfall variability, 125
  pattern of native and non-native, 158, 181
  temporal pattern, 134
specific leaf area, 102, 178, 180, 222, 223
  successional pattern, 222
Spiranthes, 116
stability, 84, 122
stochastic processes, 141, 153, 155, 167, 246
stomatal conductance, 55
storage effect, 38
stress, 39, 57, 238
stress tolerance, 57
succession
  and climate change, 252
  and conservation, 44
  and land management, 237
  as a population process, 109
  assumptions of, 47
  compositional changes, 77
  definition of, 6, 46
  ecosystem patterns, 34
  endpoint of, 7
  importance of, 4, 251
  population patterns, 88
  primary vs. secondary, 49
  rate of, 79, 172, 192
  theory, 30

Tainter, J., 44
temperature, 19
Tilman, D., 39
tolerance, 34
total cover, 69, 125, 195
  and rainfall, 117
  drought response, 119
Toxicodendron radicans, 91, 155, 200
tradeoffs
  life history, 97
  population dynamics, 99
tree, 161
tree regeneration, 81, 85, 161, 195, 199, 241
  and edges, 164, 194
trees, 69, 92, 203
Trifolium pratense, 91, 107

vegetative reproduction, 52, 104, 127, 179, 216, 242
Vitis spp., 91, 174, 199
Vitousek, P. M., 36

Warming, J. E. B., 30
Watt, A. S., 33, 190
Whittaker, R. C., 87
Wilson, E. O., 38